Problem Symbols

 Social and Behavioral Science

 Government and Public Affairs

 Business and Consumer Affairs

 Sports and Games

 Natural Science

 Agriculture

 Technology

 Education

 Medicine

MATT BINNIE 82
PH 527 8469

ELEMENTS OF STATISTICAL INFERENCE

ALLYN AND BACON, INC. BOSTON · LONDON · SYDNEY · TORONTO

Elements of Statistical Inference

FIFTH EDITION

DAVID V. HUNTSBERGER
IOWA STATE UNIVERSITY

PATRICK BILLINGSLEY
THE UNIVERSITY OF CHICAGO

Chapter opening photo credits:

1. Ellen Shub/The Picture Cube
2. Mike Mazzaschi/Stock, Boston
3. Bobbi Carrey/The Picture Cube
4. Frank Siteman/Stock, Boston
5. Peter Vandermark/Stock, Boston (also appears on title page)
6. Frank Siteman/The Picture Cube
7. Courtesy Sperry New Holland
8. Peter Menzel/Stock, Boston
9. Terry McKoy/The Picture Cube
10. Alex Novae/Stock, Boston
11. Steven M. Stone/The Picture Cube
12. George Cohen/Stock, Boston
13. Irene Shwachman/The Picture Cube

Series Editor: Carl Lindholm

Production Editor: Barbara Willette

Copyright © 1981, 1977, 1973, 1967, 1961 by Allyn and Bacon, Inc., 470 Atlantic Avenue, Boston, Massachusetts 02210. All rights reserved. No part of the material protected by this copyright notice may be reproduced or utilized in any form or by any means, electronic or mechanical, including photocopying, recording, or by any information storage and retrieval system, without written permission from the copyright owner.

Library of Congress Cataloging in Publication Data

Huntsberger, David V
    Elements of statistical inference.

    Includes index.
    1. Statistics. I. Billingsley, Patrick, joint author. II. Title.
QA276.12.H86   1981   519.5'4   80-25330
ISBN 0-205-07305-0

ISBN 0-205-07380-8 (International)

Printed in the United States of America.

10 9 8 7 6 5 4 3 2   86 85 84 83 82 81

Contents

Preface xi

1 Introduction 1

    1.1 What Statistics Is 2
    1.2 The Objectives of This Book 6
    1.3 A Word about Computation 7
    Summary and Keywords 8

2 Empirical Frequency Distributions 10

    2.1 Frequency Distributions 12
    2.2 Cumulative Frequency Distributions 21
    2.3 Graphic Presentation 24
    2.4 Stem-and-Leaf Plots 35
    2.5 Samples and Populations 37
    Chapter Problems 37
    Summary and Keywords 40

3 Descriptive Measures 42

    3.1 Introduction 44
    3.2 Symbols and Summation Notation 44
    3.3 Measures of Location 49
    3.4 The Mean 50

	3.5	The Weighted Mean	53
	3.6	The Median	56
	3.7	The Mode	59
	3.8	Selecting a Measure of Location	61
	3.9	Measures of Variation	62
	3.10	The Range	63
	3.11	The Variance and the Standard Deviation	63
	3.12	Coding, or Change of Scale	70
	3.13	Means and Variances for Grouped Data	76
	3.14	Quick Measures of Location and Spread	81
	3.15	Index Numbers	83
		Chapter Problems	89
		Summary and Keywords	92

4 Probability 94

4.1	Introduction	96
4.2	The Meaning of Probability	97
4.3	Computing Probabilities	102
4.4	Counting	115
4.5	Repeated Independent Trials	133
4.6	Finite Sampling	141
4.7	Subjective Probability and Bayes' Theorem	144
	Chapter Problems	147
	Summary and Keywords	147

5 Populations, Samples, and Distributions 150

5.1	Populations	152
5.2	Samples	153
5.3	Random Variables	155
5.4	Distributions of Random Variables	157
5.5	Expected Values of Random Variables	169
5.6	Sets of Random Variables	176

	5.7	Normal Distributions	178
	5.8	The Binomial Distribution	189
	5.9	The Normal Approximation to the Binomial	194
	5.10	The Poisson Distribution	202
		Chapter Problems	204
		Summary and Keywords	205

6 Sampling Distributions — 208

6.1	Sampling and Inference	210
6.2	Expected Values and Variances	212
6.3	Sampling from Normal Populations	217
6.4	The Standardized Sample Mean	219
6.5	The Central Limit Theorem	221
	Chapter Problems	225
	Summary and Keywords	226

7 Estimation — 228

7.1	Introduction	230
7.2	Estimation	230
7.3	Confidence Intervals for Normal Means: Known Variance	234
7.4	The t-Distribution	241
7.5	Confidence Intervals for Normal Means: Unknown Variance	243
7.6	Sample Size	249
7.7	Estimating Binomial p	252
	Chapter Problems	258
	Summary and Keywords	260

8 Tests of Hypotheses — 262

8.1	Introduction	264
8.2	Hypotheses on a Normal Mean	267

8.3	Hypotheses on a Binomial p	275
8.4	p-Values	278
8.5	The Theory of Testing	281
	Chapter Problems	288
	Summary and Keywords	289

9 Two-Sample Techniques and Paired Comparisons — 292

9.1	Introduction	294
9.2	Samples from Two Normal Populations	294
9.3	Confidence Intervals for $\mu_1 - \mu_2$ and μ_i	300
9.4	Testing the Hypothesis $\mu_1 = \mu_2$	304
9.5	Which Assumptions Can Be Relaxed?	311
9.6	Paired Comparisons	312
9.7	Groups versus Pairs	318
9.8	Two Samples from Binomial Populations	321
	Chapter Problems	328
	Summary and Keywords	330

10 Approximate Tests: Multinomial Data — 332

10.1	The Multinomial Distribution	334
10.2	A Hypothesis about Multinomial Parameters	335
10.3	Binomial Data	341
10.4	A Test for Goodness of Fit	344
10.5	Contingency Tables	350
	Chapter Problems	358
	Summary and Keywords	360

11 Regression and Correlation — 362

11.1	Linear Regression	364

	11.2	Least Squares	367
	11.3	Variance Estimates	372
	11.4	Inferences about *A* and *B*	374
	11.5	Some Uses of Regression	378
	11.6	Partitioning the Sum of Squares	382
	11.7	Multiple Regression	386
	11.8	Correlation	392
	11.9	Correlation and Cause	395
	11.10	The Least-Squares Estimates	398
		Chapter Problems	399
		Summary and Keywords	402

12 Analysis of Variance 404

	12.1	Introduction	406
	12.2	The Role of Randomization	406
	12.3	The Completely Randomized Design	407
	12.4	The Model	408
	12.5	Constructing the Analysis-of-Variance Table	409
	12.6	A Numerical Example	413
	12.7	Estimation of Effects	416
	12.8	The Hypothesis of Equal Means	419
	12.9	The Effects of Unequal Group Sizes	424
	12.10	Principles of Design	430
	12.11	The Randomized Complete Block Design	432
		Chapter Problems	440
		Summary and Keywords	443

13 Nonparametric Methods 444

	13.1	Introduction	446
	13.2	The Wilcoxon Test	446
	13.3	The Sign Test	451

	13.4	The Rank Sum Test	452
	13.5	Association	456
		Chapter Problems	461
		Summary and Keywords	463

Answers to Odd-Numbered Problems 465

Appendix 479

	Table 1	Cumulative Binomial Probabilities	480
	Table 2	Areas of the Standard Normal Distribution	481
	Table 3	Values of $e^{-\mu}$	482
	Table 4	Values of t for Given Probability Levels	483
	Table 5	95% Confidence Intervals (Percent) for Binomial Distributions	484
	Table 6	Percentage Points for the Chi-Square Distribution	485
	Table 7	Relationship between z and r (or μ_z and ρ)	486
	Table 8	Percentage Points of the F-Distribution	487
	Table 9	Random Numbers	495

Index 499

Preface

Purpose

Every year statistics plays a more important role in sociology, business and economics, agriculture, engineering, ecology, psychology, education, medicine—and many other disciplines besides. Every year, therefore, students of these subjects need a better understanding of statistics. The purpose of this book is to teach the language, principles, and practice of statistics and to do it so that the students will be able to

- understand statistical analyses they meet in the literature of their subject,
- handle many statistical problems themselves,
- talk with statisticians if the need arises.

Audience

The book is written for college students in all the areas where statistics is a requirement. Since their programs are already crowded, they need a course that is self-contained and brief—ordinarily a single quarter or semester. They need a course that presupposes no technical background and that uses no mathematics beyond some high school algebra. The book is designed for such a course.

Approach

What is hard in teaching a course of that kind is to find the right balance between applications and the statistical principles behind them. Since students want to *use* statistics, an excessively theoretical approach leaves them feeling rudderless and unmotivated. But an exclusive concentration on examples and applications, in the how-to-do-it style of a cookbook, turns statistics from a coherent subject into a loose collection of unrelated fragments, impossible to organize in the mind and soon forgotten. Our solution is to

- introduce each statistical method and technique with some example that shows why it is needed and what question it answers,
- explain the rationale underlying the method—its theoretical basis—and how it connects with other methods and fits into the framework of the subject as a whole,
- apply each method to a variety of problems to show explicitly how it is carried out and to strengthen the student's understanding of the method itself by showing that it does work in more than one area of application.

Sometimes statistics is taught as an adjunct to some one area of application, from which all the examples are then taken, an approach a friend sums up in the parody book title, *Statistics for Midwestern Dentists*. Our approach in this book reflects our conviction that students of any subject in which statistics is used understand statistics *better* if they see it applied not only to their own subject but to others as well.

Organization

The book starts off with an introductory chapter describing an investigation the Public Health Service made into the effects of drugs that control high blood pressure. This investigation, obvious in its importance to society, nicely illustrates statistical issues that are resolved in the course of the book. Then come two chapters on descriptive statistics. Important formulas such as the mean are given in standard mathematical notation, which the student gradually learns. (To us it seems unwise to slight the mathematical notation; it is the accepted language of statistics and will be needed in all of the student's future statistical work.) The fourth chapter provides flexible coverage of probability; the instructor can go into the subject in detail or move quickly on to inference. The next four chapters, the core of the book, cover sampling, distributions, and statistical inference in the form of estimation and of testing. The material up to this point in the book will ordinarily be taken up more or less in sequence, with variations to suit individual needs and preferences. Material from the last five chapters can then be selected in a variety of ways, depending on the length of the course and the interests of the class.

The Fifth Edition

For this fifth edition, we have added new features, increased the number of problems, and strengthened and emphasized the discussions of how statistical theory ties in with applications in the real world. For example:

- The introductory chapter on the **hypertension study** is new. The same study is used in the chapter on probability to show more clearly why the study of chance is important for statistics and its applications.
- This chapter has also been rearranged for greater **flexibility**: the general principles of probability now come before the treatment of combinations and permutations, which can be skipped over if the instructor wants.
- We have emphasized the general theory of testing less and *p*-**values** more.
- There is now a **binomial table,** and the normal table has been reduced to one page. A **new diagram and explanation** accompany each table.
- We have added **chapter summaries.**
- Important terms are printed in boldface where they are defined and in color in the margin beside the definition. Each of these **keywords** again appears in the chapter summary, together with a reference to the page on which it is defined. The summary, instead of serving as a substitute for the chapter, thus leads the student back into it.
- Other **new design features** should make the volume easier to use. Symbols next to problems indicate their areas of application. Subsidiary material in the text is printed in smaller type and set off by bracket rules. Three appendix tables are also on the endpapers (inside covers) for easy reference.
- There are now **more than 650 problems**—a 65% increase over the fourth edition. A section is now followed by problems keyed to that section, and a chapter ends with a set of problems requiring the use of all the methods covered in the chapter.

We have in a number of places tried to ease the exposition without making it less precise or in any way less accurate. All of these changes are the result of users' suggestions. Above all, we have heeded the voices of many users who, when asked for suggestions, told us not to sacrifice in this edition the clarity and precision they said they found in the previous ones.

Available from the publisher are an accompanying Study Guide and an Instructor's Manual.

Acknowledgments

We are grateful to the literary executor of the late Sir Ronald A. Fisher, F.R.S., to Dr. Frank Yates, F.R.S., and to Longman Group Ltd. London, for permission to reprint Table III from their book *Statistical Tables for Biological, Agricultural and Medical Research* (6th edition, 1974); to Iowa State University Press for permission to reprint Table 1.3.1 from *Statistical Methods* (6th edition, 1967), by George W. Snedecor and William G. Cochran; and to the Biometrika Trustees for permission to use the material presented here in Tables 6 and 8.

We are grateful to Dr. W. McFate Smith and William Mark Krushat, of the U.S. Public Health Service, for the data on the hypertension study; to Paul Meier and Theodore Karrison for discussions of that study; to Ronald Thisted for the random number table; to James Higgins for the new problems; to David Stones, Dale Napier, and Susan Shott for checking the new edition; and to Carl Lindholm of Allyn and Bacon for general help and encouragement.

1
Introduction

1.1 What Statistics Is

Hypertension, or high blood pressure, is often called the silent killer. In recent decades drugs have been developed to control blood pressure, and in 1966 the United States Public Health Service began a ten-year study to find out how effective they are in cases of mild hypertension.*

Normal blood pressure is around 80. (This refers to diastolic blood pressure, and the units of measurement, which don't matter here, are millimeters of mercury.) The benefits of these drugs for patients with severe hypertension—blood pressure over 115—were known, but the benefits for those in the range 90 to 115 weren't at all clear at the time. Hence the study, whose design and interpretation involved statistical analysis in important ways.

In this study, 389 patients with mild hypertension (blood pressure 90 to 115) were divided into two groups. Those in one group, the **active group**, were given pills containing an active drug that lowers blood pressure. (It was chlorothiazide plus rauwolfia serpentina, but the names and the chemistry needn't concern us.) Those in the other group, the **control group**, were given a **placebo**—that is, they were given "sugar pills," pills identical in appearance to the active ones but containing no drug at all and so having no effect on blood pressure. There were 193 patients in the active group and 196 in the placebo or control group. Over the next ten years the 389 patients were examined periodically by physicians and their histories recorded. The question was, did those in the active group do better than those in the control group? Did they experience fewer of the damaging effects of hypertension? What did the **data** say?

<small>active group</small>

<small>control group
placebo</small>

<small>data</small>

During the course of the study, the blood pressures of 24 of the patients increased to 130 or more, a dangerously high level. The 24 patients who suffered this runaway high blood pressure were all in the control or placebo group; none were on the active drug (see Table 1.1). Obviously, the drug is effective in preventing patients with mild hypertension from developing this extreme and dangerous elevation of blood pressure. The 24 who did develop it were then at such high risk of damage from their hypertension that they were removed from the placebo group and given medication to reduce their blood pressure.

An additional 126 patients suffered what are called *hypertensive events*. These are certain illnesses that high blood pressure has long been suspected of causing. Illnesses in this class include stroke, enlargement of the heart, rupture of delicate blood vessels in the eye, and congestive heart failure. A major objective of the study was to see whether the incidence of

---

*The full technical details of the study can be found in Reference 1.1 at the end of this chapter.

Table 1.1 Cases of Extreme Elevation of Blood Pressure (130 or More)

	Active	Placebo
Number of patients	193	196
Number of cases where blood pressure increased to 130 or more	0	24

these hypertensive events could in fact be reduced by lowering blood pressure. The 126 hypertensive events observed were not evenly divided between the active group and the placebo group. Of the 196 patients on the placebo, 89 experienced hypertensive events, but of the 193 on the active drug, only 37 did (see Table 1.2). Although the evidence here is not as overwhelming as that in the first table, Table 1.2 does make it quite clear that the drug helps in the prevention of hypertensive events.

But what about heart attacks and other coronary heart disease? These events are not associated directly with high blood pressure, but instead are associated with the deposit of fatty substances in the arteries (atherosclerosis). There were 73 of these severe events: 38 in the placebo group and 35 in the active group (see Table 1.3). Although there were a few more cases in the placebo group than in the active drug group, the difference is so small that no one, on the basis of Table 1.3 alone, would put much confidence in the drug's ability to prevent events in this category.

A subsequent study (see Reference 1.2) showed that lowering blood pressure *does* in fact help to reduce the incidence of coronary heart disease among mild hypertensives. But the reduction, while substantial, is not great enough to be clearly demonstrated by a study as small (389 patients) as that of the Public Health Service. The later study was much larger. It was

Table 1.2 Hypertensive Events (Cases of Stroke, Enlargement of the Heart, etc.)

	Active	Placebo
Number of patients	193	196
Number of hypertensive events	37	89

Table 1.3 Cases of Heart Attack and Other Coronary Heart Disease

	Active	Placebo
Number of patients	193	196
Number of cases of heart attack and other coronary heart disease	35	38

able to show convincingly that lowering blood pressure really does affect the chance of coronary heart disease, because it traced the medical histories of many patients—over 9,000 of them. This illustrates a general phenomenon in statistical work. If a study fails to show that a treatment is effective, it may well be that a more extensive investigation *would* show that it is effective. For this reason, a negative finding—a finding that a treatment has no effect—is always provisional and subject to revision in the light of subsequent evidence.

One more case will be instructive. Let us look at enlargement of the heart, one of the hypertensive events entering into the figures in Table 1.2. There were 12 cases of it among the patients on the active drug and 20 cases among those on the placebo (see Table 1.4). There were more cases of enlarged heart among those on placebo than among those on the active drug, but the numbers are not entirely convincing. It is unclear just what conclusion is to be drawn from the evidence in Table 1.4. Does the drug help to prevent enlargement of the heart among patients with mild hypertension? Or can the difference between 12 and 20 in the table be the result of chance—a mere accident? Only a statistical analysis can tell. In this and many similar cases where the issues are serious, it will not do to say vaguely that the evidence seems inconclusive. We need a method that will enable us to extract and appraise all the information the data contains. The science of statistics provides that method.

Tables 1.1 to 1.4 can be combined into one, as we have done in Table

Table 1.4 Cases of Enlarged Heart

	Active	Placebo
Number of patients	193	196
Number of cases of enlarged heart	12	20

Table 1.5 All Events

	Active	Placebo
Number of patients	193	196
(1) Number of cases where blood pressure increased to 130 or more	0	24
(2) Number of hypertensive events	37	89
(3) Number of cases of heart attack and other coronary heart disease	35	38
(4) Number of cases of enlarged heart	12	20

1.5. Line 1 in this table is conclusive: the drug is effective in preventing the progressive rise of blood pressure from the 90–115 range to dangerous levels of 130 or more. Line 2 is quite convincing too: the drug does help reduce the incidence of so-called hypertensive events. And line 3 is convincing in a negative sense: although final judgment must be reserved, the data do not indicate that the drug helps in the prevention of coronary heart disease. Although lines 2 and 3 are fairly convincing, a statistical analysis is really needed for an accurate appraisal of the evidence. And certainly a statistical analysis is needed for line 4. Is the drug effective in preventing enlargement of the heart?

objective statistical analysis Although the need for an **objective statistical analysis** here is quite obvious, just how it's to be done is not so obvious. The method can't be explained in this Introduction in a few sentences either: it is in fact the central subject of the book itself. The method is not mysterious or inaccessible, though, and anyone who reads the first eight chapters carefully can see how it works.

design Statistical ideas entered into the **design** of the Public Health Service study as well as into its analysis. Suppose that of the 389 mild hypertensives investigated, all those with blood pressures in the range 106–115 had been put on the active drug and all those with blood pressures in the range 90–105 had been put on placebo. That would of course have biased the study against the drug, and the reverse arrangement would have biased it in favor of the drug. To keep out this and other biases, the patients were assigned to the active and placebo groups at random. It is as though their names were put on 389 cards, and the cards shuffled thoroughly and dealt into two piles. That it should help matters to introduce chance into the study this way seems paradoxical, but it is in fact a powerful method of ensuring accuracy and objectivity.

Why were the patients in the control group given a placebo? Suppose they weren't—suppose they were given no medication at all. For all we would be able to tell in that case, the patients on the active drug might have suffered less hypertensive damage than the others merely because of the pills' psychological effect and not because of their direct physical effect. Just taking a pill can in some cases have a strong psychological effect, and giving placebos, "sugar pills," to the control group evens out this effect between the two groups.

For this control to work, it is of course essential that the patient does not know which group he is in. In this study, possible bias on the part of the examining physicians was also eliminated by keeping them ignorant of which patients were on active drug and which were on placebo. Such a study is called **double blind.**

double blind

Why was there a control group at all? Isn't it unethical to deprive these patients of the benefits of the drug? One answer is that it was not known in advance how beneficial the drug was, and without a standard of comparison, without a control group, there would be no way of assessing it. Furthermore, like almost all drugs, those that reduce blood pressure sometimes have serious side effects, and the benefits must be balanced against the risks. If in the middle of a study the evidence begins to show that the control patients should no longer be deprived of the treatment, the study can be terminated at that point. The United States Veterans Administration (see Reference 1.3) began a controlled study of the effects of drugs on patients with blood pressures in the higher range 115–129, and it became clear that they were so beneficial for this class of more severely hypertensive patient that the study was terminated early and the control group put on medication.

Statistics is the science of collecting, analyzing, and interpreting quantitative data in such a way that the reliability of the conclusions can be evaluated in an objective way. The Public Health Service hypertension study illustrates the need for such a science. But medical investigations are not the only ones requiring statistical methods—they are valuable in all quantitative sciences. They can be used in engineering as well as in the biological sciences, and in the physical sciences as well as in the social sciences. The problems in these diverse areas may be very different, but the scientific method is common to them all, and statistics can be regarded as the technology of the scientific method.

1.2 The Objectives of This Book

descriptive statistics

Statistical methods are used in part for descriptive purposes, for organizing and summarizing numerical data. **Descriptive statistics** deals with the presentation of data in graphical or pictorial form and with the calculation of descriptive measures. Chapters 2 and 3 cover these topics.

statistical inference

inductive generalization

The rest of the book mostly concerns **statistical inference**. As long as you refrain from making generalizations on the basis of your descriptive statistics, you are only describing what you observed. But as soon as you make an **inductive generalization,** you have passed beyond description and have entered the domain of inference. According to Table 1.2 above, 37 of the 193 patients on the active drug experienced hypertensive events; the fraction, then, is 37/193, which is .19, or 19%. If the medical investigator reports that of the 193 patients *in the study* who were given the drug, 19% experienced hypertensive events, then he is reporting an observed fact. Suppose he concludes that if *all* mild hypertensives in the future are put on the drug, then around 19% or 20% of them will experience hypertensive events. He has then gone beyond the facts at hand, which concern only the patients in the study. He has drawn a general conclusion about future patients who may be given the drug. Of course the investigator undertook the study exactly so that he *could* draw general conclusions of this sort. But is this particular conclusion reasonable? What rules should he follow in formulating and appraising his conclusions? Statistical inference deals with just such questions. The particular question just raised—is 19% a good prediction for the future incidence of hypertensive events among mild hypertensives on the drug?—will be dealt with later in the book when the appropriate theory and general principles have been developed. Statistical inference is the theory and practice of how to pass from empirical facts to general conclusions, and it is the principal subject of this book.

It is not the purpose of the book to make a seasoned statistician of you. The purpose is to give you an understanding of the fundamental principles of the subject—including both theory and practice—and an understanding of the language of statistics so that you may meet with statisticians on common ground. You will learn to handle many common types of problems yourself, and will learn to recognize larger problems that call for consultation with a professional statistician.

1.3 A Word about Computation

The book contains many problems for solution. These have to do with the application of statistical principles to problems in psychology, engineering, sociology, economics, and so on, and most of them involve some computation. These problems are essential to an understanding of the subject. To do the computations with pencil and paper can be somewhat tedious, and your path will be easier if you have a simple, inexpensive hand-held calculator. It need only be able to add, subtract, multiply, and divide, and take square roots.

Some instructors using previous editions of this book have also introduced their students to more complex computational systems, such as Minitab. That can be useful. Other instructors prefer to concentrate on the statistics itself and to limit the computation to what can be done on a pocket calculator. In either case, these simple instruments make many of the problems easier and more instructive.

Summary and Keywords

The Public Health Service study illustrates some of the objectives and methods of statistical work. The patients in the **active group** (p. 2) were compared with those in the **control group** (p. 2). Although in some respects the study indicated only the need for further investigation, the **data** (p. 2) definitely showed that the drug under consideration helps to prevent certain disorders. A **reliable and objective statistical analysis** (p. 5) was made possible by a **design** (p. 5) involving a **placebo** (p. 2) and a **double-blind** (p. 6) procedure. Chapters 2 and 3 will cover **descriptive statistics** (p. 6), and the rest of the book will deal primarily with **statistical inference** (p. 7)—the method of passing from data to **inductive generalizations** (p. 7).

REFERENCES

1.1 Smith, W. McFate; Edlavitch, Stanley A.; and Krushat, William Mark. "U.S. Public Health Service Hospitals Intervention Trial in Mild Hypertension." In *Hypertension* (Fifth Hahnemann International Symposium), Gaddo Onesti and Christian R. Klimpt, editors. New York: Grune and Stratton, 1979.

1.2 Shapiro, Alvin P., et al. "Five-Year Findings of the Hypertension Detection and Follow-up Program." *Journal of the American Medical Association,* December 1979.

1.3 Freis, Edward D., et al. "Effects of Treatment on Morbidity in Hypertension." *Journal of the American Medical Association,* December 1967.

2

Empirical Frequency Distributions

2.1 Frequency Distributions

Statistical data are the result of a census, sample survey, medical trial, chemical experiment—some kind of empirical investigation. Usually they consist of raw, unorganized sets of numbers. Before the data can lead to general conclusions about the phenomenon under investigation—before they can be used for statistical inference—they must be organized and summarized in such a way that the pertinent information can be extracted. This, the first step in statistical analysis, is the subject of the present chapter.

The purpose of the Public Health Service hypertension study, discussed in Chapter 1, was to discover whether a blood-pressure-reducing drug helped prevent certain kinds of disease. But the very first question to be answered was, did the drug in fact reduce blood pressure? At the outset of the study, each patient's blood pressure was measured; the result is the *baseline* blood pressure. Then, recall, some patients were put on the drug; these were the patients in the active group. Other patients were put on a placebo, a pill containing no drug at all; these were the patients in the control group. The patients were then checked periodically to see whether their blood pressure went down. (All these blood pressures are diastolic.)

Tables 2.1 and 2.2 tell the story. Table 2.1 concerns the 171 patients on the active drug.* Patient No. 1 in the active group had baseline blood pressure 101, and after 12 months on the drug his blood pressure had decreased to 78; the difference (new blood pressure minus baseline) is 78 − 101, or −23, and this is the first entry in Table 2.1. The blood pressure of patient No. 2 on the active drug in fact went up from 97 (baseline) to 104 (the 12-month reading), a difference of 104 − 97, or 7; and 7 is the next entry down in Table 2.1. And so on. The first entry in Table 2.2 is −15 because the blood pressure of the first patient in the placebo group decreased (even though he was not on the drug) from 91 to 76 (76 − 91 = −15).

Tables 2.1 and 2.2 tell the story, but not in a way that is easy to understand. A minus sign represents a decrease in blood pressure, an improvement, and since there seem to be more minuses in the first table than in the second, the drug seems to be doing some good. But to get a clearer view of what the data mean, we must construct **frequency distributions.** That is, we divide the overall range of observed values into classes, or **bins**

frequency distributions
bins

---

*The tables in Chapter 1 show 193 in the active group and 196 in the control group. The numbers here, 171 and 176, are lower because some patients left the study for various reasons.

SECTION 2.1 FREQUENCY DISTRIBUTIONS

Table 2.1 Change in Blood Pressure: Active Group (171 Patients)

−23	−11	−7	−13	4	−32	−20	−18	1
7	−32	−14	−18	6	10	−4	−15	−7
−21	−10	10	−20	−15	−10	−11	−10	−5
0	−13	−14	−6	9	−19	−10	−19	−11
5	−6	−17	−6	−15	6	−8	−17	−8
−16	2	−6	−14	−22	−11	−23	−6	−5
−12	−12	0	0	−3	−14	−34	−8	−19
−30	−17	−17	−1	−30	−31	−17	−16	−5
8	−23	−12	9	−33	4	−18	−34	−2
−28	−10	−8	−20	−8	19	−12	−11	0
−19	−12	−10	−20	−11	−2	−17	−24	−18
−18	−13	25	4	−13	−1	−7	−2	
−22	−25	−19	−8	−17	−10	−27	−1	
−6	−19	4	−16	−29	4	−8	−16	
−16	1	−7	−31	−9	0	−4	−16	
−5	−6	−14	−3	0	31	−10	−23	
−14	−24	−11	−2	20	−5	−21	−1	
−2	−3	−21	−5	−10	−12	0	−5	
10	−26	−9	−10	16	−15	−26	1	
−18	−19	−16	10	0	4	−9	−4	

Blood pressure after 12 months on drug, minus baseline blood pressure. A negative entry represents a decrease.

as they are usually called, and we count the number of observations that fall into each bin.

Let us illustrate this with the data in Table 2.1. Suppose we use bins containing five values: one bin to contain the values 0 through 4, the next to contain 5 through 9, and so on. To account for negative values, we need a bin containing −5 through −1, a bin containing −10 through −6, and so on. There is, of course, the important question, how do you choose the bin size and the number of bins in the first place? Let's postpone this question a little, until we see how the procedure works.

Table 2.3 shows the result of the tally or frequency count. The first blood-pressure difference in Table 2.1 is −23, and so we make a tally in the bin labeled "−25 to −21." The next entry down in Table 2.1 is 7, and so we make a tally in the bin labeled "5 to 9." We proceed in this way through all the 171 observations in Table 2.1. This gives Table 2.3. Notice that each bin in this distribution is specified by the greatest and the least of

Table 2.2 Change in Blood Pressure: Placebo Group (176 Patients)

−15	−2	6	−9	−3	−1	9	15	17
3	10	−10	−18	2	15	−3	−1	10
5	20	−14	−6	3	10	−3	1	16
0	0	−5	6	−14	1	8	−9	10
6	−10	4	−3	−12	5	17	2	20
3	−8	−16	−12	−13	−6	5	17	8
−5	−8	−9	12	−7	−21	−10	−2	−4
−18	6	−8	−8	10	20	5	−11	−19
−13	19	−6	−10	−2	8	2	−13	11
8	0	−17	0	−16	−23	20	−21	−6
4	−5	−12	−17	−20	−6	2	−5	5
2	0	15	1	4	18	15	19	−17
−5	−10	−14	8	4	20	−1	10	12
−5	5	−5	20	7	7	−4	0	−2
1	0	1	0	0	0	−10	7	4
0	0	−26	2	6	20	23	−2	11
−4	−7	20	4	3	20	15	2	
−10	2	−16	−1	−6	3	−19	7	
6	7	1	−11	4	−6	20	−35	
−3	23	−7	−3	5	10	14	−8	

Blood pressure after 12 months on drug, minus baseline blood pressure. A negative entry represents a decrease.

bin limits   the values in that bin; these two numbers are called the **bin limits**. The bins can't overlap: each possible value must fit into one and only one bin.

bin boundaries   Bins can also be specified by the **bin boundaries**. Take the last two bins in Table 2.3, those labeled "25 to 29" and "30 to 34." The number 29.5 lies half way between the upper limit, 29, of the next-to-last bin and the lower limit, 30, of the last bin. This half-way number, 29.5, is the *upper boundary* of the next-to-last bin; it is at the same time the *lower boundary* of the last bin. The lower limit of the next-to-last bin is 25, and the upper limit of the preceding bin is 24, and so the lower boundary of the next-to-last bin is 24.5, the number half way between. This next-to-last bin can be specified by its lower and upper boundaries, "24.5 to 29.5," just as well as by its lower and upper limits, "25 to 29." The bin boundaries are given in Table 2.4. Notice that each upper bin boundary is the lower boundary of the following bin. Notice also that each boundary is an "impossible value": since the blood-pressure differences are whole numbers (positive or nega-

SECTION 2.1 FREQUENCY DISTRIBUTIONS

Table 2.3 Change in Blood Pressure: Active Group (171 Patients)

Change in Blood Pressure	Tally	Number of Observations
−35 to −31	|||| ||	7
−30 to −26	|||| ||	7
−25 to −21	|||| |||| ||	12
−20 to −16	|||| |||| |||| |||| |||| ||||		31
−15 to −11	|||| |||| |||| |||| |||| ||	27
−10 to −6	|||| |||| |||| |||| |||| ||||		31
−5 to −1	|||| |||| |||| |||| ||	22
0 to 4	|||| |||| |||| |||	18
5 to 9	|||| ||	7
10 to 14	||||	4
15 to 19	||	2
20 to 24			1
25 to 29			1
30 to 34			1
	Total	171

tive), they cannot land exactly on a bin boundary. Again, the bins don't overlap.

Class mark

bin mark   The **bin mark** is the middle value in that bin, or the average of the lower and the upper limits. For the last bin in Table 2.4, the lower and upper limits have $(30 + 34)/2$, or 32, as their average, and this is the mark for that bin. The bin mark serves as a typical value, or representative value, for that bin. The **bin interval** is the distance from one bin mark to the next. In Table 2.4 it is 5 ($5 = 32 - 27 = 27 - 22$, etc.). It is also the distance from each bin boundary to the next, and it is the distance from one lower limit to the lower limit of the following bin.

bin interval

bin frequency   The number of observations in any bin is the **bin frequency**. It is some-
relative   times convenient to use a **relative frequency** distribution. The relative bin
frequency   frequencies are the bin frequencies divided by the total number of observations. A *percentage distribution* is obtained by multiplying the relative bin frequencies by 100 to convert them to percentages. For example, the bin "0 to 4" in Table 2.4 contains 18 values, and so the relative frequency is $18/171$, or .105, and the percentage is $.105 \times 100$, or 10.5.

Table 2.5 gives the frequency distribution of the changes in blood pressure for the 176 patients in the placebo group. It now becomes clearer that

Table 2.4 Change in Blood Pressure: Active Group (171 Patients)

Bin Limits	Bin Boundaries	Bin Mark	Bin Frequency	Relative Frequency	Percentage of Observations
−35 to −31	−35.5 to −30.5	−33	7	.041	4.1
−30 to −26	−30.5 to −25.5	−28	7	.041	4.1
−25 to −21	−25.5 to −20.5	−23	12	.070	7.0
−20 to −16	−20.5 to −15.5	−18	31	.181	18.1
−15 to −11	−15.5 to −10.5	−13	27	.158	15.8
−10 to −6	−10.5 to −5.5	−8	31	.181	18.1
−5 to −1	−5.5 to −0.5	−3	22	.129	12.9
0 to 4	−0.5 to 4.5	2	18	.105	10.5
5 to 9	4.5 to 9.5	7	7	.041	4.1
10 to 14	9.5 to 14.5	12	4	.023	2.3
15 to 19	14.5 to 19.5	17	2	.012	1.2
20 to 24	19.5 to 24.5	22	1	.006	0.6
25 to 29	24.5 to 29.5	27	1	.006	0.6
30 to 34	29.5 to 34.5	32	1	.006	0.6
		Totals	171	1.000	100.0

the drug does tend to reduce blood pressure. In the active group, most patients showed a negative blood-pressure change—a decrease. In the placebo group, there is about an even split between positive and negative changes—between increases and decreases. In Section 2.3 it is shown how to make the information in the data still clearer by means of graphs called histograms.

Many kinds of data are sorted into frequency distributions. In constructing a frequency distribution, the first question is the *number of bins* to use. One rule that has been advanced is that of Sturges (see Reference 2.3 at the end of the chapter). The rule is this: enclose the number of observations between successive powers of 2, and take the number of classes to be the exponent of the larger power of 2 (see Table 2.6). Suppose, for example, that there are 57 observations; since $2^5 = 32 < 57 \leq 2^6 = 64$, the rule says to use about six bins.

This is *not* a hard-and fast rule. It is only a guide, a rule of thumb. Since the number of patients in the active group was 171, the rule says to use 8 bins, but we actually used 14 of them. For more on how to choose the number of bins, see Section 2.3.

The next problem is to choose the *bin interval*. Remember that this is

SECTION 2.1 FREQUENCY DISTRIBUTIONS

Table 2.5 Change in Blood Pressure: Placebo Group (176 Patients)

Bin Limits	Bin Boundaries	Bin Mark	Bin Frequency	Relative Frequency	Percentage of Observations
−35 to −31	−35.5 to −30.5	−33	1	.006	0.6
−30 to −26	−30.5 to −25.5	−28	1	.006	0.6
−25 to −21	−25.5 to −20.5	−23	3	.017	1.7
−20 to −16	−20.5 to −15.5	−18	11	.062	6.2
−15 to −11	−15.5 to −10.5	−13	12	.068	6.8
−10 to −6	−10.5 to −5.5	−8	25	.142	14.2
−5 to −1	−5.5 to −0.5	−3	25	.142	14.2
0 to 4	−0.5 to 4.5	2	38	.216	21.6
5 to 9	4.5 to 9.5	7	24	.136	13.6
10 to 14	9.5 to 14.5	12	12	.068	6.8
15 to 19	14.5 to 19.5	17	12	.068	6.8
20 to 24	19.5 to 24.5	22	12	.068	6.8
25 to 29	24.5 to 29.5	27	0	.000	0.0
30 to 34	29.5 to 34.5	32	0	.000	0.0
		Totals	176	1.000	100.0

Table 2.6 Sturges's Rule for the Number of Bins

Number of Observations	Approximate Number of Bins
Between $2^3 =$ 8 and $2^4 =$ 16	4
$2^4 =$ 16 and $2^5 =$ 32	5
$2^5 =$ 32 and $2^6 =$ 64	6
$2^6 =$ 64 and $2^7 =$ 128	7
$2^7 =$ 128 and $2^8 =$ 256	8
$2^8 =$ 256 and $2^9 =$ 512	9
etc.	etc.

the distance from the lower limit of one bin to the lower limit of the next bin. To get a preliminary idea of what bin interval to use, find the largest and the smallest observations, difference them, and divide by the number of bins. In Table 2.1, the largest observation is 31 and the smallest is −34; the distance from −34 to 31 is 65 (because 31 − (−34) = 31 + 34 = 65), and 65 divided by the number of bins, 14, is 65/14, or 4.64. If the 14 bins are to cover the range from −34 to 31, the distance from one lower boundary to the next should be about 4.64. But a bin interval of 4.64 would be a nuisance, and it is much easier to use a bin interval of 5. And now, if one bin goes from 0 through 4, the next will go from 5 through 9 (0 + 5 = 5 and 4 + 5 = 9), the one after that will go from 10 through 14 (5 + 5 = 10 and 9 + 5 = 14), and so on. (You could use one bin from 3 through 7, the next from 8 through 12, and so on, but the bins used in Table 2.3 fit in more naturally with a bin interval of 5.)

A further example will help to fix these ideas. Suppose that we have 87 numerical observations from an agricultural experiment. According to Sturges's rule, we need something like 7 bins. Suppose that the observations are the amounts, in parts per million, of a nitrogen compound in 87

Table 2.7 Waiting Times between Failures of Equipment

Times between Failures (hours)	Number of Failures
0 to 49	3
50 to 99	7
100 to 149	13
150 to 199	18
200 to 249	22
250 to 299	21
300 to 349	12
350 to 399	8
400 to 449	0
450 to 499	0
500 to 549	0
550 to 599	0
600 to 649	0
650 to 699	1
Total	105

SECTION 2.1 FREQUENCY DISTRIBUTIONS

Table 2.8 Waiting Times between Failures of Equipment

Times between Failures (hours)	Number of Failures
0 to 49	3
50 to 99	7
100 to 149	13
150 to 199	18
200 to 249	22
250 to 299	21
300 to 349	12
350 to 399	8
400 or more	1
Total	105

soil samples, and suppose that the smallest is 2.76 and the largest is 5.63. The distance from smallest to largest is 5.63 − 2.76, or 2.87, and this difference divided by 7, the number of bins, is 0.41. A convenient bin interval near to 0.41 is 0.50, and convenient bins are

2.50 to 2.99, 3.00 to 3.49, 3.50 to 3.99, 4.00 to 4.49, 4.50 to 4.99, 5.00 to 5.49, 5.50 to 5.99.

These figures are the bin *limits,* and they are given to two decimal places because the observations are accurate to two decimal places. The *boundary* between the first and second bins is the number half way between 2.99 and 3.00, which is 2.995. Since this boundary has three decimal places, it can't itself occur as the value of an observation.

Sometimes it happens that the data contain a few observations whose numerical values are much smaller or much larger than the rest. If we include these values in the ordinary way, we may find that some of the bins are empty. The frequency distribution for the waiting times between failures of an airborne radio device is given in Table 2.7. Because of one extreme value, five of the bin frequencies are equal to zero.

We can get a clearer picture of the time to failure if we use an *open-ended interval,* one that has no upper limit, for the last class. Table 2.8 gives the resulting distribution. Open-ended intervals permit the inclusion of a wide range of extreme values within a relatively small number of bins, but the actual numerical values of the extreme observations are lost unless indicated in a footnote or elsewhere. Open-ended intervals also present difficulties in the calculation of some of the descriptive measures of the next chapter.

Problems

1. Use Sturges's rule for the approximate number of classes (or bins).
   a. Find the number of classes appropriate for a frequency distribution for 60 observations.
   b. Do the same for 500 observations.
   c. Find the range of the number of observations that would correspond to ten classes.
2. If the class limits in a frequency distribution are -10 to -3, -2 to 5, 6 to 13, and 14 to 21, what are the boundaries of the classes? What are the class marks?
3. Given 260 numbers such that the smallest is 103.3 and the largest is 190.7, set up a table that would be suitable for grouping the data. Give the class marks as well as the boundaries.
4. A set of measurements to the nearest hundredth of an inch is summarized in a table with these class boundaries: 74.995, 124.995, 174.995, 224.995, 274.995, and 324.995. What are the limits for the five classes of this distribution?
5. What are the class limits and the class boundaries of a distribution of integer numbers if the class marks are 8, 33, 58, 83, and 108?

6. For each of 66 prisoners the table below gives the observed time lapse in months between prison admission and the filing of a motion to review sentence. Construct the frequency distribution.

11	25	0	14	22	32
36	0	10	35	14	20
24	18	19	5	15	4
2	4	34	36	19	16
17	21	14	16	0	30
13	29	11	6	25	9
9	22	32	29	28	44
31	16	52	5	9	7
24	27	19	14	58	29
3	19	25	32	16	26
29	5	17	20	8	14

SECTION 2.2 CUMULATIVE FREQUENCY DISTRIBUTIONS 21

7. The number of automobiles per day entering an airport parking lot was recorded for a period of 30 days. Construct the frequency distribution of the data:

1172	1289	1024	1004	1143
830	1176	1051	995	817
1142	1031	949	1217	998
941	1037	1166	796	1122
1005	913	895	819	947
1049	893	1138	848	748

8. An *outlier* is an observation that is far removed from the rest of the values in a set of data. Outliers usually are the result of errors of measurement or observation. What are some reasons why they might occur in actual data?

9. The following data are the average September temperatures of a midwestern city for the last 29 years. Using a frequency distribution as an aid, find the three outliers in the data (see Problem 8):

58.9	50.7	53.2	60.7	67.1	59.1	54.5	47.8
33.5	53.3	52.7	57.0	59.6	53.6	55.3	51.0
54.2	34.1	58.2	49.0	55.4	51.8	56.2	53.8
55.4	56.6	52.6	58.8	54.0			

2.2 Cumulative Frequency Distributions

cumulative frequency distribution

Frequency distributions as described in Section 2.1 are valuable for organizing and summarizing sets of data. Sometimes we require the number of observations less than a given value; this information is contained in the **cumulative frequency distribution.**

Table 2.3 shows that no blood-pressure change was below −35, 7 of the changes were less than −30, 14 of them (7 + 7) were less than −25, 26 of them (7 + 7 + 12) were less than −20, and so on. To get the cumulative frequency of observations less than the lower limit or boundary of a specified bin, we add the bin frequencies of the preceding bins. Table

Table 2.9 Cumulative Distribution of Change in Blood Pressure: Active Group (171 Patients)

Change Less than	Number of Observations
−35	0
−30	7
−25	14
−20	26
−15	57
−10	84
−5	115
0	137
5	155
10	162
15	166
20	168
25	169
30	170
35	171

2.9 gives the completed cumulative distribution for the patients in the active group.

This table contains additional information: The number of observations equal to or greater than 5 is 171 − 155, or 16. The number of observations that are at least 5 but are less than 20 is 168 − 155, or 13.

Cumulative distributions may be constructed for relative frequencies and percentages as well as for absolute frequencies. The procedures are the same except that we add the relative frequencies or percentages, as the case may be, instead of the absolute frequencies.

Problems

10. Given the following frequency distribution,
    a. Construct the percentage distribution.

SECTION 2.2 CUMULATIVE FREQUENCY DISTRIBUTIONS

b. Construct the cumulative frequency distribution.
c. Construct the cumulative percentage distribution.

Class	Frequency
10–24	15
25–39	25
40–54	42
55–69	50
70–84	38
85–99	30

11. The table below gives the cumulative percentage distribution for 400 observations.
   a. Derive the cumulative frequency distribution.
   b. Derive the percentage distribution.
   c. How many observations were there in the 41–55 class?

Less than	Percentage of Observations
6	0
16	13
26	35
41	56
56	72
76	90
101	100

12. The points scored by the Lincoln High and Jefferson High basketball teams are given below.
   a. Construct the cumulative frequency distributions.
   b. Which team is generally the higher scoring?

Lincoln High					Jefferson High				
44	50	38	43	48	44	50	59	44	48
44	49	48	46	50	47	55	49	45	44
49	34	39	49	39	43	53	58	48	54

24　CHAPTER 2　EMPIRICAL FREQUENCY DISTRIBUTIONS

13. The daily sales of a small business, in hundreds of dollars, are given below for a 20-day period. Construct a cumulative frequency distribution.

8.2	11.5	12.1	7.3	14.8	9.4	8.3	6.1
9.7	8.4	10.1	12.3	10.2	9.5	8.7	10.3
13.4	9.8	10.9	15.1				

14. Construct a cumulative frequency distribution for the data in Problem 9. Reconstruct the cumulative frequency distribution with the outliers removed from the data set.

15. The instructor in Computer Science 101 recorded the amount of computer time (to the nearest tenth of a second) needed by each student to complete the first assignment. The data are summarized in the cumulative frequency distribution below. Construct the frequency distribution.

Less than	Cumulative Frequency
.4	3
.8	13
1.2	29
1.6	38
2.0	43

2.3　Graphic Presentation

Frequency distributions are informative, but pictorial representations of the same information often make important characteristics more immediately apparent. Here we take up the two most basic pictorial forms—histograms and ogives. Horizontal bar charts, compound bar charts, pictographs, and pie diagrams are mostly adaptations of these basic forms (see Reference 2.2).

histogram　　A **histogram** is a graphic representation of a frequency distribution. It is constructed by erecting bars or rectangles over the bins. Along the horizontal scale we mark off the bin boundaries. Along the vertical scale we mark off frequencies. We erect over each bin a rectangle whose area is proportional to the frequency for that bin.

Figures 2.1 and 2.2 show the histograms of the blood-pressure-change data for the active and the placebo groups. They show clearly that the distribution for the active group is shifted to the left in comparison with the distribution for the placebo group. The distribution for the active group, as compared with that for the placebo group, is shifted in the negative direction, the direction of decreased blood pressure. In Chapter 3 we study how to describe this shift in the histogram by means of numerical characteristics of the frequency distribution.

The bin interval used in constructing a frequency table has a marked effect on the appearance of the histogram. In Table 2.4 the blood-pressure data in Table 2.1 are sorted into bins of width 5. If a bin interval of 2 or of 15 is used instead, the frequency distributions in Table 2.10 result.

The top and bottom histograms in Figure 2.3 come from the two frequency distributions in Table 2.10, and the middle histogram is the same as Figure 2.1. The horizontal scale is the same in each case. When the raw data are grouped into bins, a certain amount of information is lost, since no distinction is made between observations falling into the same bin. The larger the bin interval is, the greater is the amount of information lost. For the blood-pressure data a bin interval of 15 is so large that the corresponding histogram gives very little idea of the shape of the distribution. A class interval of 2, on the other hand, gives a ragged-looking histogram. Little information has been lost in the bottom histogram in Figure 2.3, but the presentation of information is somewhat misleading because conspicuous irregularities in the histogram merely reflect the accidents of sampling. A bin interval of 5 represents a reasonable compromise in this case. In the choice of a class interval there is a trade-off between information content and smoothness of the histogram.

Figure 2.4 shows the histogram for the times between failures given in Table 2.8. Observe that although the open class in this distribution can't be represented by a bar in the histogram, we indicate by a note in the histogram that there was an extreme value.

It is not absolutely necessary that the bins all have the same width. If the last seven bins in Table 2.4 are combined, Table 2.11 results. The first seven bins are as before, and there is one large bin, with limits 0 and 34, which contains 34 of the observations. One advantage of using equal bin intervals is that in that case the areas of the bars in the histogram are proportional to the heights of the bars. We can therefore draw the bars so that height is proportional to bin frequency. This is not the correct way to draw the histogram if the bin intervals are unequal. The rule for drawing a histogram is that the *area* of the bar over a bin must be proportional to the frequency of the bin. In the case of equal bin intervals, this is the same

Figure 2.1

Histogram of change in blood pressure for 171 patients on active drug (from Table 2.4)

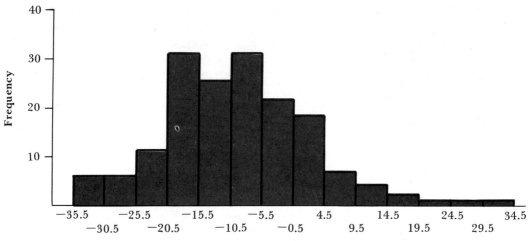

Figure 2.2

Histogram of change in blood pressure for 176 patients on placebo (from Table 2.5)

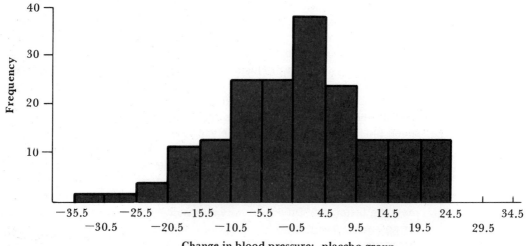

SECTION 2.3 GRAPHIC PRESENTATION

Table 2.10 Change in Blood Pressure: Active Group (171 Patients)

Bin Interval = 2				Bin Interval = 15	
Bin Limits	Number of Observations	Bin Limits	Number of Observations	Bin Limits	Number of Observations
−34 to −33	3	0 to 1	11	−35 to −21	26
−32 to −31	4	2 to 3	1	−20 to −6	89
−30 to −29	3	4 to 5	7	−5 to 9	47
−28 to −27	2	6 to 7	3	10 to 24	7
−26 to −25	3	8 to 9	3	25 to 39	2
−24 to −23	6	10 to 11	4	Total	171
−22 to −21	5	12 to 13	0		
−20 to −19	11	14 to 15	0		
−18 to −17	13	16 to 17	1		
−16 to −15	11	18 to 19	1		
−14 to −13	10	20 to 21	1		
−12 to −11	13	22 to 23	0		
−10 to −9	13	24 to 25	1		
−8 to −7	11	26 to 27	0		
−6 to −5	14	28 to 29	0		
−4 to −3	6	30 to 31	1		
−2 to −1	9	Total	171		

thing as drawing the height of the bar proportional to the bin frequency. Not so if the bin intervals are unequal. If the histogram for the distribution in Table 2.11 were drawn with height proportional to frequency, we would obtain the result shown in Figure 2.5. Here, because we have not followed the area rule, the thirty-four values in the range 0 to 34 are given an undue emphasis, and the histogram gives the impression that a large proportion of the values (blood-pressure changes) are positive. The correct representation is that in Figure 2.6, where the areas of the bars are proportional to the frequencies: the height of the rightmost bar is one-seventh what it was in Figure 2.5 because it is seven times as wide as the other bars.

ogive   Just as a frequency distribution can be represented graphically by a histogram, a cumulative frequency distribution can be represented graphically by an **ogive**. To construct an ogive we first lay out the bin boundaries on the horizontal scale, just as for a histogram. Above each bin boundary we plot a point at a vertical distance proportional to the cumulative fre-

Figure 2.3

Histograms of blood pressure change: active drug

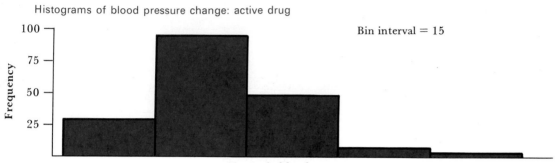

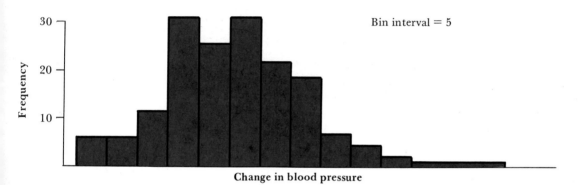

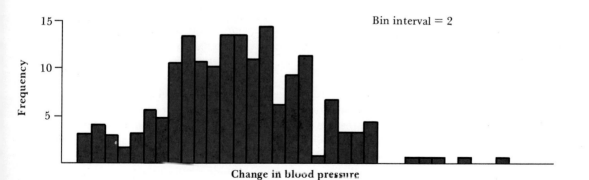

SECTION 2.3 GRAPHIC PRESENTATION   29

Figure 2.4

Histogram for times between failures of airborne radio equipment

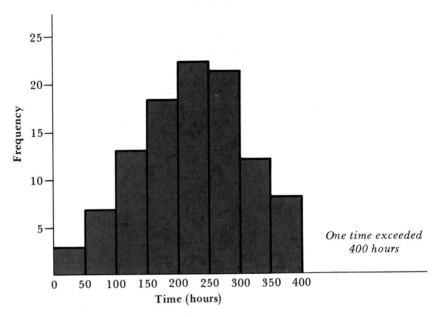

One time exceeded 400 hours

Table 2.11 Distribution of Change in Blood Pressure (Active Group) with the Last Seven Bins Combined

Bin Limits	Number of Observations
−35 to −31	7
−30 to −26	7
−25 to −21	12
−20 to −16	31
−15 to −11	27
−10 to −6	31
−5 to −1	22
0 to 34	34
Total	171

Figure 2.5

Incorrectly drawn histogram for Table 2.11

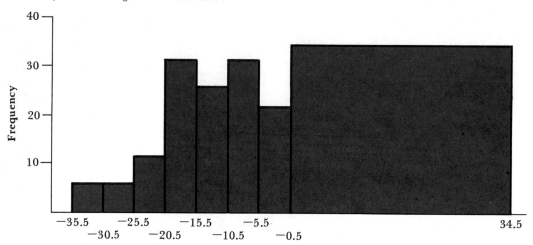

quency—proportional, in other words, to the number of observations whose numerical value is less than that bin boundary. These points are then connected by straight lines.

Using Table 2.9 in this way leads to Figure 2.7. Since there are 7 observations less than -30.5, the point above -30.5 on the horizontal

Figure 2.6

Correctly drawn histogram for Table 2.11

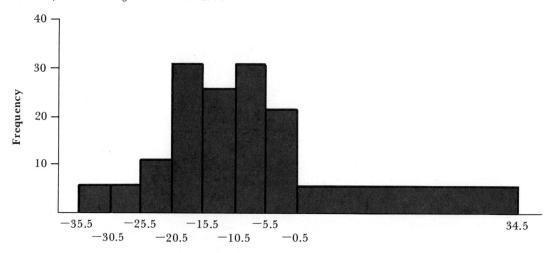

SECTION 2.3 GRAPHIC PRESENTATION

Figure 2.7

Ogive for the cumulative distribution of blood-pressure changes: active group (from Table 2.9)

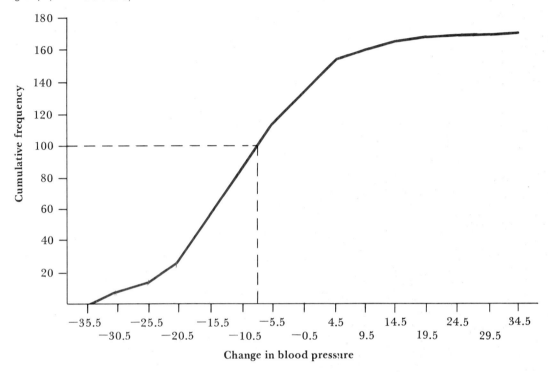

scale is at height 7 on the vertical scale. Since there are 14 observations less than -25.5, the point above -25.5 on the horizontal scale is at height 14 on the vertical scale. And so on.

We can interpolate graphically on an ogive. From Figure 2.7 we get an approximation to the number of observations less than -8 by finding the height of the curve over that point—see the dashed line in the figure. Thus we estimate that about 100 (or about 58%) of the blood-pressure changes in the active-drug group are less than -8.

Problems

16. As an experiment, toss five coins twenty times, record the number of heads obtained each time, and construct the frequency distribution for the twenty resulting numbers. Draw the histogram and the ogive.

17. For the data in Problem 6,
    a. Construct the cumulative distribution.
    b. Draw the histogram and ogive.
18. Draw the histogram and ogive for
    a. The data in Problem 7.
    b. The distribution in Problem 10.
    c. The distribution in Problem 11.
19. For the following histogram (the numbers in the bars are the frequencies),
    a. What are the boundaries and class marks?
    b. Draw the ogive.

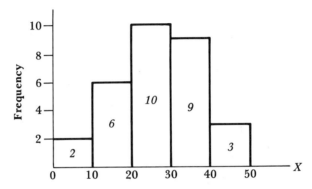

20. In a survey of legal needs, 2064 respondents were asked the number of different legal problems they had encountered in the previous five years. The following table gives the distribution of the number of problems.
    a. What percentage of people encountered no legal problems?
    b. What percentage of people encountered more than one problem?
    c. What percentage of people encountered exactly one problem?
    d. Draw the histogram and ogive.

Number of Problems	Frequency	Number of Problems	Frequency
0	498	7	14
1	645	8	5
2	375	9	5
3	262	10	2
4	161	11	2
5	56	12	1
6	38		

SECTION 2.3 GRAPHIC PRESENTATION

21. The following data are the differences in cutoff voltages between the two halves of 48 dual-triode electron tubes of a given type:

−.30	.24	.25	−.10	.53	−.16
−.37	−.83	.48	1.25	.67	.22
.28	−.15	−2.02	−1.22	.65	−.08
−.34	−.52	1.55	−.18	.15	−.12
.86	.06	.56	−.88	−1.16	−1.02
2.07	1.02	−1.43	.72	.08	1.10
−.75	−.32	.08	−.61	−.25	−1.35
.02	−.05	.61	−.80	.04	−.22

a. Construct a frequency distribution. Between −1 and 1 use intervals equal to ½ volt. Below −1 and above 1 use intervals equal to 1 volt.
b. Construct the relative frequency distribution and the percentage distribution.
c. Construct the cumulative relative frequency distribution.
d. Draw the histogram and the ogive.
e. From the ogive, estimate the proportion of tubes of this type with difference in cutoff less than zero. Greater than .50. Less than −1.

22. In a prison study 45 prisoners were found to have spent at least one day in isolation. The following is a computer printout of the number of days in isolation.
a. Summarize the data in a five-class frequency distribution.
b. Derive the cumulative percentage distribution.

Number of Days	Frequency	Number of Days	Frequency	Number of Days	Frequency
1	2	14	6	27	1
2	1	15	4	34	1
3	2	17	1	37	2
4	1	18	2	39	1
5	2	21	2	42	1
6	3	22	1	50	1
7	1	23	1	53	2
8	2	25	1	55	1
9	1	26	2		

c. Draw the histogram and ogive.
d. Among prisoners who spent at least one day in isolation what proportion spent more than ten days in isolation?

23. The following data are the amounts of impurities in percent found in 96 samples of a certain compound.
   a. Construct a frequency distribution.
   b. Construct the cumulative percentage distribution.
   c. Sketch the histogram and the ogive.

7.78	6.02	9.68	6.14	7.28	6.23	4.64	7.41
6.75	4.51	8.42	7.92	6.84	5.81	4.67	7.38
5.43	11.28	7.96	6.38	5.92	4.80	10.72	6.17
6.17	9.54	7.82	5.33	5.79	4.76	9.78	5.84
6.00	5.23	6.48	6.53	5.46	4.83	9.51	6.82
7.26	10.90	10.82	9.81	10.73	10.46	10.63	7.79
6.54	9.72	9.74	10.62	4.87	5.23	4.68	7.19
5.28	10.62	7.24	8.38	5.05	4.98	8.62	6.76
5.03	8.41	8.65	8.37	8.46	7.93	7.19	6.53
4.29	8.62	8.50	8.83	7.78	7.64	5.39	7.21
5.31	7.76	6.42	5.13	6.49	6.31	7.72	4.15
4.97	6.13	5.29	6.34	7.82	6.68	6.44	4.68

24. In a market survey conducted at a community college, 45 students were asked to give the number of times in the last month they visited the Burger Ranch restaurant. Make a histogram of the data:

3	1	3	1	3	1	0	1	2
3	1	4	4	10	7	0	4	4
2	2	5	0	1	4	0	2	2
3	2	0	1	5	0	0	1	2
1	2	5	2	5	2	0	4	2

25. Occupancy rates of the 28 largest apartment complexes in Sycamore County are given below.
   a. Draw the histogram and ogive.

91	83	86	87	90	91	86	90	94	91
89	90	96	87	90	91	89	93	86	95
87	83	88	94	84	93	84	81		

SECTION 2.4 STEM-AND-LEAF PLOTS

b. What percentage of the apartment complexes have more than 90% occupancy?

26. Thirty-five volunteers were asked to complete a task related to hand-eye coordination (times are in seconds). Make a histogram of the data using unequal intervals (see Figure 2.6).

8	6	18	39	7	10	5	8	6	14	11	10	8	60
6	6	14	15	6	7	6	5	8	11		8	15	8	8
7	8	8	6	29	5	7							

27. Levels at Lake Fritch were measured monthly for a 27-year period. Data are summarized by the ogive pictured below.
  a. What percentage of the time was the lake level above 24 feet?
  b. What percentage of the time was the lake level in the environmentally safe range of 23 to 27 feet?

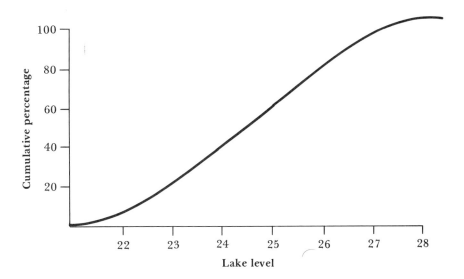

2.4 Stem-and-Leaf Plots

stem-and-leaf plot

A **stem-and-leaf plot** is a quick, simple, and often effective way to display data. Suppose 27 students in an examination got the scores shown in Table 2.12.

Table 2.12 Examination Scores

83	69	82	72	63	88	92	81	54
57	79	84	99	74	85	71	94	71
80	51	68	81	84	92	63	99	91

The stem-and-leaf plot, Table 2.13, is obtained by sorting the data according to the following rules. The digits 0 through 9 are first listed to the left of a vertical line. The first score in Table 2.12 is 83, so a 3 is entered in the 8-row (to the right of the vertical line, opposite the 8); the second score is 69, so a 9 is entered in the 6-row; the third score is 82, so a 2 is entered to the right of the 3 in the 8-row; and so on.

Table 2.13 Stem-and-Leaf Plot for the Data in Table 2.12

```
0 |
1 |
2 |
3 |
4 |
5 | 471
6 | 9383
7 | 29411
8 | 328145014
9 | 294291
```

The stem-and-leaf plot is in effect a histogram for a class interval of 10. If the plot is rotated 90 degrees counterclockwise, the result is a histogram with stacks of digits in the role of the usual vertical bars. No information has been lost, of course.

Problems

28. Draw a stem-and-leaf plot of the first 33 observations in Problem 6.
29. Draw a stem-and-leaf plot of the data in Problem 22.

SECTION 2.5 SAMPLES AND POPULATIONS

30. Grade-point averages for 32 freshmen are given below. Make a stem-and-leaf plot.

3.0	3.5	3.3	3.4	3.4	2.4	1.5	1.4
1.6	3.1	2.6	3.6	3.0	2.5	3.2	2.5
3.1	3.8	1.6	2.2	2.7	2.5	3.0	2.5
2.4	2.8	3.2	1.9	2.5	3.5	2.9	4.0

2.5 Samples and Populations

It must be stressed that the material of this and the next chapter concerns *empirical data*—data obtained by selecting a sample or performing an experiment. The actual data in hand are usually only a very small part of some much larger whole called a *population*. The 105 failure times recorded in Table 2.8, for example, constitute only a fraction of the totality of failure times for all radio equipment of that kind; they are a sample of 105 failure times from the population of all failure times.

Conceptually, we could construct a frequency distribution for the totality of possible values which might have been obtained. Such conceptual distributions we refer to as *theoretical* or *population* distributions; they are to be distinguished from the *empirical* distributions of this chapter and the next. Theoretical distributions are discussed in detail in Chapter 5. An empirical distribution *estimates* the corresponding theoretical distribution.

Chapter Problems

31. Select 100 stocks randomly from the financial section of a newspaper and make a histogram of the change in prices. On the basis of your data, would you say that it was a good day for investors?

32. After an advertising campaign was conducted to promote the Burger Ranch restaurant (Problem 24), another survey was conducted to determine the frequency with which students visit the restaurant. Comparing the data from the second survey, which are given below, with the data from the first, would you say that the advertising campaign was effective?

Less than	Cumulative Percentage
1	10
2	26
3	52
4	72
5	86
6	96
7	99
8	99
9	100
10	100

33. Studying the tendency of males to gain weight between the ages of 25 and 35, a physician examined the medical records of 24 patients and found the following data. Use descriptive methods of your choice to present a statistical summary of the data.

No.	Age 20	Age 35	No.	Age 20	Age 35
1	144	160	13	150	163
2	142	153	14	144	151
3	163	169	15	145	154
4	173	184	16	166	176
5	139	154	17	148	163
6	146	157	18	155	160
7	142	147	19	165	179
8	140	150	20	149	157
9	164	172	21	143	159
10	173	183	22	142	147
11	155	163	23	171	180
12	124	139	24	146	160

34. The cranial widths in millimeters of 106 male skulls unearthed at a site in London are given below.
    a. Make a histogram of the data.
    b. The cranial widths of 4 male skulls unearthed at another site are 120, 115, 124, 113. Are these data typical of those given below?
    c. What might we infer about the differences in physical characteristics of the former inhabitants of the two sites?

CHAPTER PROBLEMS

```
146  141  139  140  145  141  142  131  142  140
144  140  138  139  147  139  141  137  141  132
140  140  141  143  134  146  134  142  133  149
140  143  143  149  136  141  143  143  141  140
138  136  138  144  136  145  143  137  142  146
140  148  140  140  139  139  144  138  146  153
148  142  133  140  141  145  148  139  136  141
140  139  158  135  132  148  142  145  145  121
129  143  148  138  149  146  141  142  144  137
153  148  144  138  150  148  138  145  145  142
143  143  148  141  145  141
```

35. A survey of twenty-five economists was taken to get their opinions on the chance of a recession next year. The survey was repeated after three months. Each opinion was expressed as a percentage indicating the perceived chance of a recession.

  a. Provide a statistical summary of the data.

  b. Had the collective opinion changed in the time between the two surveys?

First Survey					Second Survey				
20	10	40	50	10	10	20	40	40	50
40	20	30	10	10	20	30	40	50	50
20	30	30	20	10	30	40	50	20	30
10	20	20	30	20	40	30	40	30	40
20	30	20	10	30	10	30	50	40	20

36. Data were gathered from homeowners on the percentage of income spent on housing. Data are given below in "percentage more than."

Percentage of Income: More than	Percentage of Homeowners
15	100
20	95
25	50
30	24
35	10
40	0

a. What percentage of homeowners spend between 26 and 35% of their income on housing?
b. Construct a cumulative percentage distribution.

Summary and Keywords

To make the information in statistical data stand out clearly, the values are usually sorted into classes or **bins** (p. 12) to form a **frequency distribution** (p. 12). A bin is specified by the **bin limits** (p. 14) or the **bin boundaries** (p. 14) and is represented by the **bin mark** (p. 15); the distance from one bin mark to the next is the **bin interval** (p. 15), and the number of values falling in a bin is the **bin frequency** (p. 15). The **relative frequency** (p. 15) is the frequency divided by the total number of observations. The frequency distribution is represented graphically by the **histogram** (p. 24). The sums of the bin frequencies up to the various boundaries form the **cumulative frequency distribution** (p. 21), which has the **ogive** (p. 27) as its graphical counterpart. Some data sets are very easily sorted by means of a **stem-and-leaf plot** (p. 35).

REFERENCES

2.1 Huff, Darrell. *How to Lie with Statistics*. New York: Norton, 1965. Chapters 5 and 6.
2.2 Neter, John, and Wasserman, William. *Fundamental Statistics for Business and Economics*. 2d ed. Boston: Allyn and Bacon, 1961. Chapters 4 and 5.
2.3 Sturges, H. A. "The Choice of a Class Interval." *Journal of the American Statistical Association,* March 1926.
2.4 Tukey, John W. *Exploratory Data Analysis*. Reading, Mass.: Addison-Wesley, 1977.

3

Descriptive Measures

3.1 Introduction

In the last chapter we saw how tabular and graphical forms of presentation may be used to summarize and describe quantitative data. Though these techniques make clear important features of the distribution of the data, statistical methods for the most part require concise numerical descriptions. These are arrived at through arithmetic operations on the data which yield **descriptive measures,** or **descriptive statistics.** The basic descriptive statistics are the measures of central tendency (location) and the measures of dispersion (or scatter).

descriptive measures
descriptive statistics

Discussion of these descriptive measures requires some familiarity with the mathematical shorthand used to express them. Since we are concerned with masses of data and since the operation of addition plays a large role in our calculations, we need a way to express sums in compact and simple form. The summation notation meets this requirement.

3.2 Symbols and Summation Notation

Our data consist of measurements of some characteristic of a number of people or items. They might be blood pressures of patients, as in the Public Health Service clinical trials, or they might be the annual incomes of a group of families, or the weights of a number of pigs. We designate the characteristic of interest by some letter or symbol, say, X. If we have measured two or more characteristics, we use a different letter or symbol for each. If in addition to obtaining annual income we also record the age of each person interviewed, we could represent income by X and age by Y. A third characteristic, such as educational level, we could represent by the letter Z, and so on.

In order to differentiate between the same kind of measurements made on different items or individuals, or between similar repeated measurements made on the same element, we add a subscript to the corresponding symbol; thus X_1 stands for the income of the first person interviewed, X_2 for that of the second, X_{23} for that of the twenty-third, and so on. In general, any arbitrary observed value would be represented by X_i, where the subscript i is variable in the sense that it represents any one of the observed items and need only be replaced by the proper number in order to specify a particular observation. The income, age, and educational level of the ith, or general, individual would be represented by X_i, Y_i, and Z_i, respectively.

44

SECTION 3.2 SYMBOLS AND SUMMATION NOTATION

Given a set of n observations which we represent by $X_1, X_2, X_3, \ldots, X_n$, we can express their sum as

$$\sum_{i=1}^{n} X_i = X_1 + X_2 + X_3 + \cdots + X_n$$

summation symbol

where Σ (uppercase Greek sigma) is the **summation symbol,** the subscript i is the *index of summation,* and the 1 and n that appear respectively below and above the symbol Σ designate the *range* of the summation. The combined expression says, "add all Xs whose subscripts are between 1 and n, inclusive." In place of i we sometimes use j as the summation index; any letter will do.

If we want the sum of the squares of the n observations, we write

$$\sum_{i=1}^{n} X_i^2 = X_1^2 + X_2^2 + X_3^2 + \cdots + X_n^2$$

which says, "add the squares of all observations whose subscripts are between 1 and n, inclusive." The sum of the products of two variables X and Y would be written as

$$\sum_{i=1}^{n} X_i Y_i = X_1 Y_1 + X_2 Y_2 + X_3 Y_3 + \cdots + X_n Y_n$$

If we want a partial sum, the sum of some but not all the quantities involved, the range of the summation is adjusted accordingly. For example,

$$\sum_{i=4}^{8} Y_i^2 = Y_4^2 + Y_5^2 + Y_6^2 + Y_7^2 + Y_8^2$$

$$\sum_{i=2}^{5} Y_i^2 f_i = Y_2^2 f_2 + Y_3^2 f_3 + Y_4^2 f_4 + Y_5^2 f_5$$

and

$$\sum_{i=1}^{3} (Y_i - X_i) = (Y_1 - X_1) + (Y_2 - X_2) + (Y_3 - X_3)$$

In those cases where the context makes it clear that *the sum is to be taken over all the data* we sometimes simplify the summation notation by omitting the range of summation. Thus ΣX_i is the sum of all the numbers and ΣX_i^2 is the sum of the squares of all the numbers.

We now look at some of the algebraic rules that apply to summations.

Rule 1 The summation of a sum (or difference) is the sum (or difference) of the summations:

$$\sum_{i=1}^{n}(X_i + Y_i - Z_i) = \sum_{i=1}^{n} X_i + \sum_{i=1}^{n} Y_i - \sum_{i=1}^{n} Z_i \qquad 3.1$$

Rule 2 The summation of the product of a variable and a constant is the product of the constant and the summation of the variable:

$$\sum_{i=1}^{n} cY_i = c \sum_{i=1}^{n} Y_i \qquad 3.2$$

Rule 3 The summation of a constant is the constant multiplied by the number of terms in the summation:

$$\sum_{i=1}^{n} c = nc \quad \text{(there are } n \text{ terms)} \qquad 3.3$$

Going through numerical examples will show how these rules work. Suppose that $n = 3$ and that

$$X_1 = 7, \quad X_2 = -5, \quad X_3 = 7,$$
$$Y_1 = 9, \quad Y_2 = 11, \quad Y_3 = 1.5$$

Note that $X_1 = X_3$ in this case; the X_i need not be distinct from each other (or from the Y_i). Now

$$\sum_{i=1}^{3} X_i = 7 + (-5) + 7 = 9$$

and

$$\sum_{i=1}^{3} Y_i = 9 + 11 + 1.5 = 21.5$$

Further,

$$X_1 + Y_1 = 16, \quad X_2 + Y_2 = 6, \quad X_3 + Y_3 = 8.5$$

and so

$$\sum_{i=1}^{3}(X_i + Y_i) = 16 + 6 + 8.5 = 30.5$$

Since $30.5 = 9 + 21.5$, we see that ΣX_i, ΣY_i, and $\Sigma(X_i + Y_i)$ are re-

SECTION 3.2 SYMBOLS AND SUMMATION NOTATION

lated as Rule 1 says. Suppose that $c = 2$. Then

$$cY_1 = 18, \quad cY_2 = 22, \quad cY_3 = 3$$

and so

$$\sum_{i=1}^{3} cY_i = 18 + 22 + 3 = 43$$

Since $43 = 2 \times 21.5 = c \times 21.5$, we see that ΣcY_i and $c\Sigma Y_i$ are related as Rule 2 says.

If $n = 3$ and $c = 2$, then

$$\sum_{i=1}^{3} c = 2 + 2 + 2 = 6 = 2 \times 3 = nc$$

which illustrates Rule 3. Notice that

$$\sum_{i=6}^{9} c = c + c + c + c = 4c$$

—as i runs from 6 through 9 (inclusive), it takes on 4 values, not 3 (that is, $9 - 6 + 1$ values, not $9 - 6$ values).

Each summation must be read with care so that its meaning is understood exactly. If one reads summations from the inside out, so to speak, many common errors can be avoided. Consider the summation

$$\sum_{i=1}^{n} (X_i - c)^2$$

Read from the inside out, it tells us to

(1) subtract the constant c from each X,
(2) square each of the differences obtained in step 1,
(3) sum the squares obtained in step 2.

Two important but sometimes misread sums are the *sum of squares,*

$$\Sigma X_i^2 = X_1^2 + X_2^2 + \cdots + X_n^2$$

and the *square of the sum,*

$$(\Sigma X_i)^2 = (X_1 + X_2 + \cdots + X_n)^2$$

The former is found by squaring each X and then adding the squares, the latter by adding up the Xs and then squaring the sum. These two expressions are usually unequal and often occur together in the same mathematical statement.

Problems

1. Given that

$$X_1 = 1 \quad X_4 = -2 \quad X_7 = 5$$
$$X_2 = 3 \quad X_5 = 4 \quad X_8 = 2$$
$$X_3 = 0 \quad X_6 = -1 \quad X_9 = 10$$

find

a. $\sum_{i=1}^{9} X_i$ b. $\sum_{i=3}^{5} X_i$ c. $\sum_{j=1}^{4} X_j^3$

d. $\left(\sum_{i=2}^{4} X_i\right)^2$ e. $\sum_{k=5}^{9} 3$ f. $\sum_{j=1}^{9} (7 - X_j)$

g. $3\sum_{i=1}^{5} X_i - \sum_{i=6}^{9} X_i$ h. $\sum_{k=2}^{6} X_k - 10$ i. $\sum_{i=2}^{6} iX_i$

2. Write out the following summations as sums of the terms involved and reduce the answer to the extent possible.

a. $\sum_{i=1}^{4} f_i X_i^2$ b. $\sum_{i=3}^{6} V_i^2 (Y_i - i)$

c. $\sum_{j=1}^{3} (-1)^j W_j$ d. $\sum_{j=15}^{20} (X_j - Y^j)$

e. $\sum_{i=1}^{3} X_i + \sum_{i=4}^{6} X_i - 2\sum_{i=2}^{4} X_i$ f. $\sum_{i=2}^{3} \frac{X_i}{n_i}$

g. $\sum_{k=1}^{3} kX^{k+1}$ h. $\sum_{i=2}^{3} (a^i + b^i) + 3\sum_{i=1}^{2} a^i b^{3-i} + 2ab$

3. Express the following in summation notation.
   a. $P_1 X_1 + P_2 X_2 + P_3 X_3 + P_4 X_4$ b. $a_1 + a_1 q + a_1 q^2 + a_1 q^3 + a_1 q^4$
   c. $X_1 + \dfrac{X_2}{2} + \dfrac{X_3}{3} + \dfrac{X_4}{4}$ d. $\dfrac{Y_1 + 4Y_2 + 9Y_3}{Y_1 + 8Y_2 + 27Y_3}$
   e. $U_1 - U_2^2 + U_3^3 - U_4^4$ f. $(a_1 + a_2 + a_3)^4$
   g. $m_1 X_1 + m_2 X_2 + m_3 X_3 - n_1 Y_1 - n_2 Y_2 - n_3 Y_3$

4. Given that

$$X_1 = 6 \quad X_2 = 5 \quad X_3 = 1 \quad X_4 = 4$$
$$Y_1 = 7 \quad Y_2 = 4 \quad Y_3 = 3 \quad Y_4 = 6$$

SECTION 3.3 MEASURES OF LOCATION

find

a. $\sum_{i=1}^{4} X_i Y_i$

b. $\left(\sum_{i=1}^{4} X_i\right)\left(\sum_{i=1}^{4} Y_i\right)$

c. $\sum_{i=1}^{4} (X_i - Y_i)$

d. $\sum_{i=1}^{4} (X_i - 4)(Y_i - 5)$

e. $\sum_{i=1}^{4} X_i Y_i - \frac{1}{4}\left(\sum_{i=1}^{4} X_i\right)\left(\sum_{i=1}^{4} Y_i\right)$

f. $\sum_{i=1}^{4} X_i^2 Y_i$

5. Show that
   a. $\sum_{i=1}^{n} (X_i + K)^2 = \sum_{i=1}^{n} X_i^2 + 2Kn\bar{X} + nK^2$, where $\bar{X} = n^{-1}\sum_{i=1}^{n} X_i$.
   b. $\sum_{j=1}^{m} (X_j - \bar{X})^2 = \sum_{j=1}^{m} X_j^2 - m\bar{X}^2$, where $\bar{X} = m^{-1}\sum_{j=1}^{m} X_j$.
   c. $\sum_{i=1}^{n} (X_i - \bar{X})(Y_i - \bar{Y}) = \sum_{i=1}^{n} X_i Y_i - n\bar{X}\bar{Y}$, where $\bar{X} = n^{-1}\sum_{i=1}^{n} X_i$ and $\bar{Y} = n^{-1}\sum_{i=1}^{n} Y_i$.

6. a. Show by giving a numerical example that $\sum_{i=1}^{n} X_i^2$ is not necessarily equal to $(\sum_{i=1}^{n} X_i)^2$.
   b. Similarly show that $\sum_{i=1}^{n} X_i Y_i$ is not necessarily equal to

   $$\left(\sum_{i=1}^{n} X_i\right)\left(\sum_{i=1}^{n} Y_i\right)$$

   c. Show that if $p + q = 1$, then

   $$\sum_{i=1}^{3} p^i + pq(3p + 3q + 2) + \sum_{i=1}^{3} q^i = 3$$

3.3 Measures of Location

When we work with numerical data and their frequency distributions, it soon becomes apparent that in most sets of data there is a tendency for the observed values to group themselves about some interior value; some central value seems to be characteristic of the data. This phenomenon, referred to as *central tendency,* may be used to describe the data in the sense that the central value locates the "middle" of the distribution. The blood-pressure changes for the active and the placebo groups in Tables 2.1 and 2.2 have different "middles." This is clear from the frequency distributions (Tables 2.4 and 2.5), and it is even more apparent in the histograms (Figures 2.1 and 2.2). But we need to describe this difference numerically.

measures of location

The statistics we calculate for this purpose are **measures of location**, also called measures of central tendency. For a given set of data the measure of location we use depends upon what we mean by *middle*, different definitions giving rise to different measures. We consider here three such measures and their interpretations: the mean, the median, and the mode.

3.4 The Mean

All of us are familiar with the concept of the mean, or average value. We all speak of batting averages, grade-point averages, mean annual rainfall, the average weight of a catch of fish, and the like. In most cases the term *average* used in connection with a set of numbers refers to their *arithmetic mean*. For the sake of simplicity we call it the *mean*.

mean

For a given set of n values, $X_1, X_2, X_3, \ldots, X_n$, the **mean,** or the *sample* mean, is their sum divided by n, the number of values in the set. It is denoted by $\bar{X}$ and may be expressed as

$$\bar{X} = \frac{1}{n} \sum_{i=1}^{n} X_i \qquad 3.4$$

If, for example, you catch three fish and their lengths are 10, 13, and 16 inches, their mean length is

$$\bar{X} = \frac{10 + 13 + 16}{3} = \frac{39}{3} = 13 \text{ inches}$$

Notice that the unit for the mean is the same as the unit for the observations themselves (inches, millimeters, dollars, etc.).

The blood-pressure changes in Table 2.1 (active group, page 13) have as their average the number

$$\frac{1}{171}(-23 + 7 - 21 + 0 + \cdots - 2 + 0 - 18) = -9.85.$$

The blood-pressure changes in Table 2.2 (placebo group, page 14) have as their average

$$\frac{1}{176}(-15 + 3 + 5 + 0 + \cdots - 2 + 4 + 11) = 0.23.$$

Among patients on the active drug, blood pressure went down 9.85 points on the average. Among patients on placebo, it rose slightly, by 0.23 points

SECTION 3.4 THE MEAN

Figure 3.1

The mean as a center of gravity

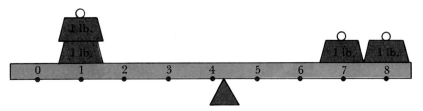

on the average. This is a numerical confirmation of the fact that the histogram in Figure 2.1 is further to the left than the one in Figure 2.2.

To find these two mean values requires adding up 171 numbers and 176 numbers. This is tedious if it is done by hand. Large sets of data are now usually handled automatically on computers, and then the additions present no problem. There is a fairly easy method of calculating a mean from a frequency distribution on a pocket calculator. This is taken up in Section 3.13. Since automatic computing equipment is now very widely available, these methods of hand computation have lost much of their importance. In this book, they are relegated to a special section (Section 3.13), which can be omitted.

The mean is the middle of a set of data in the sense that it is the center

Figure 3.2

The mean as the center of gravity of a histogram

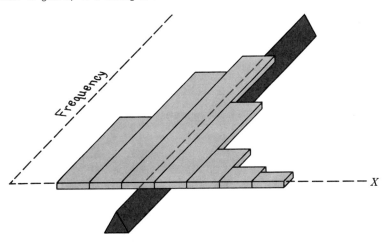

of gravity. This physical way of looking at the mean is illuminating. Suppose we have four observations—1, 8, 7, and 1—with a mean of (1 + 8 + 7 + 1)/4, or 4.25. Imagine a seesaw with a scale marked off along its edge, and imagine for each of the four observations a one-pound weight positioned according to this scale, as shown in Figure 3.1. The mean 4.25 is the point at which the fulcrum of the seesaw must be placed in order to make it balance.

Suppose we sort the data into a frequency distribution and from that make a histogram. If we draw the histogram on some stiff material of uniform density, such as plywood or metal, cut it out, and balance it on a knife edge arranged perpendicular to the horizontal scale, as shown in Figure 3.2, the X-value corresponding to the point of balance is the mean $\bar{X}$.

Problems

7. Find the means of the following sets of numbers.
    a. $-1, -3, 0, 2, -5$.
    b. 2, 2, 3, 10, 100, 1000.
    c. .001, 3×10^{-3}, -2×10^{-3}, 40×10^{-4}.

8. The final examination scores of 15 students in a statistics course are given below. Find the mean.

73	74	92	98	100	72	74	85	76	94
89	73	76	74	99				

9. The points allowed by two football teams in games against common opponents are given below. On the basis of means, which team do you think has the better defense?

Team A:	10	15	12	25	40
Team B:	9	15	20	23	21

10. The following data are the daily low temperatures in degrees Celsius in Casper, Wyoming, for the first week in December. Compute the mean.

0.6	1.1	-1.7	-3.9	-4.4	-3.3	-1.1

SECTION 3.5 THE WEIGHTED MEAN 53

11. Seven items were purchased at each of three clothing stores (the same items at each store). Based on average prices, which store has the least expensive clothing?

Item	1	2	3	4	5	6	7
Store A	15.95	33.89	65.75	22.99	27.67	38.65	74.95
Store B	14.98	40.98	66.75	27.50	29.35	42.32	77.52
Store C	19.25	33.99	67.98	24.50	31.50	41.15	75.95

12. One measure of the complexity of a piece of writing is the average number of words per sentence. Select two articles from a newspaper, perhaps a letter to the editor and a syndicated column, and use this measure to determine which article is the more complex.
13. The average of 23 numbers is 14.7. What is the sum of these numbers?
14. A coin is tossed n times. When "tails" is tossed a "0" is recorded, and when "heads" is tossed a "1" is recorded. What interpretations may be given to the sum and the mean of this sequence of "0's" and "1's"?

3.5 The Weighted Mean

When we compute the simple arithmetic mean of a set of data, we assume that all the observed values are of equal importance and we give them equal weight in our calculations. In situations where the numbers are not equally important we can assign to each a weight which is proportional to its relative importance and calculate the **weighted mean.**

weighted mean

Let $V_1, V_2, \ldots, V_k$ be a set of k values, and let $w_1, w_2, \ldots, w_k$ be the weights assigned to them. The weighted mean is found by dividing the sum of the products of the values and their weights by the sum of the weights; that is,

$$\bar{V} = \frac{w_1 V_1 + w_2 V_2 + w_3 V_3 + \cdots + w_k V_k}{w_1 + w_2 + w_3 + \cdots + w_k}$$

In summation notation, this is

$$\bar{V} = \frac{\sum w_i V_i}{\sum w_i} \qquad 3.5$$

Every student is familiar with the concept of the weighted mean; the grade-point average is such a measure. It is the mean of the numerical values of the letter grades weighted by the numbers of credit hours in which the various grades are earned. Suppose these numerical values are A = 4, B = 3, C = 2, and D = 1. If a student takes a 3-credit course and makes an A in it ($w_1 = 3$, $V_1 = 4$), a 5-credit course and makes a B ($w_2 = 5$, $V_2 = 3$), another 3-credit course and makes a C ($w_3 = 3$, $V_3 = 2$), and a 2-credit course and makes another A ($w_4 = 2$, $V_4 = 4$), then the grade-point average for the term is

$$\bar{V} = \frac{(3 \times 4) + (5 \times 3) + (3 \times 2) + (2 \times 4)}{3 + 5 + 3 + 2} = \frac{41}{13} = 3.15$$

The weighting procedure is also used to find the mean when several sets of data are combined. Suppose we have three sets of data consisting of n_1, n_2, and n_3 observed values and having means $\bar{X}_1$, $\bar{X}_2$, and $\bar{X}_3$, respectively. The mean for the combined data is the weighted average of the individual means, the respective weights being the sample sizes n_1, n_2, and n_3:

$$\bar{X} = \frac{n_1\bar{X}_1 + n_2\bar{X}_2 + n_3\bar{X}_3}{n_1 + n_2 + n_3}$$

Failure to weight the means when combining data is not an uncommon error. Imagine the male student body of a school split into two groups. In the first group the mean height is 75 inches, and in the second group the mean height is 69 inches. The average $(75 + 69)/2 = 72$ is not the mean height of male students in the school if the first group consists of the 15 members of the basketball team and the second group consists of the remaining 568 male students in the school. The true mean then is

$$\frac{(15 \times 75) + (568 \times 69)}{583} = 69.15 \text{ inches}$$

in accordance with the equation above for two (rather than three) means.

Problems

15. Three sets of data had means of 25, 20, and 22; the numbers of observations were 20, 25, and 30, respectively. What is the mean if these sets are combined?

16. The hourly rate and number of employees at each salary level in a machine shop are given as follows. Find the mean hourly rate.

SECTION 3.5 THE WEIGHTED MEAN

Salary Level	1	2	3	4	5
Hourly rate	4.50	5.00	5.50	6.00	6.50
Number of employees	25	20	15	10	5

17. The students using a college placement service were classified according to age. What is the average age of this group?

Age	20	21	22	23	24	25
Number	2	20	35	19	10	6

18. A conservation officer asked 50 fishermen at the Sandy Shores Pier to record the number of fish caught during a 5-hour period. Find the mean catch.

Number of fish	0	1	2	3	4
Frequency	18	13	10	5	4

19. A legal case may be tried before a jury or before a judge alone. The following table gives the percentage of jury trials in each of eight samples of trials. If the eight samples are combined into one large sample, what is the percentage of jury trials in this large sample?

Sample Size	Percent Jury Trials
30	9.0
10	3.5
80	7.0
100	6.0
50	8.5
75	8.0
40	5.5
150	6.5

20. The number of imperfections in each of 25 sheets of fiberboard was noted by a quality-control inspector. Find the mean number of imperfections per sheet.

number	0	1	2	3
frequency	14	8	2	1

3.6 The Median

median

The **median** of a set of data is a number such that about half the observations are less than that number and about half the observations are greater than that number. To be specific, suppose there are n observed values and that n is an odd number. The array is formed in which the values are lined up in increasing order, and the median is by definition the observation in the middle position. For example, in the array

$$1 \quad 1 \quad 2 \quad 3 \quad 3 \quad \mathbf{8} \quad 11 \quad 14 \quad 19 \quad 19 \quad 20$$

the median is 8 (there are five observations to its left and five to its right). If n is even, there is no one observation in the middle position. In this case, we take the median to be the average of the *pair* of observations occupying the two central positions. In the array

$$2 \quad 5 \quad 5 \quad 6 \quad 7 \quad \mathbf{10} \quad \mathbf{15} \quad 21 \quad 21 \quad 23 \quad 23 \quad 25$$

the observations 10 and 15 occupy the two center positions (there are five observations to the left of 10 and five to the right of 15), and so the median is (10 + 15)/2, or 12.5.

The median can be found without difficulty from a stem-and-leaf plot if the set of data is not too large. For the 27 scores plotted in Table 2.13 the median is the fourteenth in numerical order. Rows 5, 6, and 7 contain a total of 12 scores, the smallest 12 scores. Hence the median must be the second smallest score in the 8-row, namely 81 (it happens to occur twice).

Since we often have our data in the form of a grouped frequency distribution, we need a way of estimating the median for grouped data. Recall that when raw data are grouped into bins, information is lost because the distinction between the various observations in each individual bin is lost. The following method assumes that the observations in each bin are more or less evenly spread over that bin, so that this loss of information is of little importance.

The method is most easily understood and carried out with the ogive. Look at the ogive in Figures 2.7 and 3.3 for the blood-pressure changes of the 171 patients in the active group in the hypertension study. The height of the curve over a given value on the horizontal scale represents approxi-

SECTION 3.6 THE MEDIAN

mately the number of observations numerically less than that given value. (See the end of Section 2.3, where the approximate number of observations less than -8 is found by graphical interpolation in Figure 2.7.) The number of observations is 171, and half of this is 85.5; make it 85 to get a whole number. We want the value such that half of the observations are smaller than it is, and so we want the value such that 85 of the observations are smaller than it is. We should therefore look for the value on the horizontal scale such that the height of the curve above it is 85. The way to do this is to find the cumulative frequency 85 on the vertical scale, trace horizontally over to the curve and then vertically down to the scale at the bottom (see the line labeled "median" in Figure 3.3), and read off the value of the point where the traced line intersects the horizontal scale. The result is about -10.5: the median is about -10.5. The actual median according to the definition above is -10. We rounded 171/2, or 85.5, down to 85, and found this value on the vertical scale. If we had instead rounded 85.5

Figure 3.3

Ogive for the cumulative distribution of blood-pressure change: active group (from Table 2.9)

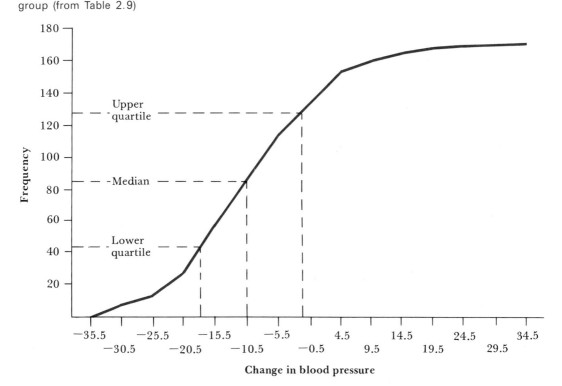

upward to 86 and located this on the vertical scale, or if we had located 85.5 itself on the vertical scale, the median we read off on the bottom scale would have been practically the same as before.

The upper and lower quartiles shown in Figure 3.3 will be defined in Section 3.14.

Problems

21. Find the median of the following sets of numbers.
    a. 2, 1, 5, −1, 4, 6.
    b. 17.1, 16.4, 17.5, 16.9, 17.2.
    c. 1, 1, 2, 1, 3, 1, 1, 1, 2, 2.
22. Find the median of the students' scores in Problem 8.
23. Find the median age of the students using the college placement service referred to in Problem 17.
24. A survey of records of a retail business was taken to determine the number of absences per employee in January. Find the median number of absences.

Number of absences	0	1	2	3	4	5	6	7
Frequency	5	15	23	22	17	10	6	3

25. A set of data should consist of the numbers 28, 15, 17, 24, and 19, but the digits in 28 were transposed so that the data actually recorded were 82, 15, 17, 24, and 19. If the error were not detected, which would be the better measure of central tendency, the median or the mean?
26. Compare the defenses of the two teams referred to in Problem 9 using medians instead of means. Which do you think is the better measure for comparing the points scored against the two teams, the mean or the median?
27. For two countries, A and B, the following table gives information concerning the distribution of annual income per capita (in U.S. dollars) in each.

	A	B
Median	4000	1250
Mean	3750	4750

SECTION 3.7 THE MODE

a. Which of the two countries has, in your opinion, a higher standard of living, assuming all other factors (e.g., prices, taxes, etc.) are similar?

b. Assuming the two countries have equal population size, what can you say about the median and mean of the distribution of annual income per capita of the two countries taken together?

c. If a 10% income tax is imposed uniformly in each country, and if their population sizes are equal, which country will collect more?

28. The cumulative frequency distribution of the IQ scores of 100 high-school students is given below. Draw the ogive and use the graphical method shown in Figure 3.3 to find the median.

IQ Less than	Cumulative Frequency
80	4
90	18
100	39
110	67
120	90
130	100

3.7 The Mode

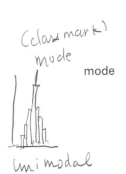

When data are grouped we often find that there is one class that has maximum frequency and that the frequencies of the other classes tend to fall away continually as we move away from the maximum class in either direction. This class is called the *modal class*, and we define the **mode** as the class mark of that class.

For example, for the distribution of blood-pressure differences in Table 2.5 the bin with greatest frequency is 0 to 4, the frequency being 38. Moreover, as a glance at the histogram in Figure 2.2 shows, the frequencies taper off on either side of this class. This is therefore the modal class, and its class mark (midpoint) 2 is the mode M:

$$M = 2$$

If there is such a class with maximum frequency, then the distribution is said to be *unimodal*. If there is no such class, the distribution is said to be *multimodal*, and the mode is undefined. In a histogram like that in Figure 3.4, the distribution is multimodal; A is the major mode, and B and C are

Figure 3.4

A histogram showing major and minor modes

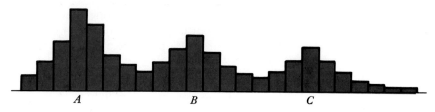

minor modes. The concentration of values is greatest around A, but there are secondary concentrations around B and C.

The histogram in Figure 2.1 is almost unimodal. There is a slight dip at the bin from -15 to -11, but that is a minor matter.

Problems

 29. Find the mode of the data in Problem 18.

 30. Find the mode of the number of absences in Problem 24. Compare the mode with the median and mean.

 31. Make a histogram of the grade-point averages in Problem 30 of Chapter 2, and find the mode.

 32. Electric motors were placed under extremely high stress until they failed. The cumulative frequency distribution of the number of hours to failure is given below. Find the mode and median.

Hours-to-Failure Less than	Cumulative Frequency
10	5
20	22
30	36
40	47
50	53

 33. Forty applicants for a secretarial job were timed in typing a letter. The following data give the times in minutes.
  a. Make a histogram of the data and find the major and minor modes (use five intervals).
  b. If the amount of time needed to correct errors was included in the typing time, how might this explain the two modes?

0.8	0.9	0.7	1.4	1.5	1.0	1.0	1.5
1.0	1.3	1.4	1.6	1.5	0.9	0.9	0.9
1.0	1.7	1.0	1.1	1.1	0.7	1.0	0.9
1.3	1.4	1.0	1.1	0.9	0.9	1.2	1.0
1.1	1.4	1.2	0.8	1.2	1.3	1.4	0.9

3.8 Selecting a Measure of Location

In deciding which measure of location to report for a set of data, we first of all consider the use to which the results are to be put. In addition, we need to know the advantages and disadvantages of each measure of location as regards its calculation and interpretation.

If the distribution of the data is symmetric and unimodal, then the mean $\bar{X}$, the median m, and the mode M will all coincide; but as the distribution becomes more and more skewed (lopsided), the differences among these measures will become greater. This is illustrated in Figure 3.5.

The mean is sensitive to extreme values. If in a small town the average annual income of the 100 heads of household is reported as $5,990, this figure is correct but very misleading if one head of household is a multimil-

Figure 3.5

Mean, median, and mode for symmetrical and asymmetrical distributions

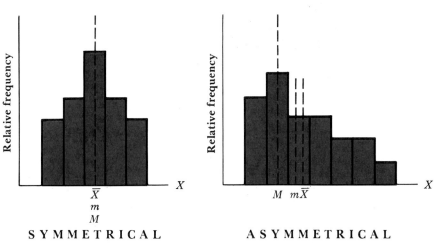

SYMMETRICAL   ASYMMETRICAL

lionaire with an income of $500,000 and the remaining 99 are paupers with incomes of $1000. On the other hand, a few extreme values have little or no effect on the median and the mode. For the numbers

$$1, 3, 4, 6, 6, 9, 13$$

we have $\bar{X} = 6$, $m = 6$, and $M = 6$. If we add the number 70 to this set, the mean will be equal to 14, a shift of 8 units, but the median and mode remain unchanged. For this reason, the median or mode may be better to use than the mean if the data contains a few observations far removed from the others. For other measures of location useful in this circumstance, see Section 3.14.

If we want to combine measures for several sets of data, the algebraic properties of the mean give it a distinct advantage. We have seen that we can use the weighted mean for this purpose. The median and the mode are not subject to this type of algebraic treatment.

The calculation of a measure of location is usually a first step toward making inferences about the source of the data. In this case the mathematical and distributional properties of the mean give it a distinct advantage. In statistical inference a primary consideration is statistical stability. It can be shown that if a large number of sets of data are taken from the same source and all three measures are calculated for each set, there will be less variation among the means than among the medians or among the modes; hence the mean is more stable. This, coupled with the fact that it is more amenable to mathematical and theoretical treatment, makes the mean an almost universal choice for all but purely descriptive purposes.

3.9 Measures of Variation

The measures of location discussed in the preceding sections serve to describe one aspect of numerical data; but they tell us nothing about another aspect of equal importance—the amount of variation or scatter among the observed values. In any set of statistical data the numerical values will not be identical but will be scattered or dispersed to a greater or lesser degree. The statistics we calculate to measure this characteristic of the data we refer to as **measures of variation** or **measures of dispersion**. Since several sets of data could have the same, or nearly the same, mean, median, and mode, but vary considerably in the extent to which the individual observations differ from one another, a more complete description of the data results when we evaluate one of the measures of variation in addition to one or more of the measures of location. Several such measures are considered in the following sections.

measures of variation
measures of dispersion

3.10 The Range

range
The **range** is the difference between the largest and smallest values in the data. In Table 2.1 the largest value is 31 and the smallest value is -34; hence the range is $31 - (-34)$, or 65.

Although the range is easy to calculate and is commonly used as a rough-and-ready measure of variability, it is not generally a satisfactory measure of variation for several reasons. In the first place, it uses only a fraction of the available information concerning variation in the data and reveals nothing about the way in which the bulk of the observations are dispersed within the interval bounded by the smallest and largest values. Second, as the number of observations is increased the range generally tends to become larger; therefore, it is not proper to use the ranges to compare the variation in two sets of data unless they contain the same number of values. Finally, the range is the least stable of our measures of variation for all but the smallest sample sizes; that is, in repeated samples taken from the same source the ranges will exhibit more variation from sample to sample than will the other measures.

The range differs from most of our statistical measures in that it is a relatively good measure of variation for small numbers of observations but becomes less and less reliable as the sample size increases. Because it is easy to calculate and is reasonably stable in small samples, it is commonly used in statistical quality control, where samples of four or five observations are often sufficient. See Section 3.14 for related measures of variation.

3.11 The Variance and the Standard Deviation

Since the disadvantages of the range limit its usefulness, we need to consider other measures of variation. Suppose we have n numbers, $X_1, X_2, X_3, \ldots, X_n$, whose mean is $\bar{X}$. If we were to plot these values and $\bar{X}$ on the X-axis as in Figure 3.6 (only five values are shown) and measure the distance of each of the Xs from the mean $\bar{X}$, it seems reasonable that the average, or mean, of these distances should provide a measure of variation. These distances, or *deviations from the mean,* are equal to $X_1 - \bar{X}, X_2 - \bar{X}, \ldots, X_n - \bar{X}$, or, in general, $X_i - \bar{X}$, and their mean is

$$\frac{1}{n} \sum_{i=1}^{n} (X_i - \bar{X})$$

Figure 3.6

Deviations from the sample mean

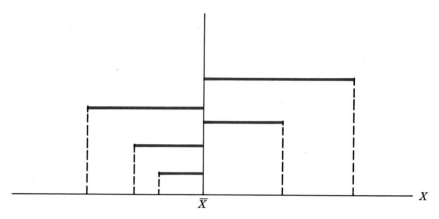

But this quantity is always equal to zero, since the algebraic sum of the deviations of a set of numbers from its mean is always equal to zero.

This mathematical fact can be seen as follows:

$$\Sigma(X_i - \bar{X}) = \Sigma X_i - n\bar{X}$$

but

$$\bar{X} = \frac{1}{n}\Sigma X_i \quad \text{or} \quad n\bar{X} = \Sigma X_i$$

Therefore

$$\Sigma(X_i - \bar{X}) = \Sigma X_i - \Sigma X_i = 0 \qquad 3.6$$

The trouble is that we have used not the distances of the X_i to the mean $\bar{X}$, but the signed distances, and one remedy is just to ignore their algebraic signs. We have then the *mean deviation*

$$\frac{1}{n}\sum_{i=1}^{n}|X_i - \bar{X}| \qquad 3.7$$

where $|X_i - \bar{X}|$, the *absolute value* of $X_i - \bar{X}$, is just $X_i - \bar{X}$ with the sign converted to $+$ if it happens to be $-$.

The mean deviation, though it appears to be simple, is not of great interest. It is not particularly easy to calculate, and the absolute values

SECTION 3.11 THE VARIANCE AND THE STANDARD DEVIATION

make theoretical developments difficult. There are theoretical reasons why one should use squares instead of absolute values. Suppose we do use the square instead of the absolute value as a measure of deviation. In the squaring process the negative signs will disappear; hence the sum of the squares of the deviations from the mean will always be greater than zero unless all the observations have the same value. In symbols the sum of squares of deviations from the mean is

$$\sum_{i=1}^{n} (X_i - \bar{X})^2 \qquad 3.8$$

sum of squares

This quantity, for simplicity, will be referred to as the **sum of squares**. Hereafter, whenever the phrase *sum of squares* appears it should be taken to mean *sum of squares of deviations from the sample mean*. To avoid ambiguity, if the sum of squares of the observations or the sum of squares for any other quantities is meant, the phrase will be qualified accordingly.

It is clear that the sum of squares provides a measure of dispersion. If all the observed values are identical, the sum of squares is equal to zero. If the values tend to be close together, the sum of squares will be small, but if they scatter over a wide range, the sum of squares will be correspondingly large. To illustrate this characteristic and, incidentally, to show how the sum of squares is calculated according to the definition given by equation (3.8), we use the heights in inches of ten plants:

$$12, 6, 15, 3, 12, 6, 21, 15, 18, 12$$

Their mean is

$$\bar{X} = \frac{1}{n} \sum X_i = \frac{120}{10} = 12$$

To find the deviations from the mean we subtract $\bar{X} = 12$ from each of the X values. We get

$$0, -6, 3, -9, 0, -6, 9, 3, 6, 0$$

We can use the fact that the sum of the deviations from the mean is equal to zero to check our arithmetic:

$$(3 + 9 + 3 + 6) - (6 + 9 + 6) = 0$$

To complete our calculations we square each of the deviations and sum them all:

$$0 + 36 + 9 + 81 + 0 + 36 + 81 + 9 + 36 + 0 = 288$$

Thus the sum of squares for these ten numbers is

$$\sum_{i=1}^{10} (X_i - \bar{X})^2 = 288$$

The calculations are summarized in the following table:

X_i	$X_i - \bar{X}$	$(X_i - \bar{X})^2$
12	0	0
6	−6	36
15	3	9
3	−9	81
12	0	0
6	−6	36
21	9	81
15	3	9
18	6	36
12	0	0
120	0	288

$n = 10 \qquad \bar{X} = 12$

In contrast to the preceding example, which displays a fair amount of variation, suppose the ten plants have heights

12, 10, 12, 14, 10, 13, 12, 11, 14, 12

We find the sum of squares as before:

X_i	$X_i - \bar{X}$	$(X_i - \bar{X})^2$
12	0	0
10	−2	4
12	0	0
14	2	4
10	−2	4
13	1	1
12	0	0
11	−1	1
14	2	4
12	0	0
120	0	18

$n = 10 \qquad \bar{X} = 12$

SECTION 3.11 THE VARIANCE AND THE STANDARD DEVIATION

As could be expected, since the variation in the second set of numbers is less than that in the first set, the sum of squares, 18, is considerably smaller than that calculated for the first set. Notice that the mean is the same as that in the preceding example; the sum of squares measures the spread *about* the mean, whatever the mean may be.

variance   The **variance** is the sum of squares divided by $n - 1$:

$$s^2 = \frac{1}{n-1} \sum_{i=1}^{n} (X_i - \bar{X})^2 \qquad 3.9$$

The variance is an average for the squares of the deviations. In calculating the mean $\bar{X}$ we divide $\sum_{i=1}^{n} X_i$ by n, the number of observations (see (3.4)). But notice that in calculating the variance by the formula (3.9) we divide not by n but by $n - 1$. The reason for this will be discussed in Chapter 7. To divide by $n - 1$ strikes everyone as strange at first, and the simplest thing to do is to accept it temporarily until the justification comes along in Chapter 7.

standard deviation   The **standard deviation** is the positive square root of the variance. The variance is usually denoted by s^2 and the standard deviation by s; they are related by

$$s = \sqrt{s^2} \qquad 3.10$$

The standard deviation s is the most widely used measure of the spread in the data. Notice that its units are those of the original data. If the X_i are measurements in inches, then $\bar{X}$ has inches as its unit. The variance s^2, however, being an average of squares of deviations, has inches squared as its unit; but the square root s of s^2 has inches as its unit again, which is what we want. The spread should be measured in the same units as the data.

For the data in Table 2.1 (the changes in blood pressure for the 171 patients on the active drug), the mean was -9.85; the standard deviation comes out to 11.55. For the data in Table 2.2 (the changes in blood pressure for the 176 patients on placebo), the mean was 0.23; the standard deviation comes out to 11.13. The two sets of data have very different means, but their standard deviations are about the same. This is reflected graphically in the histograms in Figures 2.1 and 2.2: The histogram for the active group is situated further left on the horizontal scale than the one for the placebo group is, but the amount of spread in the two histograms looks about the same.

For the ten plant heights (in inches)

12, 6, 15, 3, 12, 6, 21, 15, 18, 12,

the sum of squares was 288. Since n is 10, $n - 1$ is 9; dividing 288 by 9 gives 288/9, or 32.00 for the variance: $s^2 = 32.00$. And $s = \sqrt{32.00} = 5.66$: the standard deviation is 5.66 inches.
For the ten plant heights

$$12, 10, 12, 14, 10, 13, 12, 11, 14, 12,$$

the sum of squares was 18, and so $s^2 = 18/9 = 2.00$ and $s = \sqrt{2.00} = 1.41$. The standard deviation in this case, 1.41 inches, is less than in the first case, reflecting the fact that the ten measurements are less widely spread out.

If n is large, it is very laborious to compute the sum of squared deviations by the formula (3.8). There is another method of computing the sum of squares that does not require that we find the individual deviations from the mean. The method is based on the formula

$$\sum_{i=1}^{n}(X_i - \bar{X})^2 = \sum_{i=1}^{n} X_i^2 - \frac{1}{n}\left(\sum_{i=1}^{n} X_i\right)^2 \qquad 3.11$$

There is a companion formula for the variance:

$$s^2 = \frac{1}{n-1}\left[\sum_{i=1}^{n} X_i^2 - \frac{1}{n}\left(\sum_{i=1}^{n} X_i\right)^2\right] \qquad 3.12$$

machine formulas These are called the **machine formulas** because they are convenient to use on hand calculators.

For the ten plant heights

$$12, 6, 15, 3, 12, 6, 21, 15, 18, 12$$

considered earlier, the calculations via (3.11) and (3.12) go like this: First,

$$\begin{aligned}\sum_{i=1}^{10} X_i^2 &= (12)^2 + (6)^2 + (15)^2 + (3)^2 + (12)^2 \\ &\quad + (6)^2 + (21)^2 + (15)^2 + (18)^2 + (12)^2 \\ &= 144 + 36 + 225 + 9 + 144 \\ &\quad + 36 + 441 + 225 + 324 + 144 = 1728\end{aligned}$$

As before,

$$\begin{aligned}\sum_{i=1}^{10} X_i &= 12 + 6 + 15 + 3 + 12 + 6 \\ &\quad + 21 + 15 + 18 + 12 = 120\end{aligned}$$

SECTION 3.11 THE VARIANCE AND THE STANDARD DEVIATION

Using (3.11) gives

$$\sum_{i=1}^{10} (X_i - \bar{X})^2 = 1728 - \frac{1}{10}(120)^2 = 1728 - \frac{14400}{10}$$

[handwritten: variance =]

$$= 1728 - 1440 = 288,$$

which checks with the previous answer. And (3.12) gives, as before, $s^2 = 288/9 = 32.00$.

For a derivation of (3.11) we rewrite the sum of squares as

$$\Sigma(X_i - \bar{X})^2 = \Sigma(X_i^2 - 2\bar{X}X_i + \bar{X}^2)$$

If we now apply Rules 1, 2, and 3 of Section 3.2, we get

$$\Sigma(X_i - \bar{X})^2 = \Sigma X_i^2 - 2\bar{X}\Sigma X_i + n\bar{X}^2$$

But

$$\bar{X} = \frac{1}{n}\sum X_i \quad \text{and} \quad \bar{X}^2 = \frac{1}{n^2}\left(\sum X_i\right)^2$$

Therefore

$$\Sigma(X_i - \bar{X})^2 = \Sigma X_i^2 - 2\frac{1}{n}\left(\sum X_i\right)^2 + \frac{1}{n}\left(\sum X_i\right)^2$$

which is the same as (3.11) above.

[handwritten: rule of thumb: S = 1/6 (range) usually.]

Problems

34. For the following empirical spring constants of ten wood beams find
 a. The mean.   b. The variance.
 c. The range.   d. The standard deviation.

| .0135 | .0190 | .0185 | .0201 | .0173 |
| .0218 | .0270 | .0234 | .0156 | .0181 |

35. The lengths in centimeters of 10 leaves of a house plant are 6.8, 10.0, 10.2, 6.5, 7.0, 7.8, 10.8, 6.1, 5.9, 8.9. Find the variance and range.

36. The station-to-station rates for the initial 3 minutes on telephone calls from Springfield to the five largest cities in the state are $1.95, $2.10, $1.95, $2.00, $2.30. Find the mean and standard deviation.

37. Seven different thermometers were used to measure the temperature of a substance. The readings in degrees Celsius are −4.10, −4.13, −4.09, −4.08, −4.10, −4.09, −4.12. Find the variance and standard deviation.

38. Use both the definition (3.9) and the machine formula (3.12) to compute the variances of the following sets of numbers.
   a. −1, 2, 3, −4.
   b. .01, .05, .08, .02, .03.

39. A weight lifter's maximum bench press (in pounds) in each of six successive weeks was

$$280, 295, 275, 305, 300, 290$$

Find the variance by (3.9) and by (3.12). Find the standard deviation.

40. An amateur astronomer recorded the number of meteorites appearing in the night sky for ten evenings in September. Find the mean and standard deviation of the data.

Number	Frequency
0	1
1	2
2	4
3	2
4	1

41. Compute the standard deviation of the two sets of scores in Problem 9. On the basis of standard deviations, which team has the more consistent defense?

42. The six houses most recently sold by a real-estate agent spent 3.5, 7.0, 2.5, 4.0, 4.5, and 13.0 months on the market. Find the mean, median, standard deviation, and range.

43. a. What can be said about the data values when the variance is zero?
    b. If a computed value of the variance is negative, what does this mean?

3.12 Coding, or Change of Scale

Now that we have seen how certain measures of location and of variation are calculated, we consider a principle whose application will, in many

SECTION 3.12 CODING, OR CHANGE OF SCALE

instances, lead to a substantial reduction in the amount of arithmetic required to obtain some of these measures. As already noted, computational simplicity becomes less important as the use of computers increases. On the other hand, an understanding of the material in this section will clarify some of the facts about random variables developed and used from Chapter 5 onward.

change of scale The computational simplicity mentioned is achieved by applying to the data a **change of scale,** or a *linear transformation,* as it is often called. A linear transformation may consist of a translation of the data to a new origin, or it may take the form of an expansion or contraction of the scale of measurement. A translation results when the same constant is added to or subtracted from every value in a set of data. An expansion or contraction of the scale of measurement is obtained by multiplying every observed value by the same suitably chosen constant. In most applications the transformation employed will involve both a translation and a change in the units of measurement.

For a set of n numbers, $X_1, X_2, \ldots, X_n$, a translation alone will simplify the computations if these n values are such that the addition to each of the same constant will give us a set of values containing fewer digits than the original numbers. For example, given the numbers

$$381, 386, 383, 389, 393, 384, 386$$

if we subtract 380 from each (add minus 380), we get the new set

$$1, 6, 3, 9, 13, 4, 6$$

and it is obvious that it would be easier to work with the latter. Formally, we have used a transformation whose equation is

$$U_i = X_i - 380$$

where the U_i are the *transformed values.*

Suppose we have the numbers

$$.003, .008, .011, .015, .006, .008, .005$$

Here we are inclined to ignore the decimal point and to work with the integers

$$3, 8, 11, 15, 6, 8, 5$$

In other words, we transform the original numbers by multiplying each by 1000. The corresponding equation is

$$U_i = 1000 X_i$$

The transformations used in the preceding examples are both special cases of the general linear transformation

$$U_i = AX_i + B \qquad 3.13$$

where the X_i are the observed values, the U_i are the transformed values, and A and B are constants. The additive constant B may be positive or negative. It may also be zero. In the latter case we would have only an expansion or contraction. The multiplicative constant A may have any value greater than zero. If A is equal to 1 we have a simple translation. Keep in mind that division by A is the same as multiplication by the reciprocal $1/A$.

After a set of raw data has been transformed, we find the mean, variance, and standard deviation of the Us in the usual way; that is,

$$\bar{U} = \frac{1}{n} \sum U_i$$

$$s_U^2 = \frac{1}{n-1} \sum (U_i - \bar{U})^2$$

$$s_U = \sqrt{s_U^2}$$

After these measures have been calculated for the transformed values, it will be necessary to retransform them, or convert them back, to the original units and origin. The first formula is

$$\bar{U} = A\bar{X} + B \qquad 3.14$$

This can be solved for $\bar{X}$ in terms of $\bar{U}$:

$$\bar{X} = \frac{1}{A}(\bar{U} - B) \qquad 3.15$$

The standard deviations are related by

$$s_U = As_X \qquad 3.16$$

and

$$s_X = \frac{1}{A} s_U \qquad 3.17$$

We see that if every X_i is multiplied by a positive constant A, then the mean and standard deviation are multiplied by the same constant and the variance is multiplied by the square of the constant. If a constant B is added to every X_i, it is also added to the mean but has no effect on the variance and standard deviation.

The truth of these formulas can be grasped visually by imagining a city 6000 feet above sea level and four mountains with peaks 1200, 1800, 3600, and 4200 feet above the city (see Figure 3.7). The peaks average

$$\frac{1200 + 1800 + 3600 + 4200}{4} = 2700 \text{ feet}$$

SECTION 3.12 CODING, OR CHANGE OF SCALE

Figure 3.7

Change of scale

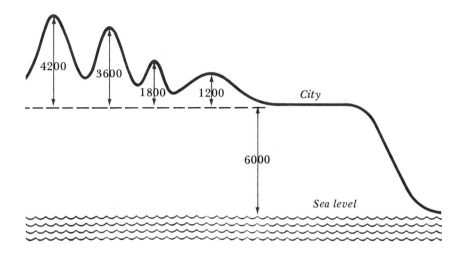

above the city. To find the average of their heights above sea level we could convert all three to measurements from sea level by adding 6000 and then average the results, but we know intuitively that the average height above sea level is 2700 + 6000, or 8700, feet. This is Equation (3.14) with A equal to 1 and B equal to 6000.

Similarly, to find the average height above the city in yards instead of feet, we could convert all four heights to yards by multiplying by $\frac{1}{3}$ and then averaging, but intuitively we know that the average height above the city in yards is $\frac{1}{3} \times 2700$, or 900. This is Equation (3.14) with A equal to $\frac{1}{3}$ and B equal to 0.

Finally, the standard deviation of 1200, 1800, 3600, and 4200 turns out to be 1428.3. And if the altitudes of the mountains are spread out by an amount 1428.3 feet, they ought to be spread out an amount $\frac{1}{3} \times 1428.3$, or 476.1, yards. This is a case of Equation (3.16) with A equal to $\frac{1}{3}$.

Example 1

Find the mean and standard deviation of the nine measurements

2.43, 2.46, 2.42, 2.45, 2.43, 2.48, 2.46, 2.47, 2.45

Procedure: A helpful transformation is $U_i = 100X_i - 240$. Here $A = 100$, $B = -240$.

X_i	$U_i = 100X_i - 240$	$U_i - \bar{U}$	$(U_i - \bar{U})^2$
2.43	3	−2	4
2.46	6	1	1
2.42	2	−3	9
2.45	5	0	0
2.43	3	−2	4
2.48	8	3	9
2.46	6	1	1
2.47	7	2	4
2.45	5	0	0
	45	0	32

$$\bar{U} = \frac{45}{9} = 5$$

$$s_U^2 = \frac{32}{8} = 4$$

$$s_U = \sqrt{4} = 2$$

We convert to the original scale as follows, by using (3.15) and (3.17):

$$\bar{X} = \frac{\bar{U} + 240}{100} = \frac{5 + 240}{100} = 2.45$$

$$s_X = \frac{1}{100} s_U = \frac{2}{100} = .02$$

The relationships (3.14) through (3.17) can be derived as follows. If we sum both sides of Equation (3.13) we get

$$\Sigma U_i = A \Sigma X_i + nB$$

Dividing both sides of this equation by n gives (3.14), and solving (3.14) for $\bar{X}$ gives (3.15). The relationship between the standard deviations is derived as follows:

$$U_i = AX_i + B$$
$$\bar{U} = A\bar{X} + B$$
$$U_i - \bar{U} = A(X_i - \bar{X})$$
$$(U_i - \bar{U})^2 = A^2(X_i - \bar{X})^2$$

SECTION 3.12 CODING, OR CHANGE OF SCALE

$$\Sigma (U_i - \bar{U})^2 = A^2 \Sigma (X_i - \bar{X})^2$$

$$\frac{1}{n-1} \Sigma (U_i - \bar{U})^2 = \frac{A^2}{n-1} \Sigma (X_i - \bar{X})^2$$

This last equation is $s_U^2 = A^2 s_X^2$, and taking square roots gives (3.16). Finally, dividing each side of (3.16) by A gives (3.17).

Problems

44. If X has a mean equal to 100 and a standard deviation equal to 5, what are the mean and standard deviation of U, where
  a. $U = X + 10$?   **b.** $U = 40X$?   **c.** $U = 40X + 10$?

45. If the mean of X is equal to 6 and the variance of X is equal to 10, what are the mean and variance of U, where
  a. $U = X + 5$?   **b.** $U = 3X$?   **c.** $U = 3X + 5$?

46. If $Y = 10X - 25$, $\bar{Y} = 5$, and $s_Y^2 = 100$, find the mean, variance, and standard deviation of X.

47. If $U = 5X + 20$, $\bar{U} = 100$, and $s_U^2 = 40$, what are the mean and variance of X?

48. Find the mode, median, and range of U in Problem 45 if the mode, median, and range for X are respectively 8, 9, and 25.

49. For each of the following sets of numbers use a linear transformation to find the mean, variance, and standard deviation.
  a. 97, 106, 94, 103, 97, 112, 106, 109, 103.
  b. 532, 535, 531, 534, 532, 537, 535, 536, 534.

50. For each of the following sets of numbers use a linear transformation to find the mean, variance, and standard deviation.
  a. 6876, 6878, 6881, 6880, 6881, 6878.
  b. 4.32×10^{-6}, 3.35×10^{-6}, 4.39×10^{-6}, 4.41×10^{-6}, 4.43×10^{-6}.

51. Given the following set of ten measurements of a physical constant,

2.481×10^{-6}	2.475×10^{-6}
2.478×10^{-6}	2.477×10^{-6}
2.481×10^{-6}	2.480×10^{-6}
2.483×10^{-6}	2.488×10^{-6}
2.485×10^{-6}	2.491×10^{-6}

use a linear transformation to find the mean, variance, and standard deviation. Give these measures on both the U-scale and the X-scale.

52. If F represents Fahrenheit temperature and C represents Celsius, the formula relating F and C is C = $\frac{5}{9}$(F − 32). The mean and standard deviation of the January temperature in Detroit are 26°F and 3°F. Express these in Celsius.

3.13 Means and Variances for Grouped Data

grouped data

To illustrate the calculation of the mean for **grouped data** we apply the procedure to the blood-pressure data for the active group as summarized in the frequency distribution given in Table 2.4.

To find the mean of this distribution we operate as if each observed value in a given bin were exactly equal to the bin mark for that bin. Recall that the bin mark is the center point of the bin—the average of the upper and lower bin limits. Now the observations are $X_1, X_2, \ldots, X_n$, where $n = 171$ for this set of data, and they are grouped into k bins, where $k = 14$ in this case. Let v_j denote the bin mark for the jth bin, and let f_j denote the bin frequency for that bin. These are given in the second and third columns of Table 3.1; they come from the third and fourth columns of Table 2.4. If each observation X_i falling into bin j had value exactly v_j (instead of having a value near v_j), the sum of all the observations in bin j would be exactly $v_j f_j$. In Table 3.1 the bin mark for the fourth bin (−20 to −16) is $v_4 = -18$, and the bin frequency is $f_4 = 31$. If each observation in the fourth bin were −18, the sum of the values for these 31 observations would be exactly $v_4 f_4 = (-18) \times (31) = -558$, and this will approximate the sum even though the 31 values are spread over the range from −20 to −16. The sum of all 171 values is approximated by summing the individual bin totals:

$$\sum_{j=1}^{k} v_j f_j = v_1 f_1 + v_2 f_2 + \cdots + v_k f_k$$

Notice that since each of the n observations goes into exactly one of the k bins, the bin frequencies must add to the total number, n, of observations:

$$\sum_{j=1}^{k} f_j = f_1 + f_2 + \cdots + f_k = n$$

The mean for the grouped data is thus given by

$$\bar{X} = \frac{1}{n} \sum_{j=1}^{k} v_j f_j \qquad 3.18$$

Table 3.1 shows the calculation for the blood-pressure frequency distri-

SECTION 3.13 MEANS AND VARIANCES FOR GROUPED DATA

Table 3.1 Calculating the Mean for Grouped Data

Bin	Bin Mark v_j	Frequency f_j	Product $v_j f_j$
−35 to −31	−33	7	−231
−30 to −26	−28	7	−196
−25 to −21	−23	12	−276
−20 to −16	−18	31	−558
−15 to −11	−13	27	−351
−10 to −6	−8	31	−248
−5 to −1	−3	22	−66
0 to 4	2	18	36
5 to 9	7	7	49
10 to 14	12	4	48
15 to 19	17	2	34
20 to 24	22	1	22
25 to 29	27	1	27
30 to 34	32	1	32
		$\Sigma f_j = 171$	$\Sigma v_j f_j = -1678$

$$\bar{X} = \frac{-1678}{171} = -9.81$$

bution. The mean comes out to −9.81. The original calculation according to the formula $\bar{X} = \Sigma X_i/n$, as in Section 3.4, gave a mean of −9.85. The difference between −9.85 and −9.81 is due to the fact that the values in a bin are not all exactly at the bin mark. The two answers are close enough for all practical purposes.

The mean as center of gravity was illustrated in Figure 3.1 for the four observations 1, 8, 7, and 1. Now suppose the observations are grouped as follows:

Bin	Mark	Frequency
0–2	1	2
3–5	4	0
6–8	7	2

And now imagine weights of 2, 0, and 2 pounds placed at the bin marks 1, 4, and 7, respectively. The grouped mean

$$\frac{1 \times 2 + 4 \times 0 + 7 \times 2}{4} = 4$$

Figure 3.8

The mean as a center of gravity

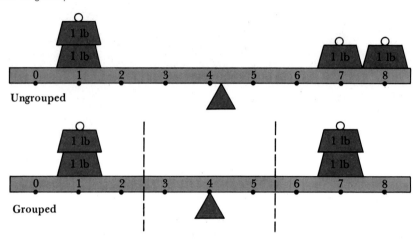

is the balance point for this arrangement of weights, as shown in the bottom half of Figure 3.8. The rightmost weight, originally at 8, has been moved to the bin mark 7, changing the balance point from 4.25 to 4. The same kind of arrangement of weights can be imagined for more complicated sets of data. The point is that if the bin interval is small, no weight travels very far when it is moved from its original position to its bin mark, which means that the balance point cannot travel very far either.

The variance and standard deviation can also be calculated for grouped data. As when computing the mean, we proceed as though each value in a given bin were equal to the bin mark. To find the sum of squares we calculate the mean by the formula (3.18). Then we find the deviations of the bin marks from this mean. The deviation for bin j is $v_j - \bar{X}$, its square is $(v_j - \bar{X})^2$, and the f_j observations in bin j contribute $(v_j - \bar{X})^2 f_j$ to the total sum of squares. Therefore the total sum of squares is the sum of the contributions of the various bins, that is, $\Sigma_{j=1}^{k} (v_j - \bar{X})^2 f_j$. And the variance for the grouped data is

$$s^2 = \frac{1}{n-1} \sum_{j=1}^{k} (v_j - \bar{X})^2 f_j \qquad 3.19$$

As before, the standard deviation, s, is the square root of this:

$$s = \sqrt{s^2} \qquad 3.20$$

Table 3.2 shows the calculations for the blood-pressure data, and the standard deviation comes out to 11.78. In Section 3.11 the standard devia-

SECTION 3.13 MEANS AND VARIANCES FOR GROUPED DATA

Table 3.2 Calculating the Standard Deviation for Grouped Data

v_j	f_j	$v_j - \bar{X}$	$(v_j - \bar{X})^2$	$(v_j - \bar{X})^2 f_j$
−33	7	−23.19	537.78	3764.43
−28	7	−18.19	330.88	2316.13
−23	12	−13.19	173.98	2087.71
−18	31	−8.19	67.08	2079.36
−13	27	−3.19	10.18	274.75
−8	31	1.81	3.28	101.56
−3	22	6.81	46.38	1020.27
2	18	11.81	139.48	2510.57
7	7	16.81	282.58	1978.03
12	4	21.81	475.68	1902.70
17	2	26.81	718.78	1437.55
22	1	31.81	1011.88	1011.88
27	1	36.81	1354.98	1354.98
32	1	41.81	1748.08	1748.08
	171			23588.00

$$s^2 = \frac{1}{n-1} \sum (v_j - \bar{X})^2 f_j = \frac{23588.00}{170} = 138.75$$

$$s = \sqrt{s^2} = \sqrt{138.75} = 11.78$$

tion was computed for the ungrouped data by means of the formulas (3.9) and (3.10), and the answer was 11.55. The discrepancy between 11.55 and 11.78 shows the effect of grouping—the effect is small.

The machine formula (3.19) for the variance can be put in a form analogous to (3.12):

$$s^2 = \frac{1}{n-1} \left[\sum_{j=1}^{k} v_j^2 f_j - \frac{1}{n} \left(\sum_{j=1}^{k} v_j f_j \right)^2 \right]$$

Problems

53. Apples for market are graded by weight in ounces. A sample of the harvest has the following frequency distribution. Find the grouped mean and standard deviation.

Weight	Frequency
.21–.30	31
.31–.40	45
.41–.50	36
.51–.60	23
.61–.70	11

54. For the distribution given in Problem 10 of Chapter 2, find
    a. The mean.   b. The median.
    c. The variance.   d. The standard deviation.
55. Find the mean and median.

Class	Relative Frequency
20 but less than 40	.12
40 but less than 60	.28
60 but less than 80	.36
80 but less than 100	.24

56. Find the mean and median.

Class	Relative Frequency
5 but less than 10	.2
10 but less than 15	.5
15 but less than 20	.3

57. The following data are the March contributions to a local charity. Find approximate values for the mean and standard deviation

Amount (Dollars)	Frequency
1–5	151
6–10	213
11–19	125
20–49	74
50–99	52
100–200	31

SECTION 3.14 QUICK MEASURES OF LOCATION AND SPREAD 81

58. For each of the following sets of data find
  a. The mean and median without grouping the data.
  b. The variance and standard deviation without grouping.
  c. The mean, median, and mode after grouping the data.
  d. The variance and standard deviation after grouping.
    (i) The 66 time-lapse observations of Problem 6 of Chapter 2.
    (ii) The differences in cutoff voltages of the 48 electron tubes of Problem 21 of Chapter 2.
    (iii) The 96 determinations of Problem 23 of Chapter 2.

3.14 Quick Measures of Location and Spread

upper quartile
lower quartile

The median for ungrouped data was defined in Section 3.6. The observations are arranged in an array in order of size from smallest to largest. The median is the middle value in this array. The **upper quartile** and the **lower quartile** are the observations lying one-quarter of the way along, counting respectively from the top and bottom of the array. These definitions must be made precise, just as it was necessary in defining the median to distinguish the cases where the number n of observations is odd or even.

To find the lower quartile first calculate one-quarter of n and add $\frac{1}{2}$ to obtain $\frac{1}{4}n + \frac{1}{2}$. Round this to the nearest integer and count up that many observations from the bottom of the array; the observation in that position is the lower quartile. Consider as an example the array

$$1 \quad 1 \quad \mathbf{2} \quad 3 \quad 3 \quad 8 \quad 11 \quad 14 \quad \mathbf{19} \quad 19 \quad 20$$

the first one in Section 3.6. Here $n = 11$, and so $\frac{1}{4}n + \frac{1}{2} = \frac{1}{4}(11) + \frac{1}{2} = 3.25$, which rounds off to 3; therefore the lower quartile is the third observation from the bottom, namely 2.

There is a case that the rule as formulated does not yet cover. Suppose $\frac{1}{4}n + \frac{1}{2}$ stands midway between two integers; should it be rounded up or rounded down? The rule is not to round at all, but to take the two integers that $\frac{1}{4}n + \frac{1}{2}$ stands between, take the observations in these two positions in the array, and take their average as the lower quartile. For the second array in Section 3.6,

$$2 \quad 5 \quad \mathbf{5} \quad \mathbf{6} \quad 7 \quad 10 \quad 15 \quad 21 \quad \mathbf{21} \quad \mathbf{23} \quad 23 \quad 25$$

we have $\frac{1}{4}n + \frac{1}{2} = \frac{1}{4}(12) + \frac{1}{2} = 3.5$. According to the rule the lower quartile is the average of the observations in positions 3 and 4, which is $(5 + 6)/2$, or 5.5.

The upper quartile is defined the same way, but the observations are counted from the top of the array instead of from the bottom. For the first array the upper quartile is the observation in position 3 from the top, namely 19. For the second array it is the average of the observations in positions 3 and 4 from the top, namely $(21 + 23)/2 = 22$.

A rough idea of the location and spread of the data can be obtained quickly from the median and the two quartiles. For the first array above, they are

$$\text{lower quartile} = 2, \quad \text{median} = 8, \quad \text{upper quartile} = 19$$

interquartile range
trimean

A common measure of spread is the **interquartile range,** the difference of the two quartiles, in this case $19 - 2$, or 17. And a measure of location increasingly coming into use is the **trimean,** computed from the formula

$$\text{trimean} = \frac{\text{lower quartile} + (2 \times \text{median}) + \text{upper quartile}}{4}$$

In our example, the trimean is $[2 + (2 \times 8) + 19]/4 = 37/4 = 9.25$. Recall (see Section 3.8) that the mean can be unrepresentative if there are wild observations (excessively large or small). The trimean has the advantage that it is not affected by a few extreme observations; it has the advantage over the median that it takes some account of the pattern in which the observations fall on either side of the middle value.

For small n all these measures can be found from a stem-and-leaf plot. For the 27 examination scores in Table 2.12, $\frac{1}{4}n + \frac{1}{2} = \frac{1}{4}(27) + \frac{1}{2} = 7.25$, which rounds to 7. The seventh smallest observation is the largest in the 6-row of Table 2.13, and the seventh largest is the largest in the 8-row. The median was earlier found to be 81. Hence,

$$\text{lower quartile} = 69, \quad \text{median} = 81, \quad \text{upper quartile} = 88$$

$$\text{interquartile range} = 88 - 69 = 19$$

$$\text{trimean} = \frac{69 + (2 \times 81) + 88}{4} = \frac{319}{4} = 79.75$$

These measures are for ungrouped data and are used mostly for small values of n. The median for *grouped* data was defined by graphical interpolation on the ogive, as in Figure 3.3. The quartiles for grouped data can be defined similarly. The lower quartile is a number that exceeds about one-quarter of the observations, and for the 171 blood-pressure changes $\frac{1}{4}n = \frac{1}{4}(171) = 42.75$. Find this number on the vertical cumulative frequency scale in Figure 3.3 and trace horizontally over to the curve and down. The point arrived at on the bottom scale, about -17.0 in this case, is the approximate lower quartile.

The upper quartile is found similarly. Find $\frac{3}{4}n = \frac{3}{4}(171) = 128.25$ on the cumulative frequency scale, trace over and down as indicated in Figure 3.3, and read off the approximate upper quartile—about -2.0 in this case.

SECTION 3.15 INDEX NUMBERS

There are more generally defined quantities of which the median and quartiles are special cases. These are the *percentiles*. If p is a number between 0 and 100, the pth percentile is the value that exceeds about $p\%$ of the observations. To find the 40th percentile from Figure 3.3, calculate 40% of 171, getting $.40 \times 171 = 68.4$; find this on the vertical scale, and then trace over to the curve and down in the usual way, arriving at the number -13.0 (approximately). This number, -13.0, is the 40th percentile. In this terminology, the lower quartile is the 25th percentile, the median is the 50th percentile, and the upper quartile is the 75th percentile.

Problems

59. Find for the data in Problem 22 of Chapter 2
  a. The median.   **b.** The interquartile range.
  c. The trimean. **d.** The 20th and 80th percentiles.

60. Problem 18 of Chapter 2 involved the ogives for the data in three earlier problems. From each of these three ogives find approximate values for
  a. The median.   **b.** The lower quartile.
  c. The upper quartile. **d.** The interquartile range.
  e. The 10th and 90th percentiles.

61. For the following data find the lower quartile, the upper quartile, the median, the trimean, and the interquartile range.
  a. $-2, 4, -1, 3, 5, 7, 0$.
  b. $1, 3, 5, 7, 9, 11, 13, 15, 17, 19, 21, 23, 25, 27, 29, 31$.
  c. $5, 1, 2, 5, 4, 5, 3, 3$.

62. Find the trimean and the interquartile range for the lake levels in Problem 27 of Chapter 2.

63. The largest value in a data set of size 100 was incorrectly recorded in such a way that the incorrect value is larger than the correct value. Which of the following quantities are affected by this error: the mean, the median, the trimean, the variance, the range, the interquartile range?

3.15 Index Numbers*

Index numbers give quantitative descriptions of change over time. The change they describe usually has to do with business or economics or social

*The material in this section is not required in the rest of the book.

Table 3.3 U.S. Population

	Year		
	1940	1960	1980
Population (millions)	132	180	220
Population index (1940 base)	100	136	167

conditions. Table 3.3 gives the population of the United States for 1940, 1960, and 1980. The ratio of the population for 1960 to the population for 1940 is 180/132, or 1.36; multiplying by 100 gives 136, the second entry in the bottom row of the table. We can say that in 1960 the population was 136% what it was in 1940, or that the population was up 36%. The ratio of the population for 1980 to that for 1940 is 220/132, or 1.67; multiply by 100 to get the third entry, 167, in the bottom row. Between 1940 and 1980 the population rose 67%.

An index number is a ratio multiplied by 100 to put it in the form of a percentage.

Table 3.4 shows for the years 1976 to 1979 the amounts of oil imported into the United States from OPEC countries. Dividing 2,260,482 by 1,849,017 gives 1.223, and this multiplied by 100 is 122.3, the second entry in the second row. Imports in 1977 were 122.3% what they were in 1976—up 22.3%. The other two indices in the second row are computed in the same way. Imports in 1978 were below what they were in 1977, but still they were 11.3% above the 1976 level.

The indices in the second row of Table 3.4 all have for their denomina-

Table 3.4 U.S. Oil Imports from OPEC Sources

	Year			
	1976	1977	1978	1979
Oil imported (thousands of barrels)	1,849,017	2,260,482	2,057,468	2,023,341
Import index: 1976 base	100.0	122.3	111.3	109.4
Import index: 1978 base	89.9	109.9	100.0	98.3

SECTION 3.15 INDEX NUMBERS

Table 3.5 Simple Aggregate Price Index

Item	1970 Price p_{0i}	1975 Price p_{1i}	1980 Price p_{2i}
1. Milk (dollars/qt)	$0.27	$0.31	$0.49
2. Steak (dollars/lb)	1.08	1.17	1.71
3. Butter (dollars/lb)	0.76	0.85	1.89
4. Pepper (dollars/lb)	2.50	2.10	2.20
Total	$4.61	$4.43	$6.29
Price index computed by formulas (3.21)	100.0	96.1	136.4

tor 1,849,017, the import figure for 1976, which is thus the *base year* and serves as the standard of comparison. If, instead, we divide the import figures by 2,057,468, the figure for 1978, we obtain the import indices with 1978 as the base year; these are given in the last row of the table. For example, imports of OPEC oil in 1979 were 98.3% of what they were in 1978.

An index number gives a clear picture of trend. The theory and computation of an index number are ordinarily more complicated than are those for the population index and the import index above. This is because an index number ordinarily represents the total combined effect of changes in a number of varying quantities. The best-known index is the U.S. Bureau of Labor Statistics Consumer Price Index, or CPI, usually called the cost-of-living index. The CPI is a statistical measure of changes in prices of goods and services bought by urban wage earners and clerical workers. Its importance lies partly in the fact that union contracts often contain cost-of-living escalation clauses.

The prices of some 400 items go into the CPI. To make clear its structure, we construct a simple illustrative example involving the four items whose prices are listed in Table 3.5. We take 1970 as the base year, and the problem is to construct a sensible index for 1975 compared with 1970 and for 1980 compared with 1970. The first procedure that comes to mind is simply to add the prices for a year and divide by the total 4.61 for 1970, which gives the indices

$$100 \times \frac{4.61}{4.61} = 100.0, \quad 100 \times \frac{4.43}{4.61} = 96.1,$$

$$100 \times \frac{6.29}{4.61} = 136.4$$

In other words, if p_{0i}, p_{1i}, and p_{2i} represent the respective prices of item i ($i = 1, 2, 3, 4$) for each of the three years, we compute

$$100 \times \frac{\Sigma_i p_{0i}}{\Sigma_i p_{0i}} = 100, \qquad 100 \times \frac{\Sigma_i p_{1i}}{\Sigma_i p_{0i}}, \qquad 100 \times \frac{\Sigma_i p_{2i}}{\Sigma_i p_{0i}} \qquad 3.21$$

But the index 96.1 for 1975 makes it appear that prices have dropped below the 1970 level. A glance at the table shows that although the prices of milk, steak, and butter have risen considerably, this increase is more than offset in our computation by the decrease in the price of pepper, which costs a lot per pound. Since pepper is surely a negligible item in the average budget, we should get a clearer idea of food costs if we leave it out, which leads to new totals and indices:

	1970	1975	1980
Total (pepper omitted)	$2.11	$2.33	$4.09
Price index	100.0	110.4	193.8

Now the indices reflect an increase in prices. Pepper played a disproportionate role. But perhaps one of the other items plays a disproportionate role, too, in some less obvious way.

This shows that the prices must somehow be weighted according to their importance. The problem of units also suggests that prices should be weighted. The milk price is given in dollars per quart, but why not use gallons instead? Why is a quart rather than a gallon comparable to a pound, the unit for the other three items? If the milk cost is expressed in dollars per gallon, it becomes four times what it was and exerts proportionally greater influence in the computation given by the formulas (3.21).

Suppose the typical family of four in 1970 purchased the four items in the amounts shown in Table 3.6. The four food items, together with the quantities in which they were purchased, are called a *market basket*. It leads to a reasonable and widely used index. For item i, the 1970 price was p_{0i} and the 1970 quantity was q_{0i}; hence, $p_{0i}q_{0i}$ was the total spent on item i in the base year. For example, 728 quarts of milk at $0.27 per quart (Table 3.6) comes to 0.27 × 728, or $196.56, for the year's expenditure on milk. The remaining figures in the 1970 cost column of Table 3.6 are computed the same way; their sum, $576.07, is the total spent on the market basket in 1970.

Now in 1975 the price for item i was p_{1i}; if the quantity were still q_{0i} as in the base year, the total spent on item i in 1975 would be $p_{1i}q_{0i}$. Thus 728 quarts of milk at $0.31 per quart comes to $225.68. The rest of the

Table 3.6 Price Index With Base-Year Quantity Weights (The Laspeyres Index Number)

Item	1970 Quantity q_{0i}	1970 Cost $p_{0i}q_{0i}$	1975 Cost $p_{1i}q_{0i}$	1980 Cost $p_{2i}q_{0i}$
1. Milk	728 qt	$196.56	$225.68	$356.72
2. Steak	312 lb	336.96	365.04	533.52
3. Butter	55 lb	41.80	46.75	103.95
4. Pepper	0.3 lb	0.75	0.63	0.66
	Total	$576.07	$638.10	$994.85
Price index computed by Equation (3.22)		100.0	110.8	172.7

1975 column is computed the same way; its sum, $638.10, is what the 1970 market basket would cost in 1975.

In point of fact, buying patterns change. Nonetheless, we use the 1970 quantities q_{0i} in 1975 also—we hold the market basket constant. If I switch from beer to champagne, my increased expenses are not to be entirely attributed to inflation or the diminished purchasing power of the dollar. To measure that, we must imagine that living standards are constant. The total array of foods which in 1970 cost $576.07 in 1975 cost $638.10, and the increase reflects a genuine increase in cost, not an increase in standard of living (which may have taken place also). The ratio 638.10/576.07, or 1.108, an index of 110.8, measures the increase relative to the base year; in 1975 prices were up 10.8% over their 1970 level. The index for 1980 being 172.7, prices then were up 72% over their 1970 level.

Notice that pepper now plays a negligible role, as it should. The 1975 pepper expenditure was down 12 cents, but this hardly affects the total picture.

The actual CPI is computed as in our example, although it involves not four food items, but around 400 items concerning food, transportation, medical costs, and so on. The general description of a price index is this: Let the index i refer to the item; i ranges from 1 to 4 in our example and from 1 to about 400 for the real CPI. Let the index n refer to the time period; n ranges over 0, 1, and 2 in our example, but any number of times can be considered and the period could be a month, say, rather than five years. The price of item i in period n is p_{ni} and the quantity is q_{ni}. The amount spent on item i in the base period is $p_{0i}q_{0i}$, for a grand total of $\Sigma p_{0i}q_{0i}$. If in period n the quantities were q_{0i} as in the base period, the total for item i

would be $p_{ni}q_{0i}$ and the grand total expended would be $\Sigma p_{ni}q_{0i}$. The index

$$I_{n:0} = 100 \times \frac{\Sigma_i p_{ni}q_{0i}}{\Sigma_i p_{0i}q_{0i}} \qquad 3.22$$

measures the change in cost for a fixed buying pattern. This is called the *Laspeyres index number*.

The Laspeyres index has the algebraically equivalent form

$$I_{n:0} = 100 \times \frac{\Sigma_i p_{0i}q_{0i}(p_{ni}/p_{0i})}{\Sigma_i p_{0i}q_{0i}}$$

In other words, $I_{n:0}$ is 100 times the weighted sum (see Equation (3.5), page 53) of the price ratios p_{ni}/p_{0i} for the various items, the weights being the amounts $p_{0i}q_{0i}$ spent in the base period. The *Paasche index number* is defined by

$$P_{n:0} = 100 \times \frac{\Sigma_i p_{ni}q_{ni}}{\Sigma_i p_{0i}q_{ni}} = 100 \times \frac{\Sigma_i p_{0i}q_{ni}(p_{ni}/p_{0i})}{\Sigma_i p_{0i}q_{ni}}$$

The Laspeyres index measures the relative costs of maintaining base-period standards in the base period and in period n; the Paasche index measures the relative costs of maintaining period-n standards in the base period and in period n. The Laspeyres index is the more convenient to use on a continuing basis, because the weights remain fixed. The CPI is a modified Laspeyres index computed monthly, with the weights revised from time to time to account for changes in buying patterns.

Hosts of indices in some form related to (3.22) are computed to measure changes in regional and national retail and wholesale prices, industrial and agricultural production, and so on. The rationale for each is the same as that for the CPI.

Problems

64. A family spent $75.10 on 12 grocery items in 1980. The very same items would have cost only $68.70 in 1979. The quantity 75.10/68.70 = 1.09 is an example of which index, the Laspeyres or the Paasche?

65. For these retail sales volumes in millions of dollars

1977	66,978
1978	65,810
1979	68,352
1980	63,409

CHAPTER PROBLEMS

find the sales indices (a) with 1977 as the base year, and (b) with 1978 as the base year.

66. In 1978 the CPI was 106.7; in 1979 it was 108.1. If a worker in 1978 earned $3.90 per hour and in 1979 earned $4.00 per hour, did his buying power increase or decrease?

67. For the following data find the Laspeyres index and Paasche index with 1975 as the base year.

	1975		1980	
Item	Unit Price	Units Sold	Unit Price	Units Sold
A	$ 1.50	350	$ 2.00	600
B	15.00	100	12.50	125
C	7.50	200	10.00	300

68. For the following data find the retail price index for each year, using 1970 as the base.

Retail Price of Selected Electrical Appliances

	Average Unit Price			Thousands of Units Sold
Appliance	1970	1975	1980	1970
A	$295	$305	$308	3650
B	334	340	345	1025
C	250	261	270	1300
D	43	44	47	1275

CHAPTER PROBLEMS

69. For each of the following sets of numbers find
    a. The mean
    b. The median.
    c. The mode or modes.
    d. The range.
    e. The variance. Calculate this two ways—by the defining formula (3.9) and by the machine formula for the sum of squares.

f. The standard deviation.
  (i) 3, 5, 7, 9, 11
  (ii) 1, 1, 2, 3, 4, 4
  (iii) −3, −1, 0, 0, 1, 3
  (iv) 3, 10, 8, 5, 4, 1, 8, 2, 4
  (v) 2, 2, 4, 2, 3, 4, 0, 1, 3
  (vi) −2, −8, 3, −4, −1, 3, −5
  (vii) 20, 23, 19, 22, 25, 15, 23
  (viii) −1, 3, 3, −3, 0, 3, 2, 1

70. For the following yields of corn (grams) obtained from each of thirty hills find
   a. The mean.
   b. The median.
   c. The range.
   d. The variance.
   e. The standard deviation.
   f. The trimean.

982	1205	258	927	620	1023
395	1406	1012	762	840	960
1056	793	713	736	1582	895
1384	862	1152	1230	1261	624
862	1650	368	358	956	1425

71. The following is the distribution for the number of defective items found in 404 lots of manufactured items. Find the mean, median,

Number of Defective Items	Number of Lots
0	53
1	110
2	82
3	58
4	35
5	20
6	18
7	12
8	9
9	3
10	1
11	2
12	1

CHAPTER PROBLEMS

mode, variance, and standard deviation of the number defective per lot.

72. The medians of two samples of sizes 20 and 40 were 41 and 52, respectively. The smaller sample contained 7 observations larger than 52, and the larger sample contained 16 observations smaller than 41. Find the median of the combined samples.

73. Let $d_i = X_i - Y_i$ represent the difference between two observations on the same individual, $i = 1, 2, \ldots, n$. Show that

$$s_d^2 = s_X^2 + s_Y^2 - \frac{2}{n-1} \sum_{i=1}^{n} (X_i - \bar{X})(Y_i - \bar{Y})$$

where

$$\bar{X} = n^{-1} \sum_{i=1}^{n} X_i \quad \text{and} \quad \bar{Y} = n^{-1} \sum_{i=1}^{n} Y_i$$

74. Let X_i and Y_i represent the income in two consecutive years for the ith person in a group of n. Let $d_i = X_i - Y_i$. Which of the following relations, if any, are correct?
   a. $M_d = M_X - M_Y$.
   b. $m_d = m_X - m_Y$.
   c. $\bar{d} = \bar{X} - \bar{Y}$.
   d. $s_d = s_X - s_Y$.

75. In a study of the impact of legal periodicals the following distribution of number of citations was observed.
   a. Find the median and the mode of the number of citations.
   b. Using a change of scale, find the mean and standard deviation.

Number of Citations	Number of Publications
$0 \leq X < 1{,}000$	5
$1{,}000 \leq X < 2{,}000$	18
$2{,}000 \leq X < 3{,}000$	30
$3{,}000 \leq X < 4{,}000$	43
$4{,}000 \leq X < 5{,}000$	71
$5{,}000 \leq X < 6{,}000$	80
$6{,}000 \leq X < 7{,}000$	63
$7{,}000 \leq X < 8{,}000$	30
$8{,}000 \leq X < 9{,}000$	10
$9{,}000 \leq X < 10{,}000$	4
$10{,}000 \leq X < 11{,}000$	1
	355

Summary and Keywords

Descriptive measures or **statistics** (p. 44) summarize in compact form the main features of a data set. The principal **measures of location** (p. 50) are the **mean** (p. 50), the **weighted mean** (p. 53), the **median** (p. 56), and the **mode** (p. 59). The principal **measures of variation** or **dispersion** (p. 62) are the **range** (p. 63), the **variance** (p. 67), and the **standard deviation** (p. 67). The mean and the standard deviation, the most important measures, are conveniently defined in terms of the **sum of squares** (p. 65) by means of the **summation symbol** (p. 45). The variance is most easily calculated from the **machine formula** (p. 68); often a **change of scale** (p. 71) makes the computations simpler, and special formulas are required for **grouped data** (p. 76). The **quartiles** (p. 81), the **interquartile range** (p. 82), and the **trimean** (p. 82) are useful descriptive statistics that can be computed very quickly.

REFERENCES

3.1 Coxton, Frederick E., Cowden, Dudley J., and Bolch, Ben W. *Practical Business Statistics*. 4th ed. Englewood Cliffs, N.J.: Prentice-Hall, 1969. Chapters 3 and 4.

3.2 Hoel, Paul G. *Elementary Statistics*. 3d ed. New York: Wiley, 1971.

3.3 Huff, Darrell. *How to Lie with Statistics*. New York: Norton, 1965. Chapters 5 and 6.

3.4 Snedecor, George W., and Cochran, William G. *Statistical Methods*. 6th ed. Ames, Iowa: Iowa State University Press, 1967. Chapter 5.

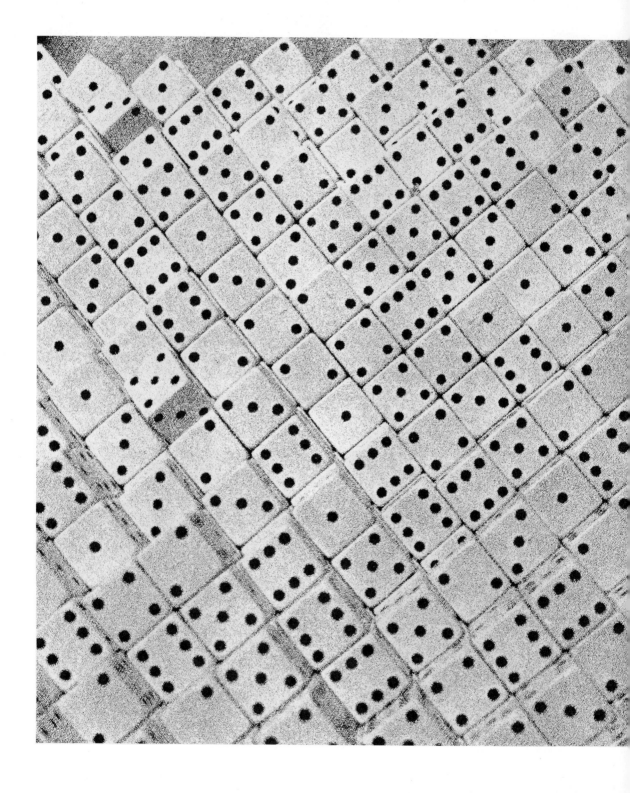

4
Probability

4.1 Introduction

Let's go back to Table 1.5 in Chapter 1, repeated here as Table 4.1. According to line 3, of the 196 patients on the placebo (the "sugar pill"), 38 suffered heart attacks or other coronary heart disease, while of the 193 on the blood-pressure-reducing drug, only 35 did. Suppose someone asserted that the drug helps prevent these events, saying, "After all, more patients in the placebo group had trouble." You would probably say, "But 38 opposed to 35—that's no evidence. It could happen by chance alone, so it's no evidence in favor of the drug."

On the other hand, most of us would find line 2 in the table pretty convincing evidence that the drug does help prevent high-blood-pressure damage. The excess of 89 over 37 is great enough that it's hard to believe that it's just an accident, that it could happen by pure chance, that the drug is in fact worthless in this department. Evaluating line 4 involves the idea of chance, too. It is not clear how strong the evidence in line 4 is. There were 20 cases of enlarged heart among the patients on placebo and only 12 among those on the active drug. Does this show that the drug helps, or could the excess of 20 over 12 be accidental—be the result of chance alone?

These problems and questions illustrate the fact that to understand statistical inference—to understand how to pass from observed facts to general conclusions—it is necessary to understand something about the science of chance, or, as it is usually called, the theory of probability. This chapter is an introduction to the elements of that subject.

Public-opinion polling provides a social-science example of the role

Table 4.1 Events Sustained by Patients in the Public Health Service Hypertension Study

	Active	Placebo
Number of patients	193	196
1. Number of cases where blood pressure increased to 130 or more	0	24
2. Number of hypertensive events	37	89
3. Number of cases of heart attack and other coronary heart disease	35	38
4. Number of cases of enlarged heart	12	20

probability plays in statistical questions. Imagine the voting population of a city split between those who intend to vote for candidate A in a coming election and those who intend to vote for candidate B. An opinion poller trying to predict the outcome of the election draws at random a sample of size 100 from the population of voters. The precise definition (to be given later on) of the term *at random* itself requires ideas from probability theory. Suppose that 72 out of the 100 voters in the sample say that they intend to vote for candidate A. In that case, the poller will with some confidence predict A as the winner, because if a majority of the voters in the population really favored B, it is very unlikely that so many as 72 voters in the sample would favor A. To understand the principles of polling it is necessary to know precisely *how* unlikely that would be. Again, to see how to pass from observed facts (in this case the composition of the poller's sample) to larger conclusions (in this case the composition of the entire voting population in the city) requires some understanding of probability theory.

Probability theory impinges on sciences other than statistics, too. Mendel's laws of heredity provide a famous example: among second-generation hybrids (say of pea plants), the probability is one in four that an individual will have a recessive characteristic rather than a dominant one (say a white blossom rather than a red one). Probability models are to be found throughout the physical, biological, and social sciences. Although there are many reasons for studying probability theory, our reason for studying it here is to use it in statistics.

Historically, probability was first studied in connection with gambling problems in the seventeenth century. It was only gradually realized how many and various its applications were. It is still true that a good way to learn the basic principles of the science of probability is through a study of problems involving cards and dice. These principles will be illustrated in this chapter mostly by such problems—by examples from gambling. The connection between probability theory and statistics will become fully apparent in the chapters that follow this one.

4.2 The Meaning of Probability

The purpose of this section is to explain the meaning of the term *probability*. Although it is worthwhile to try to arrive at a definition of probability, it is much more important to understand how probability is *used*, to see how probabilities are actually calculated.

outcomes

Imagine an operation which can result in any one of a definite set of possible **outcomes,** but which is governed by chance, so that the actual

Table 4.2 Sample Space for Rolling Dice

1-1	1-2	1-3	1-4	1-5	1-6
2-1	2-2	2-3	2-4	2-5	2-6
3-1	3-2	3-3	3-4	3-5	3-6
4-1	4-2	4-3	4-4	4-5	4-6
5-1	5-2	5-3	5-4	5-5	5-6
6-1	6-2	6-3	6-4	6-5	6-6

outcome can't be predicted with complete certainty. We have in mind the drawing of a card from a well-shuffled deck, for example, or the rolling of a pair of dice.

If the random experiment consists of drawing a card from an ordinary deck, there are 52 possible outcomes—the 52 cards in the deck. This set of all possible outcomes or results of the experiment is called the **sample space**.* Since 13 of the 52 possible outcomes in the sample space are spades, almost everyone will say that if the deck was shuffled well, the chance of drawing a spade (the chance that the outcome is a spade) is 13/52. Similarly, almost everyone will say that the chance of drawing an ace is 4/52. This is because the outcomes in the sample space are regarded as being equally likely.

For rolling a pair of dice, the sample space—the set of all 36 possible outcomes or results—is exhibited in full in Table 4.2. Here 3-1 denotes the result that the one die shows three dots and the other shows one dot. The outcome 1-3, where the order is reversed, is also listed in the table because the two dice are to be distinguished from one another. (Perhaps it is best to think of a red die and a green one, even though the dice of an actual pair are made so similar that telling them apart would require very close examination.) If the dice are fair ones (not loaded), everyone expects the 36 outcomes to be equally likely. Since the total number of dots showing is 4 for three of the outcomes (namely, for 3-1, 2-2, and 1-3), we expect the probability of rolling a 4 to be 3/36. Since the total showing is 7 for six of the outcomes (namely, for 6-1, 5-2, 4-3, 3-4, 2-5. and 1-6), we expect the probability of rolling a 7 to be 6/36.

These examples are instances of the classical conception of probability: There is a definite, finite set of possible outcomes, or results, or cases (the 52 cards, or the 36 combinations for the rolled pair of dice); this is the

---

*Sometimes it is called the *outcome space;* this is a better term, but *sample space* has become standard.

sample space. Within the sample space there is some smaller set of outcomes (the spades, the aces, the dice pairs that total 4, or the ones that total 7) whose probability we seek; this smaller set is called an **event,** or sometimes a *subset,* and the outcomes or cases making up the event are called the *favorable cases.* The probability of the event is taken to be the ratio of the number of outcomes in the event to the number of outcomes in the sample space, that is, the ratio of the number of favorable cases to the total number of cases:

event

$$\text{Probability} = \frac{\text{Number of favorable cases}}{\text{Total number of cases}} \qquad 4.1$$

$$= \frac{\text{Number of outcomes in the event}}{\text{Total number of outcomes in the sample space}}$$

In drawing a card the probability of the event "ace"—the chance of getting an ace—is 4/52 because the event "ace" comprises four of the 52 cases. In rolling dice the probability of the event "seven"—the chance of rolling a total of 7—is 6/36 because the event "seven" comprises six of the 36 cases.

The formula (4.1) is to be viewed as a *way of computing* a probability and *not* as a *definition* of the probability. Whether or not the computation leads to a result that corresponds with reality depends on the circumstances. The chance of drawing an ace is 4/52 if the deck is well shuffled, but not if a sleight-of-hand expert is in charge. The chance of rolling a 7 is 6/36 if the dice are fair, but not if they are loaded. The chance a coin will land heads may be 1/2 (there is one favorable case, heads, out of two possible cases, heads and tails), but this is not necessarily true if the coin is bent (coins even exist with heads on both sides). The formula (4.1) leads to the right answer if all the possible cases (all the outcomes in the sample space) are equally likely.

The words *right answer* in this last sentence presuppose that there is such a thing. Most of us feel that if a particular pair of dice is rolled, even a loaded pair, there does exist a certain definite probability that the result will be a seven, even though we may be entirely ignorant of the actual numerical value of that probability. This parallels the fact that most of us feel that the sixth moon of Jupiter has a certain definite mass in tons, even though we may be entirely ignorant of the actual numerical value of that mass. And most of us feel that if a particular coin (perhaps bent or unbalanced in some way) is tossed, there does exist a certain definite probability (perhaps unknown to us) that it will land heads upward.

To give a satisfactory definition of this probability is very difficult, just as it is very difficult to define with precision concepts such as mass and force in physics. As in the case of physical concepts, the fruitful procedure is to worry not about how to define probability but about how to measure it. To get an idea of the probability of rolling seven with a pair of loaded dice, the obvious procedure is to roll the dice a number of times and check the fraction of times they total seven. If 500 rolls yield 80 sevens, the probability of seven must be something like 80/500, or .16.

Experience shows that if a coin is tossed repeatedly, the relative frequency (fraction) of heads tends to stabilize at a definite value, which we take to be the probability of heads. Figure 4.1 shows the results of such an experiment. The horizontal scale in this figure represents the number of tosses of the coin, and the height of the curve over any given point on the horizontal scale represents the relative frequency of heads up to that point in the experiment. In the first few tosses there weren't many heads, and so the curve at the beginning drops down to about .4. Then there came a number of heads, and so the curve goes up to about .6: in the first dozen or so tosses, the fraction of heads was about .6. After that the curve drops down some more, and so on. The coin was tossed 600 times in all. Since the relative frequency seems to be settling down at a value near .5, we can say that the coin is well balanced.

Figure 4.1

Relative frequency of heads for a balanced coin

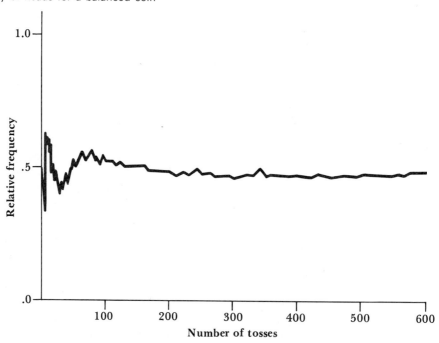

Figure 4.2

Relative frequency of heads for an unbalanced coin

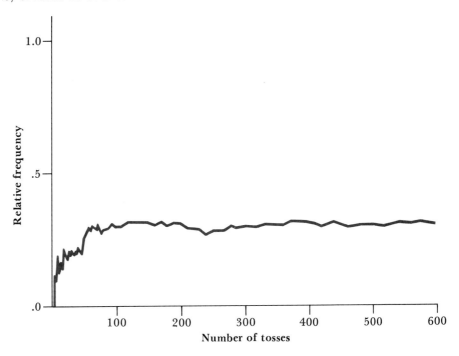

Figure 4.2 shows the same effect for a different coin, a bent one. Here the relative frequency is converging to something like .3, and the coin is seen to be unbalanced.

Experiments repeated under constant conditions show this kind of stability. Suppose we repeat over and over again some experiment, like rolling a pair of dice, and suppose we keep track of some event, like obtaining a seven. Suppose that at each repetition of the experiment we compute the relative frequency with which the event has occurred in the sequence of trials up to that point. This relative frequency will stabilize at some limit, which we take to be the **probability** of the event. This probability is to be regarded as a natural characteristic of the experiment and of the event itself, just as the mass of the sixth moon of Jupiter is a physical characteristic of that body. (A different conception of probability is briefly treated in Section 4.7.)

Now just as physical quantities like force and mass can often be computed mathematically, so can this probability. The formula applicable in many cases of interest to us is (4.1): The probability as given by (4.1) will often agree closely with the value at which the relative frequency of the event would stabilize in a long sequence of empirical trials of the experiment. This kind of agreement will exist when the outcomes of the experiment (the elements of the sample space) are equally likely or nearly so, and this in turn will be the case in the presence of an appropriate symmetry: when the coin is balanced, for example, or the dice are uniformly made, or the deck is shuffled so that no card has precedence over any other.

Probability, then, represents long-range frequency of occurrence. But, as we said at the beginning of this section, how probability is defined is much less important than how it operates. The next section sets out the laws that probabilities obey and the rules for calculating them.

4.3 Computing Probabilities

Recall that our probabilities are to be computed by the formula

$$\text{Probability} = \frac{\text{Number of favorable cases}}{\text{Total number of cases}} \qquad 4.2$$

This formula is based on the assumption that all cases are equally likely. The numerator and the denominator here can often be found by a simple counting out of the possibilities.

Example 1

If a pair of dice are rolled, what is the probability that both dice show the same face? The sample space consists of the 36 outcomes laid out in Table 4.2. The event in question consists of the outcomes along the diagonal: 1-1, 2-2, 3-3, 4-4, 5-5, 6-6. There are six of them, and so the probability that the numbers on the dice are the same is 6/36, or $\frac{1}{6}$.

Example 2

Suppose an ordinary deck of 52 cards is shuffled thoroughly and two cards are then dealt out, one to Alice and one to Bill. What is the probability that the two cards dealt are of the same suit? The dots in Table 4.3 represent the outcomes of the experiment. The row the dot is in specifies Alice's card and the column specifies Bill's card; for example, the dot in the lower left corner is the case where Alice gets the ace of spades and Bill gets the two of clubs. Table 4.3 is of course much larger than Table 4.2, and for clarity the outcomes are represented only by dots. The important difference, however, is that there are no outcomes along the diagonal. If Alice holds the king of

SECTION 4.3 COMPUTING PROBABILITIES

Table 4.3 Sample Space for Dealing Two Cards

clubs, then Bill can't hold that card, and so in the twelfth row there is no dot in the twelfth column. Since there are 52 rows and there are 51 outcomes in each row, the number of outcomes in the sample space is 52 × 51, which comes to 2,652.

What is the probability that the two cards are of the same suit? How many favorable cases are there? How many outcomes are there in this event? If this event is to occur, then whatever card Alice gets, Bill must get one of the 12 other cards of the same suit. For example, if Alice gets the ace of clubs, then Bill must get a club, but of course not the ace of clubs; in the thirteenth row of Table 4.3, the first twelve dots are the ones in the event we are talking about. Each of the 52 rows contributes 12 outcomes to the event, and so the number of favorable cases is 52×12, or 624. The probability that the two cards dealt are of the same suit is therefore $(52 \times 12)/(52 \times 51)$, or $12/51$, which is about .24.

We turn now to some general properties of probabilities. We illustrate them mostly by examples in which the outcomes are equally likely, so that probabilities are given by the formula (4.2). Each property also holds, however, even if the outcomes are not equally likely. We denote events by A and B, etc., and we denote their probabilites by $P(A)$ and $P(B)$, etc.

The first property is easy to see:

$$0 \leq P(A) \leq 1 \qquad 4.3$$

The second property has to do with the addition of probabilities. An example will make clear its meaning and truth.

Example 3

For rolling a pair of dice the sample space consists of the 36 outcomes in Table 4.2. If A is the event that the total is 4, we know that A contains three outcomes:

$$A = \{3\text{-}1,\ 2\text{-}2,\ 1\text{-}3\}$$

The event B that the total is 3 contains two outcomes:

$$B = \{2\text{-}1,\ 1\text{-}2\}$$

Now "A or B" stands for the event that the total is *either* 4 *or* 3:

$$A \text{ or } B = \{3\text{-}1,\ 2\text{-}2,\ 1\text{-}3,\ 2\text{-}1,\ 1\text{-}2\}$$

By three applications of formula (4.2) for computing probabilities,

$$P(A \text{ or } B) = \frac{5}{36} = \frac{3}{36} + \frac{2}{36} = P(A) + P(B)$$

disjoint
mutually
exclusive

Two events are **disjoint**, or **mutually exclusive**, if there is no outcome that belongs to both of them. This is true of the events A and B in Example 3. The general rule is this:

SECTION 4.3 COMPUTING PROBABILITIES

mutually exclusive

Addition of probabilities If the events A and B are <u>disjoint, then</u>

$$P(A \text{ or } B) = P(A) + P(B) \qquad 4.4$$

If events A and B are disjoint, so that they have no outcomes in common, then the number of outcomes in "A or B" must be the number in A plus the number in B, and so the equation (4.4) is a consequence of (4.2).

Events A and B may be said to be disjoint *<u>if they cannot both happen at the same time</u>*. Understanding this point will remove a common source of confusion. The sample space represents *a single trial* of the experiment, and A and B are disjoint if they cannot both occur in one trial. The dice in Example 3 cannot total both 4 and 3 at once; but none of this prevents them from totaling 4 on one trial and 3 on another trial.

Equation (4.4) can be extended to three events. If A, B, and C are disjoint in the sense that no two can happen at once, then

$$P(A \text{ or } B \text{ or } C) = P(A) + P(B) + P(C) \qquad 4.5$$

The same thing holds for four or more events, as illustrated by the next example.

Example 4

The events that dice total 2, 4, 6, 8, 10, and 12 have respective probabilities (refer to Table 4.2) 1/36, 3/36, 5/36, 5/36, 3/36, and 1/36. The probability that some one of these events occurs—the probability, that is, of an even-valued total—is therefore

$$\frac{1}{36} + \frac{3}{36} + \frac{5}{36} + \frac{5}{36} + \frac{3}{36} + \frac{1}{36} = \frac{18}{36} = \frac{1}{2}$$

If A and B are not disjoint, Equation (4.4) does not apply, but there is another formula that does apply:

$$P(A \text{ or } B) = P(A) + P(B) - P(A \text{ and } B) \qquad 4.6$$

An example from cards will show how the rule works.

Example 5

A card is drawn at random from an ordinary deck; A is the event of drawing a spade, and B is the event of drawing a face card. Here "A or B" is the event of drawing a card that is a spade *or* a face card *or both*—the word "or" does not exclude the possibility that *both* events occur. Of course, "A and B" is the event that the card is both a spade *and* a face card. The number of outcomes in A is 13, the number in B is 12, and the number in

"A and B" is 3. By Equation (4.6), the probability of drawing a spade or a face card is

$$P(A \text{ or } B) = \frac{13}{52} + \frac{12}{52} - \frac{3}{52} = \frac{22}{52}$$

Each of the two terms $P(A)$ and $P(B)$ on the right in Equation (4.6) accounts for the outcomes that are both in A and in B; subtracting $P(A$ and $B)$ compensates for this double counting. The formula (4.6) contains as a special case the rule (4.4) that applies when A and B are disjoint events: If A and B are disjoint, then the event "A and B" can't happen and hence has probability 0.

complement   The **complement** of an event A is the "opposite" event, the one that happens exactly when A does not. We denote it by $\bar{A}$. There is a simple connection between the probabilities of the two events.

Complementary events   The probabilities of A and $\bar{A}$ are related by

$$P(\bar{A}) = 1 - P(A) \qquad \qquad 4.7$$

Another dice example will illustrate this formula.

Example 6

If two dice are rolled, what is the probability that the two numbers that turn up are *different?* The probability of the opposite event, the complementary event, was calculated in Example 1. This complementary event is the event that the two numbers are the *same*, and its probability is $\frac{1}{6}$. The probability that the two numbers are different is therefore $1 - \frac{1}{6}$, or $\frac{5}{6}$.

Example 7

If a card is dealt to Alice and a card is dealt to Bill, as in Example 2, what is the chance that the two cards are of different suits? Alice can get any of the 52 cards, and if Bill is to have a card of a different suit, then he must get one of the 39 cards in the other three suits. For example, if Alice gets the king of spades, Bill can get any of the 13 clubs, or any of the 13 hearts, or any of the 13 diamonds, which gives 39 possibilities in all. Thus each of the 52 rows in Table 4.3 contributes 39 outcomes to the event that the cards are of different suits, so that there must be 52 × 39 outcomes in this event. The probability that the two cards dealt are of different suits is therefore (52 × 39)/(52 × 51), which reduces to $\frac{39}{51}$.

SECTION 4.3 COMPUTING PROBABILITIES

On the other hand, the probability that the two cards are of the *same* suit was calculated in Example 2, and it was $\frac{12}{51}$. And by (4.7), the probability of the opposite event, the event that the suits are *different*, is $1 - \frac{12}{51}$. But $1 - \frac{12}{51}$ and $\frac{39}{51}$ are equal—the two methods give the same answer. Often it is easier to get at a probability by using the formula (4.7) than it is to calculate it directly.

conditional probability

An important adjunct to the notion of probability is the notion of **conditional probability.**

Conditional probability The conditional probability of an event B given another event A, which we denote $P(B|A)$, is defined by

$$P(B|A) = \frac{P(A \text{ and } B)}{P(A)} \qquad 4.8$$

This equation represents a definition, not a fact. (If $P(A) = 0$, the conditional probability is not defined.) An example will show that the definition is a sensible one.

Example 8

In rolling dice, let A be the event that the total does not exceed 5. The triangle in the accompanying table encloses the outcomes in A, and the probability of A, $P(A)$, is $10/36$. Let B be the event that the total is even. We saw in Example 4 that $P(B)$ is $1/2$; the 18 outcomes favoring B are shown in boldface in the table. If you know that A has occurred—that the total is at most 5—what should you take as the probability of B?

1-1	1-2	1-3	1-4	**1-5**	1-6
2-1	**2-2**	2-3	**2-4**	2-5	**2-6**
3-1	3-2	**3-3**	3-4	**3-5**	3-6
4-1	**4-2**	4-3	**4-4**	4-5	**4-6**
5-1	5-2	**5-3**	5-4	**5-5**	5-6
6-1	**6-2**	6-3	**6-4**	6-5	**6-6**

It is difficult (at least for those of us untrained in Zen) to imagine rolling the dice and noticing that the total is 5 or less without also noticing whether or not the total is even. An effective way to understand the notion of conditional probability is to imagine that a friend or referee has rolled the dice out of your sight and that he reports to you that the total does not exceed 5 but does not tell what the total is. If you are in receipt of this partial information about the outcome, what should be your probability for B?

To know that the total is 5 or less is to know that the outcome is one of those in the triangle. As far as you are concerned, after the referee's report, the sample space consists of the ten outcomes in the triangle. Four of these, the ones in boldface, correspond to an even total, and so your *conditional probability,* given the partial information, should be 4/10.

This is exactly the answer that (4.8) gives. The event "*A* and *B*," the event that the total is both even and at most 5, consists of the four boldface outcomes within the triangle, and so $P(A \text{ and } B)$ is equal to 4/36; since we know already that $P(A)$ equals 10/36, the formula (4.8) gives

$$P(B|A) = \frac{\frac{4}{36}}{\frac{10}{36}} = \frac{4}{10}$$

which does check.

The ordinary probability $P(B)$ is sometimes called an *unconditional probability* to distinguish it from the conditional probability $P(B|A)$. A second illustration of conditional probability follows.

Example 9

For the experiment of drawing a single card from a deck of 52 cards, let *A* be the event "spade," and let *B* be the event "face card." Then "*A* and *B*" consists of the three face cards that are also spades, and so $P(A \text{ and } B)$ is $\frac{3}{52}$. Since $P(A)$ equals $\frac{13}{52}$, the definition (4.8) gives

$$P(B|A) = \frac{\frac{3}{52}}{\frac{13}{52}} = \frac{3}{13}$$

As before, this makes sense: If the referee draws a card and tells you it is a spade, as far as you are concerned the 13 spades are now the possible outcomes and, of these, three favor the event "face card." Notice that since $P(B) = \frac{12}{52} = P(B|A)$, the conditional and unconditional probabilities coincide in this case, which is not true in Example 8.

Example 10

As in Examples 2 and 7, the experiment consists of dealing a card to Alice and one to Bill. Let *A* be the event that Alice gets a spade, and let *B* be the event that Bill gets a club. The outcomes in *A* are represented by the dots in the last 13 rows of Table 4.3—the rows corresponding to a spade for Alice—and there are 13×51 of these (remember, 51 dots to a row). Therefore $P(A)$ is $(13 \times 51)/(52 \times 51)$, which by cancellation is the same thing as $\frac{13}{52}$. This of course makes sense: when Alice is dealt her card, there are 52 cards she might get, and 13 of these are spades.

The event "A and B" consists of those outcomes where Alice gets a spade and Bill gets a club; these outcomes correspond to the dots in the 13-by-13 block of dots in the lower left-hand corner of Table 4.3, and so there are 13 × 13 of them. Therefore $P(A$ and $B)$ is $(13 \times 13)/(52 \times 51)$. By (4.8), the conditional probability that Bill gets a club, given that Alice has got a spade, is

$$P(B|A) = \frac{P(A \text{ and } B)}{P(A)} = \frac{(13 \times 13)/(52 \times 51)}{(13 \times 51)/(52 \times 51)} = \frac{13}{51}$$

This, too, makes sense. After Alice has received her card, there are 51 left in the deck. And if her card is a spade, then the number of clubs left is 13.

In this example, interpreting $P(B|A)$ requires no conceptual referee. Two cards are dealt in succession; there is no difficulty in imagining yourself between the two, knowing the first card was a spade but ignorant of what the second will be.

Multiplying both sides of the formula (4.8) by $P(A)$ gives

$$P(A \text{ and } B) = P(A) \times P(B|A) \qquad 4.9$$

Sometimes, as in Examples 8, 9, and 10, we know $P(A)$ and $P(A$ and $B)$ in advance and we put them into (4.8) to find the value of $P(B|A)$. On the other hand, sometimes it is $P(A)$ and $P(B|A)$ that we know in advance, and we put them into (4.9) to find the value of $P(A$ and $B)$, as in the next example.

Example 11

I have just won my semifinal match in a tennis tournament, and the other semifinal match, between Al and Bob, is in progress. Let A be the event that Al will win, and suppose that $P(A) = .6$. Let B be the event that I will take the final match and hence the tournament, and suppose that $P(B|A) = .3$—my chance of beating Al, if he wins the semifinal, is .3. Then by (4.9),

$$P(A \text{ and } B) = P(A)P(B|A) = .6 \times .3 = .18;$$

the chance that Al will win the other semifinal match and that I will then beat him in the final match is .18.

Suppose that $P(B|\overline{A}) = .5$; if Bob wins instead of Al, the probability that I can beat him in the final match is .5 (Bob is not as strong a player as Al). Since by (4.7) we have $P(\overline{A}) = 1 - P(A) = 1 - .6 = .4$, the formula

(4.9) gives
$$P(\bar{A} \text{ and } B) = .4 \times .5 = .20;$$
the chance that Bob will win the other semifinal match and that I will then beat him in the final match is .20.

These probabilities can be combined. I can win the tournament by facing Al in the final and beating him (probability .18) *or* by facing Bob in the final and beating him (probability .20). These events are disjoint. My chance of winning the tournament is therefore the sum .18 + .20, which is .38.

---

independence

Perhaps the most important single concept in probability theory is that of **independence**.

Independence The events A and B are called independent if
$$P(A \text{ and } B) = P(A) \times P(B) \qquad 4.10$$

To see the idea behind this definition, divide both sides of (4.10) by $P(A)$, which gives $P(A \text{ and } B)/P(A) = P(B)$. By definition (4.8) then, (4.10) is the same as
$$P(B|A) = P(B) \qquad 4.11$$
In other words, A and B are independent if the conditional probability of B, given A, is the same as the unconditional probability of B.

In Example 8, $P(B)$ is $1/2$ and $P(B|A)$ is $4/10$, so A and B are not independent. Imagine that an adversary offers to bet you on the outcome of the experiment. Each of you is to stake $5.00; if the dice show an even total—if B occurs—you win, and if the total is odd, your adversary wins. Since $P(B)$ equals $1/2$, the bet is fair. But suppose the referee rolls the dice and announces to you and your adversary that the total was at most 5—that A occurred—and *then* your adversary offers to bet you as before. Since now your chance of winning is only $P(B|A) = 4/10$, you will reject the offer. (The bet is now fair if he puts up $6.00 to your $4.00.) The point is that A influences B—the occurrence of A decreases the chance that B will occur. In this sense A and B fail to be independent.

In Example 9, on the other hand, $P(B|A)$ is equal to $P(B)$. The probability of a face card is $3/13$, and this is so even if you know the card is a spade. In this example A does not influence B—the occurrence of A leaves unaltered the chance that B occurs. In this sense A and B are independent.

Equation (4.10) defines independence. Dividing it through by $P(A)$ gives Equation (4.11), which is perhaps more intuitive as a condition for independence. We could just as well divide through by $P(B)$, which leads from equation (4.10) to

SECTION 4.3 COMPUTING PROBABILITIES

$$P(A|B) = P(A) \qquad 4.12$$

Here the occurrence of B neither increases nor decreases the probability of A. Thus Equations (4.10), (4.11), and (4.12) all mean the same thing, and A has no influence on B if and only if B has no influence on A.

Sometimes we can compute P(A), P(B), and P(A and B) and then use Equation (4.10) to find out whether or not A and B are independent. Sometimes, on the other hand, we *assume* that A and B are independent; if we know the values of P(A) and P(B), we can then use Equation (4.10) to find the value of P(A and B). The following example concerns such a case.

Example 12

The experiment consists of rolling a pair of dice twice in succession. Let A be the event that the first roll gives a 7, and let B be the event that the second roll gives a 4. From our previous computations, we know that A should have probability 6/36 and B should have probability 3/36. Now it is reasonable to assume that A and B are independent. Since the dice have no memory, the occurrence of a 7 on the first roll should not alter the chance of a 4 on the second roll. Under this assumption the chance of a 7 followed by a 4 is

$$P(A \text{ and } B) = \frac{6}{36} \times \frac{3}{36} = \frac{1}{72}$$

Recall that A and B are disjoint if they cannot both occur *on the same trial* of an experiment. In the same way, independence of A and B refers to a single trial because A and B themselves refer to the same trial. Thus A and B are independent if knowing that A has occurred on a particular trial of an experiment does not affect your probability for B on that trial—nothing is said about the occurrence of B on some other trial of the experiment. This is so even in Example 12. There the experiment consists of rolling the dice *twice in a row*. In other words, the experiment consists of carrying through twice the simpler experiment of rolling the dice once.

The concepts *independent* and *disjoint* (or mutually exclusive) are often confused with each other. They are in fact diametrically opposed—disjoint events cannot be independent. Suppose that A and B are disjoint: that they cannot both happen. If you know that A has occurred, then you automatically know that B cannot have occurred, and so P(B|A) must be 0. Thus P(B|A) and P(B) must differ (except in the uninteresting case where P(B) equals 0, in which B is impossible in the first place), and so A and B are not independent.

Problems

1. In Table 4.2 count the number of outcomes for which
   a. Both dice are even.
   b. At least one die is even.
   c. Exactly one die is even.
   d. At most one die is even.
   e. Neither is even.

2. A pair of dice are tossed once. Use (4.2) to find the probability that
   a. The total is 10.
   b. The total is 10 or more.
   c. At least one die shows 3 dots or more.
   d. The total is 7 or 11.

3. A penny, a nickel, and a dime are tossed. List all possible outcomes. Count the number of outcomes for which
   a. The dime shows heads.
   b. Either the nickel or the dime shows heads.
   c. Either the nickel or the dime shows heads, but not both of them.
   d. The penny and the nickel do not agree.
   e. The penny agrees with the nickel but not with the dime.

4. A coin is tossed and a die is rolled. List the possible outcomes. Count the number of outcomes with
   a. Heads (on the coin) and four (on the die).
   b. Heads and an even number.
   c. Heads or an even number.
   d. Tails and a number less than four.
   e. Tails and anything.

5. A clothing store is having a sale on shirts and slacks. The shirts come in red, white, beige, and blue. The slacks come in grey, beige, and blue.
   a. List all the color combinations of shirts and slacks.
   b. How many combinations are there in which the color of the shirt does not match the color of the slacks?

6. From the text of the preceding problem as printed above, a line is selected at random. What is the probability that the word "and" appears at least once in the line?

7. A small college football conference has 6 teams in it. Two are to be selected at random to play in a special postseason game.
   a. How many possible pairings are there? List them. For convenience, label the teams A, B, C, D, E, F.
   b. If the teams are ranked from best to worst, what is the probability that one of the two best teams will play one of the two worst teams?

SECTION 4.3 COMPUTING PROBABILITIES   **113**

8. A consumer is asked to test three glasses of cola. She is told that two of them contain brand A while the third contains brand B. If she simply guesses, what is the probability that she correctly identifies the two glasses containing brand A?

9. One card is selected at random from a deck of 52. Find the probability that it is
   a. A face card. 12/52
   b. A spade. 13/52
   c. An ace or a king.
   d. An ace and a king. 0
   e. An ace or a face card.
   f. A heart or a jack.
   g. A heart or a spade or a jack.

10. Two dice are thrown, a red die and a green die. Let r be the number of dots showing on the red die and g the number on the green. Denote the total $r + g$ by t. What is the probability that
    a. r is even?
    b. r is even and g is odd?
    c. r is even or g is odd?
    d. t is even and g is odd?
    e. t is even or g is odd?
    f. $t < 6$ and g is odd?

11. A box contains six red tags numbered 1 through 6 and six white tags numbered 1 through 6. One tag is drawn at random. What is the probability that it is
    a. Red?
    b. An even number?
    c. Red and even?
    d. Red or even?
    e. Neither red nor even?

12. How many possible pairs of initials can be formed from the letters of the alphabet?

13. A man has in his pocket three keys, one of which fits the lock. If he tries them in random order, what is the chance he gets the right one on the second try?

A#2

14. A survey of women students at a college showed that 10% smoke cigarettes, 30% drink coffee, and 5% both smoke cigarettes and drink coffee.
    a. What percentage of these students neither drink coffee nor smoke cigarettes?
    b. Of those students who smoke cigarettes, what percentage drink coffee?

c. Of those students who do not drink coffee, what percentage smoke cigarettes?

15. In noting the causes of breakdowns, a television repair man finds that 50% fail because of the tuner, 15% fail because of the picture tube, 5% fail because of audio problems, 20% fail because of the color circuitry, and 10% fail because of other problems.
    a. Given that the problem is not audio, what is the probability that it is the tuner?
    b. Given that the problem is not the picture tube or audio, what is the probability that it is the color circuitry?

16. Suppose A and B are independent events with $P(A) = .2$ and $P(B) = .3$. What is the probability that
    a. Both occur?
    b. At least one occurs?
    c. Exactly one occurs?
    d. Neither occurs?

17. In a college of 1000 students and 100 faculty, 10% of the faculty are Democrats and 90% are Republicans, whereas among the students these percentages are reversed. A member of the college is drawn at random and found to be a Republican. What is the conditional probability that he is a student?

18. Two dice are rolled. Given that the two numbers appearing are different, find the conditional probability that the sum is even. (See Table 4.2.)

19. Two cards are drawn simultaneously from a deck of 52. Let A be the event they are of the same color; let B be the event they are of the same suit; let C be the event they are both aces.
    a. Find $P(A)$.
    b. Find $P(C|A)$.
    c. Are A and B independent?
    d. Are A and B mutually exclusive?
    e. Find $P(A|B)$ and $P(B|A)$.

20. A box contains nine coins of the following denominations and years:

Pennies	1974	1976	1978	1980
Nickels	1974	1978	1980	
Dimes	1978	1981		

A coin is drawn at random. Let A be the event the coin drawn is a penny, B the event it is a nickel, and C the event it bears the year 1978.

SECTION 4.4 COUNTING

    a. Find $P(C)$. ← 4/9
    b. Find $P(C|A)$.
    c. Find $P(A|C)$.
    d. Are A and C independent?
    e. Are B and C independent?

21. A certain electrical appliance is made up of three components. For the appliance to function properly, all its components must function properly. If the components fail independently of one another and if each has probability .20 of failing, what is the probability that the appliance fails?

22. A mini-submarine used in oceanographic research carries a primary air supply and a backup supply. The probability that the primary air supply functions correctly during any given dive is .95. If the primary supply fails, the probability that the backup supply functions correctly is .90. What is the probability that a dive is completed with at least one of the air supplies functioning correctly?

23. A penny, a nickel, and a dime are tossed. Let A be the event exactly two heads occur, B the event the penny comes up heads, and C the event the nickel comes up heads.
    a. Are A and C independent?
    b. Find $P(B|A)$.
    c. Are B and C mutually exclusive?

24. An urn contains three red, four white, and five blue balls. A second urn contains one red, six white, and three blue balls. One ball is selected at random from each urn. Find the probability that
    a. The balls are of the same color.
    b. One is red and one is white.
    c. At least one is red.
    d. Neither is blue.
    e. Neither is blue, given that at least one is red.

4.4 Counting

Computing a probability by the formula (4.2) requires computing the numerator (the number of favorable cases) and the denominator (the total number of cases). In the examples considered thus far, these values were found by actually counting out the cases. This cannot usually be done, however, and we need some mathematical techniques for computing these numbers. Often they can be found by what we call the *addition principle* and the *multiplication principle*.

If the professor has counted the 11 women in a class and the 13 men in the same class, she knows without counting the whole class over again that there are 24 students altogether. This is because of the addition principle. As in the preceding section, sets of objects are said to be *disjoint*, or *mutually exclusive*, if there is no object that belongs to both sets.

The addition principle If one set contains a objects and a second set contains b objects, and if the two sets are disjoint, then the two sets taken together contain $a + b$ objects.

For example, there are 26 red cards in an ordinary deck and 13 spades, so there are $26 + 13$, or 39, cards that are *either* red *or* spades. It is essential that the two sets in question be disjoint, however: There are 4 aces and 13 spades, but the number of cards that are either aces or spades is not 17 ($4 + 13$) because the set of aces and the set of spades share a card—the ace of spades. (The correct answer is one less than 17, since exactly one card has been double counted.)

Of course the addition principle is the basis of the formula (4.4) for adding the probabilities of disjoint events.

The addition principle extends in an obvious way to three or more sets: Since there are four jacks, four queens, and four kings, there are $4 + 4 + 4$, or 12, face cards all told.

We tend to use the addition principle correctly and without reflection. The multiplication principle is more complicated.

multiplication principle

A good starting point in coming to an understanding of the **multiplication principle** is the old rhyme

> As I was going to St. Ives,
> I met a man with seven wives.
> Every wife had seven sacks,
> Every sack had seven cats,
> Every cat had seven kits.
> Kits, cats, sacks, and wives,
> How many were going to St. Ives?

The answer is *one:* Only *I* was going to St. Ives.* But of course the point of the rhyme is the seeming difficulty of calculating the number of kits, cats, sacks, and wives coming *from* St. Ives. This is not so hard as it may at first appear.

The number of wives is seven. Since each wife has seven sacks, the total number of sacks is 7×7, or 49. Since each of these 49 sacks has

---

*According to *The Oxford Dictionary of Nursery Rhymes* one eighteenth-century authority gives the answer *none,* interpreting the question as, "How many Wives, Sacs, Cats, and Kittens went to St. Ives?"

SECTION 4.4 COUNTING

seven cats, the total number of cats is 49×7, or 343. Since each of these 343 cats has seven kits, the total number of kits is 343×7, or 2401.

Wives	7
Sacks	49
Cats	343
Kits	2401
Total	2800

The number of kits, cats, sacks, and wives coming from St. Ives is, therefore, 2800. If we include the man as well, the total increases to 2801, a number less round than 2800 and hence more arresting.

The successive multiplications here are instances of what we call the multiplication principle. It can be used to solve a surprising variety of counting problems.

The multiplication principle We are to make two choices, a first-stage choice followed by a second-stage choice:
(i) The number of alternatives open to us at the first stage is a.
(ii) No matter which alternative we select at the first stage, the number of alternatives open to us at the second stage is b.

In these circumstances the number of two-stage alternatives, or simply results, is $a \times b$. That is, the number of ways in which we can make the first choice and then the second is $a \times b$.

Example 1

If at the first stage we are to choose one of the St. Ives man's wives, the number of alternatives, a, is 7. If at the second stage we are to choose one of the sacks in the keeping of this wife, then no matter which wife we happen to have selected at the first stage, the number of alternatives now open to us, b, is 7. According to the principle the number of sacks is $a \times b = 7 \times 7$, or 49.

Starting afresh with another task, that of choosing a sack at the first stage and then choosing one of this sack's cats at the second stage, we know from the first computation that the alternatives, at the (new) first stage, number 49: a is 49. And no matter which sack we select at the first stage, the number of second-stage alternatives, b—the number of cats belonging in this sack—is 7. According to the principle, the number of cats is $a \times b = 49 \times 7$, or 343.

Finally, for choosing a cat and then one of the cat's kits, a is 343 (the number of cats, just derived) and b is 7 (the number of kits per cat), so there are $a \times b = 343 \times 7$, or 2401, kits in all. (To show that there are 2800 kits, cats, sacks, and wives *in toto*, we need only apply the addition principle.)

Example 2

A host has in his larder beef, trout, and capon. In his wine closet are Burgundy, Claret, Zinfandel, Chablis, Moselle, Riesling, and Champagne. Taste or fashion dictates that with the beef he may serve only Burgundy, Claret, Zinfandel, or Champagne and that with trout or capon he may serve only Chablis, Moselle, Riesling, or Champagne. How many meat-wine combinations (count trout and capon as meat) are available to him?

He first chooses the meat and then the wine. The alternatives at the first stage are beef, trout, and capon; their number, a, is 3. The set of alternatives open to the host at the second stage varies with the selection made at the first stage; the following table lists them.

Choice of Meat	Alternatives at the Second Stage			
Beef	Burgundy	Claret	Zinfandel	Champagne
Trout	Chablis	Moselle	Riesling	Champagne
Capon	Chablis	Moselle	Riesling	Champagne

Although the *set* of possible wines varies with the meat chosen, their *number* does not; this number, b, is 4.

By the multiplication principle the number of meat-wine pairs is $a \times b = 3 \times 4$, or 12. An alteration of the above table leads to an exhibit of all 12 pairs. With each second-stage alternative we simply couple the meat dish at the left of its row:

Beef-Burgundy	Beef-Claret	Beef-Zinfandel	Beef-Champagne
Trout-Chablis	Trout-Moselle	Trout-Riesling	Trout-Champagne
Capon-Chablis	Capon-Moselle	Capon-Riesling	Capon-Champagne

This exhibit shows why the multiplication principle works: To find the number of pebbles in a rectangular array we can multiply the number of rows by the number of pebbles in each row; there is no need to count the pebbles individually.

For a case in which the multiplication principle does not apply suppose the host finds Riesling and Champagne unacceptable with trout but will serve Chablis with beef:

SECTION 4.4 COUNTING

Choice of Meat	Alternatives at the Second Stage				
Beef	Burgundy	Claret	Zinfandel	Champagne	Chablis
Trout	Chablis	Moselle			
Capon	Chablis	Moselle	Riesling	Champagne	

Again a is 3, but the number of alternatives at the second stage is no longer constant. There *is* no b, and the multiplication principle does not apply. While the number of meat-wine pairs can be *counted up* (there are eleven of them), it cannot be *computed*. Of course the answer for the original set of rules, too, could be arrived at by counting up all twelve possibilities, but the idea is to derive the answer from a general principle that works even when an exhaustive count would be prohibitive.

Example 3

The multiplication principle was in fact used several times in the preceding section. In Example 2 of that section, a card is dealt to Alice and then a card is dealt to Bill. To count the number of points in the sample space, let the first-stage choice consist of choosing the card that goes to Alice. The number, a, of alternatives is 52. Let the second-stage choice consist of choosing the card that goes to Bill. Whatever card was given to Alice, the set of second-stage alternatives is the set of 51 cards remaining in the deck: $b = 51$. The number of outcomes is $a \times b$, or 52×51. Table 4.3 makes this clear.

The multiplication principle was also used to find the number of outcomes for which the two cards dealt are of the same suit. Here the first-stage choice is the same as before, and so a is still 52. But now the set of second-stage alternatives consists of the 12 cards that are different from the one given to Alice but are of the same suit as that card. Hence b is 12, and the number of outcomes where Alice and Bill get the same suit is $a \times b$, or 52×12.

These two applications of the multiplication principle give the numerator (52×12) and the denominator (52×51) needed to find the probability ((52×12)/(52×51), or 12/51) in Example 2 of the preceding section.

Example 4

A woman in New York may take a plane to Chicago or to St. Louis. From Chicago she may fly on to Seattle, to San Francisco, or to Los Angeles. From St. Louis she may fly on to San Francisco, to Los Angeles, or to San Diego. To how many West Coast cities may she go?

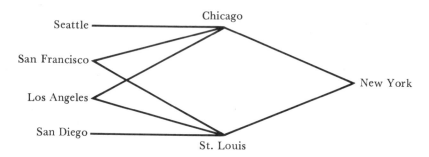

At the first stage she has $a = 2$ alternatives, and at the second she has $b = 3$. By the multiplication principle the number of destinations is $a \times b = 2 \times 3$, or 6. This answer is wrong—the number of destinations is only 4. The trouble is not far to seek: The various first-stage selections coupled with the various second-stage selections do not lead to distinct results (destinations).

Thus, to the statement of the multiplication principle we must add this proviso:

Each first-stage alternative coupled with each one of its second-stage successors leads to a distinct result.

This condition holds in Examples 1 and 3 and in the first part of Example 2, and it also holds in Example 4 if we ask not for the number of *destinations,* but for the number of *routes* to the West Coast.

The multiplication principle also works if we have a sequence of three choices instead of just two. Here the first-stage and second-stage choices are to satisfy conditions (i) and (ii) above, and the third-stage choice must satisfy the analogous condition:

(iii) No matter which alternative we select at the first stage, and no matter which alternative we select at the second stage, the number of alternatives open to us at the third stage is c.

In these circumstances the number of different results is $a \times b \times c$. (As before, we assume that different choices lead to different results.)

Example 5

Suppose the woman in Example 4 may from Seattle or San Franciso fly on to Tokyo or to Hong Kong and may from Los Angeles or San Diego fly on to

SECTION 4.4 COUNTING

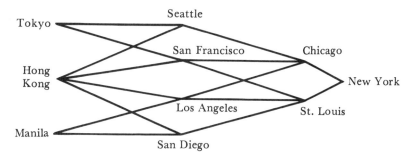

Hong Kong or to Manila. How many routes to the Orient are available to her? As before, $a = 2$ and $b = 3$. No matter which West Coast city she is in after the first two legs of the journey (and in the cases of San Francisco and Los Angeles, no matter whether she got there via Chicago or via St. Louis), she has $c = 2$ third-stage alternatives open to her. By the three-stage multiplication principle there are $a \times b \times c = 2 \times 3 \times 2$, or 12, routes all told.

The multiplication principle extends in the obvious way to a sequence of four or more choices.

Example 6

In the original St. Ives problem we found the number of cats by using the two-stage multiplication principle twice. A single application of the three-stage principle with a, b, and c equal to 7 yields the correct answer, $a \times b \times c = 7 \times 7 \times 7$, or 343, in one operation. And the four-stage principle shows at one stroke that there are $7 \times 7 \times 7 \times 7$, or 2401, kits.

With the multiplication principle in hand we can now turn to problems involving what are called *combinations* and *permutations*. Although it may not be immediately apparent why, these problems in fact carry us nearer to statistical applications.

Example 7

What is the number of three-letter "words" that can be made up from the five letters *A, B, C, D,* and *E*? The "word" need not make sense (*CDA* is counted as well as *BAD*), a letter may be repeated (*BDB* is counted), and the order in which the letters come is taken into account (*AED* and *ADE* are counted as different).

The number of such words is 5 × 5 × 5, or 125, because there are five ways to choose the first letter of the word, five ways to choose the second, and five ways to choose the third. To put it differently, we must successively choose letters to fill the three blanks __ __ __. Since in each blank we can use any of the five letters, the three-stage multiplication principle with *a, b,* and *c* each equal to 5 shows that there are indeed 5^3 possible words.

words

More generally, we can consider an alphabet of *a* letters and can make up **words** of length *n*. (In the preceding example $a = 5$ and $n = 3$.) Repeated letters are allowed, and the order of the letters in the word is taken into account. The rule is as follows:

Words There are a^n words of length *n* from an alphabet of *a* letters.

> The reasoning underlying this rule is as in Example 7. To make a word we must successively put letters into *n* blanks set out in a row. Each blank can take any of the *a* letters of the alphabet. By the *n*-stage multiplication principle the number of possibilities is $a \times a \times a \times \cdots \times a$, where *a* appears *n* times in the product, and this is just a^n, the *n*th power of *a*.

permutations

We now come to **permutations**. These are words in which repeated letters are not allowed. Consider again the alphabet *A, B, C, D, E* of five letters, and make up words of length three.

Example 8

This differs from Example 7 in that this time repeats are prohibited; for example, *BDB* is not counted. There are five choices for the first letter, but

Table 4.4 The Sixty Permutations of *A, B, C, D, E* Taken Three at a Time

ABC	ABD	ABE	ACD	ACE	ADE	BCD	BCE	BDE	CDE
ACB	ADB	AEB	ADC	AEC	AED	BDC	BEC	BED	CED
BAC	BAD	BAE	CAD	CAE	DAE	CBD	CBE	DBE	DCE
BCA	BDA	BEA	CDA	CEA	DEA	CDB	CEB	DEB	DEC
CAB	DAB	EAB	DAC	EAC	EAD	DBC	EBC	EBD	ECD
CBA	DBA	EBA	DCA	ECA	EDA	DCB	ECB	EDB	EDC

SECTION 4.4 COUNTING

since repeats are ruled out there are only four choices for the second and three choices for the third. So the number of possibilities is $5 \times 4 \times 3$, or 60. Table 4.4 is a systematic table of the possibilities.

> In more detail, the argument for the answer 60 is this. We must successively choose letters to fill the three blanks ___ ___ ___. Since we may fill the first blank with any of the five letters, the number of first-stage alternatives is $a = 5$:
>
> A ___ ___
> B ___ ___
> C ___ ___
> D ___ ___
> E ___ ___
>
> If the first blank was filled with A, the second may be filled with B, C, D, or E; if the first blank was filled with B, the second may be filled with A, C, D, or E; and so on:
>
> | First-Stage Choice | Second-Stage Alternatives | | | |
> |---|---|---|---|---|
> | A ___ ___ | AB ___ | AC ___ | AD ___ | AE ___ |
> | B ___ ___ | BA ___ | BC ___ | BD ___ | BE ___ |
> | C ___ ___ | CA ___ | CB ___ | CD ___ | CE ___ |
> | D ___ ___ | DA ___ | DB ___ | DC ___ | DE ___ |
> | E ___ ___ | EA ___ | EB ___ | EC ___ | ED ___ |
>
> Thus $b = 4$ this time; there are always four alternatives at the second stage. Finally, the third blank may be filled with any of the three unused letters; for example, from BC___ we may go on to BCA, BCD, or BCE. So $c = 3$, and the total number of words is thus $a \times b \times c = 5 \times 4 \times 3$, or 60.

Thus the five letters A, B, C, D, and E can be lined up in $5 \times 4 \times 3$ ways when taken three at a time, repeats not allowed. Such a lining up is called a *permutation*; if we denote the number of permutations of five distinct things taken three at a time by $_5P_3$, what we have shown is that $_5P_3 = 5 \times 4 \times 3$.

factorial notation Answers like this are most conveniently expressed by using the **factorial notation**. The symbol $n!$, read n-factorial, is the product of the integers from 1 to n:

$$n! = n(n-1)(n-2) \cdots (3)(2)(1) \qquad 4.13$$

For example,

$$5! = 5 \times 4 \times 3 \times 2 \times 1$$

From the definition it is clear that if r is a number between 1 and n, then

$$\begin{aligned} n! &= n(n-1)! = n(n-1)(n-2)! \\ &= n(n-1)(n-2) \cdots (n-r+1)(n-r)! \end{aligned} \qquad 4.14$$

and if we divide both sides of Equation (4.14) by $(n-r)!$, we see that

$$\frac{n!}{(n-r)!} = n(n-1)(n-2) \cdots (n-r+1) \qquad 4.15$$

Thus

$$_5P_3 = 5 \times 4 \times 3 = \frac{5!}{2!}$$

It is convenient to define $0!$ as 1:

$$0! = 1 \qquad 4.16$$

We can now state the general rule governing permutations or linings up.

Permutations The number of permutations of n distinct things taken r at a time $(1 \leq r \leq n)$ is

$$_nP_r = n(n-1) \cdots (n-r+2)(n-r+1) = \frac{n!}{(n-r)!} \qquad 4.17$$

The argument for the relation (4.17) is the same as for the special case in Example 8. We can think of the n objects as distinct symbols or letters with which we are to fill in a succession of r blanks, using a different letter in each blank. The set of alternatives at any stage is the set of letters as yet unused, and the numbers of these for the first stage, second stage, third stage, etc., are n, $n-1$, $n-2$, etc. According to the multiplication principle, the number of permutations is the product of these numbers, of which there are r (one for each blank). This gives the relation (4.17).

SECTION 4.4 COUNTING   125

Example 9

The size of the sample space in Table 4.3 is 52×51, and this illustrates permutations. To give a card to Alice and a card to Bill is, in effect, to create one of the permutations of the 52 cards taken two at a time. By (4.17) the number of these is $52 \times (52 - 1)$, or $52!/(52 - 2)!$; either way, the number is 52×51.

If r equals n, we have a special case. All n objects are to be lined up, and we suppress the phrase "taken n at a time."

Permutations The number of permutations of n distinct things is

$$_nP_n = n! = n(n - 1) \cdots (3)(2)(1) \qquad 4.18$$

(Because of (4.16), the right side of (4.17) for $r = n$ is $n!/0! = n!/1 = n!$.)

The number of permutations of the letters A, B, C is $3!$, which is 6; a list of them follows:

$$ABC \quad ACB \quad BAC \quad BCA \quad CAB \quad CBA$$

The number of permutations of the digits 0 through 9 is

$$10! = 3,628,800$$

and the number of permutations of the 26 letters of the alphabet is

$$26! = 403,291,461,126,605,635,584,000,000$$

In the last two cases the answers certainly must be arrived at by the application of general principles, not by counting out all the possibilities.

combinations We turn next to **combinations**. These are like permutations, but with the important difference that the order in which the letters come is disregarded. Consider once more the alphabet A, B, C, D, E and sets of size three.

Example 10

In Example 8 we found the number of three-letter words that can be made from A, B, C, D, and E with repetitions disallowed. We now ask how many *subsets* of size three can be made up. We still distinguish ABC from ABD, since they contain different letters, but we do not distinguish ABC from ACB, since they differ merely in the order in which the letters come. The letters in the *set* are not lined up at all; we could write them in a jumble:

$$\begin{array}{ccc} & & C \\ & C & B \\ A \quad B & \text{or} & \\ & & A \end{array}$$

The usual mathematical notation is $\{A, B, C\}$.

Such a *set*, such an unordered selection of three of the five letters, we call a *combination*; it is to be sharply distinguished from a permutation. We want the *number* of such combinations (the number of combinations of five distinct objects taken three at a time), a number we denote $_5C_3$. There are in fact ten such combinations:

$$\{A, B, C\}, \{A, B, D\}, \{A, B, E\}, \{A, C, D\}, \{A, C, E\},$$
$$\{A, D, E\}, \{B, C, D\}, \{B, C, E\}, \{B, D, E\}, \{C, D, E\}$$

To see that the list is complete, look at the 60 *permutations* in Table 4.4. There are 3!, or 6, permutations of the letters in the set $\{A, B, C\}$; the first column of the table lists them. The second column of the table lists the six permutations of the letters in the set $\{A, B, D\}$. In fact, each of the ten columns consists of the six permutations of a *set* or combination of three letters, so there must be ten such sets. Compare the top row of Table 4.4 with the list of ten combinations just above.

That there are ten of these permutations can be deduced from the multiplication principle in the following way. Let us temporarily write x in place of $_5C_3$, the number whose value we seek. Consider this two-stage procedure: At the first stage we choose one of the x combinations. At the second stage we take the three letters chosen (the combination) and line them up in some order. For example, we may choose the combination $\{A, C, D\}$ and then line up the letters in it in the order *DAC*. The number of alternatives at the first stage is x. The number of alternatives at the second stage is simply the number of permutations of the three letters chosen at the first stage, and this number we know to be 3!. So, in the notation of the multiplication principle (p. 117), $a = x$ and $b = 3!$. But after making these two choices, what we arrive at is some permutation of three of the five letters. Each permutation can arise in exactly one way from such a pair of choices. By the two-stage multiplication principle, the product $a \times b$, or $x \times 3!$, is the same as the number $_5P_3$ of permutations of five things taken three at a time. Since we already know that $_5P_3 = 5!/2!$, we can conclude that $x \times 3! = 5!/2!$. Solving for x gives our answer,

$$_5C_3 = x = \frac{5!}{3!2!}$$

which comes out to 10.

For the general case of combinations, consider a set of n distinct letters or any n distinct objects. A combination of these is a subset, or unordered subcollection, of them. Suppose the subset is to contain r of the objects. The number of such subsets is then the number of combinations of n things taken r at a time and is denoted by $_nC_r$, or more commonly by $\binom{n}{r}$. (In Example 10, $n = 5$ and $r = 3$.)

Combinations The number of combinations of n distinct objects taken r at a time (the number of subsets of size r) is

$$\binom{n}{r} = {_nC_r} = \frac{n!}{r!(n-r)!} \qquad 4.19$$

> The validity of this rule can be seen by the sort of argument in Example 10. We can choose a combination in $_nC_r$ ways and then permute the r objects in it in $r!$ ways, arriving at one of the $_nP_r$ permutations. By the two-stage multiplication principle, $_nC_r \times r! = {_nP_r}$. Since we already know that $_nP_r = n!/(n-r)!$, division by $r!$ gives the rule above for $_nC_r$.

Notice that for the case where r is equal to n, the number of combinations is $n!/n!0!$, which is 1 by the convention (4.16). This is the right answer—there is but one subset that contains all the n objects. The answer is also 1 for the case where $r = 0$. This is itself just a convention—there exists exactly one set with nothing in it.

Example 11

The number of combinations of 52 cards taken two at a time is

$$\binom{52}{2} = \frac{52!}{2!(52-2)!} = \frac{52!}{2!50!} = \frac{52 \times 51 \times 50!}{2 \times 50!} = \frac{52 \times 51}{2}$$

Suppose that in Table 4.3 we keep the dots above the diagonal and discard those below it. The number of dots remaining is then $52 \times 51/2$, and there is one dot for each combination. For example, the one in the upper right-hand corner corresponds to the unordered pair consisting of the two of clubs and the ace of spades.

It is sometimes hard to tell whether a problem requires combinations or whether it requires permutations. To answer this question, we ask ourselves whether or not the arrangement or order of the objects is relevant. If we need to take order into account, permutations are called for; if order is irrelevant, combinations are called for. Consider the following two examples.

Example 12

Problem: If there are seven horses in a race, in how many ways can three from among them finish first, second, and third?

Analysis: Here we are definitely interested in the order in which the horses cross the finish line. If we have bet on a particular horse to win, we collect nothing if he places or shows. Since order is important, we use permutations; the answer is

$$_7P_3 = \frac{7!}{4!} = 210$$

Example 13

Problem: How many five-card hands can be dealt from a 52-card deck?

Analysis: Here order is irrelevant; no player will object if in order to have a straight flush he must rearrange the cards he has been dealt. Since order is irrelevant, we use combinations; the answer is

$$\binom{52}{5} = \frac{52!}{5!47!} = 2,598,960$$

In this section we have looked into the beginnings of *combinatorial analysis*, the branch of mathematics that deals with counting. Very often the solution of a combinatorial problem requires the use of the rules for combinations or permutations together with the multiplication principle itself. The following example is typical.

Example 14

Problem: In a college with 55 faculty members and 750 students there is to be formed a committee consisting of three faculty and five students. How many such committees are possible?

Analysis: The number of ways to select the set of three faculty members for the committee is $\binom{55}{3}$. The number of ways to select the set of five students for the committee is $\binom{750}{5}$. The full committee

SECTION 4.4 COUNTING

can be made up of any of these sets of faculty together with any of these sets of students. By the multiplication principle the total number of possible committees is therefore $\binom{55}{3} \times \binom{750}{5}$.

This equation happens to work out numerically to 51,192,216,130,497,750. In such a problem, however, the important thing is not the numerical answer, but the method.

These counting principles are important for the solution of many problems in probability theory.

Example 15

Problem: What is the probability of a flush (all cards of the same suit) in a five-card hand from an ordinary deck?

Analysis: In Example 13 we saw that the total number of hands (order disregarded) is $\binom{52}{5}$. With thorough shuffling and fair dealing, these outcomes are equally likely; the denominator for (4.2) is $\binom{52}{5}$. Now a flush in spades is a combination of some five of the 13 spades, and so there are $\binom{13}{5}$ of them; similarly, there are $\binom{13}{5}$ flushes in each of the other suits. Hence the numerator is $4 \cdot \binom{13}{5}$, and so the probability of a flush is

possible straights / possible Hands

$$\frac{4 \cdot \binom{13}{5}}{\binom{52}{5}} = \frac{5148}{2,598,960} = .00198$$

Example 16

In drawing a five-card hand from an ordinary deck, what is the probability that there is more than one suit represented in the hand? It would be complicated to solve this directly, but in the preceding problem the probability of the complementary event (a flush) was found. By the formula (4.7) for complementary events, the probability we seek is

$$1 - \frac{5148}{2{,}598{,}960} = .99802$$

Problems

25. A Morse code letter is a sequence of dots and dashes. How many letters are there of length
    a. One?
    b. Two?
    c. Three?
    d. One, two, or three?

26. From a club of six boys a committee of three is chosen by lot.
    a. What is the probability Tom is on the committee and Dick is not?
    b. Given that Tom is on the committee, what is the conditional probability that Dick is also on it?
    c. Is the event that Tom is on the committee independent of the event that Dick is on it?

27. Five tags are labeled a, b, c, d, e. A set of three is drawn at random. Let A be the event that a is in the set drawn; let B be the event that b is in the set drawn; let C be the event that d is not in the set drawn.
    a. Find $P(A)$, $P(B)$, $P(C)$.
    b. Are A and C independent?
    c. Find $P(B|A)$.
    d. Find $P(\overline{C}|B)$.
    e. Find $P(A \text{ or } C)$.

28. How many different license plates are there consisting of
    a. Four digits?
    b. Four letters?
    c. Two letters followed by four digits?
    d. Four digits if the first one may not be zero?

29. How many three-letter words can be made from ten different letters if
    a. Repeats are allowed?
    b. Repeats are not allowed?
    c. A letter may be repeated only once?

30. By how many routes may the woman of Example 5, Section 4.4, travel from New York
    a. To the Orient via San Francisco?

SECTION 4.4 COUNTING   **131**

      b. To Tokyo?
      c. To Tokyo and return?
      d. To Hong Kong and return?
      e. To the Orient and return?

31. a. In how many ways can a president, a secretary, and a treasurer be selected from a club of 20 people?
    b. In how many ways can a committee of three be selected from the club?

32. A home builder offers four floor plans, brick or wood siding, a double-car or single-car garage, and three grades of roofing material. How many possibilities does he offer?

33. How many words of length five can be made from
    a. The English alphabet?
    b. The Russian alphabet (32 letters)?

34. A computer "word" consists of a sequence of 0s and 1s.
    a. How many computer words of length 5 are there?
    b. How many computer words are there of length 5 or less?

35. A chemist at a pharmaceutical laboratory investigates possible compounds for relieving allergy symptoms. If there are five chemicals to choose from and they are to be combined three at a time, how many compounds must the chemist investigate?

36. a. From a set of 30 phonograph records, in how many ways can a man select one for each of his four children?
    b. In how many ways can he select four for his wife?
    c. In how many ways can he select four for his wife and one for each child?

37. In how many ways can the 26 letters of the alphabet be lined up so that *A* and *B* are adjacent?

38. With the eight points shown in the following figure used as vertices, how many triangles can be constructed?

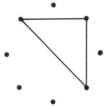

39. A manufacturer of breakfast cereal puts prizes in the boxes. There are 10 different kinds of prizes, which are distributed randomly among the boxes of cereal. A customer buys 3 boxes of cereal.

a. What is the probability that all prizes will be the same?
b. What is the probability that at least two prizes will be the same?

40. The telephone company is adding two new exchanges to its service area. The seven-digit telephone numbers for these new exchanges must begin with either a 5 or a 6. How many numbers are available to the telephone company for the new exchanges?

41. a. How many 13-card bridge hands can be dealt from a deck of 52?
    b. In how many ways can 13-card hands be dealt to North, South, East, and West?

42. How many four-letter words can be made from the letters of the word *sectional* if repeats are not allowed and
    a. The letter *s* may not be used?
    b. The letter *s* must be used?
    c. The word must start with *s*?
    d. The word must start with a vowel?

43. How many five-card poker hands consist of
    a. Two pair?
    b. A full house?
    c. A straight flush?
    d. Four of a kind?

44. In a town with a rectangular network of streets, a student lives four blocks west of her school and six blocks south of it.
    a. How many routes (ten blocks long) has she to school?
    b. How many if she wants to approach school from the south?
    c. How many if she wants to approach school from the west?

45. A "combination" for a lock with 40 positions consists of four settings, and no setting can coincide with the preceding one. How many "combinations" are there?

46. In a club of six men and six women, in how many ways can a committee of four be selected if
    a. The committee must consist of two men and two women?
    b. There must be at least one man?
    c. There must be at least one of each sex?
    d. The women must outnumber the men?

47. Six people are seated in six chairs in a circle. If everyone moves the same number of places to the left, the seating is considered the same as before. How many seatings are there?

48. Ten people, consisting of five couples, are to be seated in a row, keeping the couples together. How many such seatings are there?

49. If an entering college student must take one of five English courses, one of six history courses, one of five science courses, and one of three sociology courses, how many programs are available to him?

SECTION 4.5 REPEATED INDEPENDENT TRIALS

50. At the Medici pizzaria, the pizzas come with a choice of mushroom, pepperoni, anchovy, sausage, and tomato. Counting the pizza garnished with everything and the pizza garnished with nothing, how many possibilities are there?

51. During the next week a man plans to play bridge four of the nights and chess the other three. How many schedules are there?

52. The digits 0 through 9 are lined up at random. What is the probability that the digits 0, 1, and 2
    a. Are adjacent and in increasing order?
    b. Are in increasing order?
    c. Are adjacent?

53. A 13-card bridge hand is dealt. What is the probability that
    a. All cards are of the same suit?
    b. There are no aces?
    c. There is at least one spade?
    d. There is exactly one king?

54. Poker dice have their six faces labelled *ace, king, queen, jack, ten,* and *nine*. If we roll a poker die and draw a card at random from a standard deck of 52, what is the probability that the card and the uppermost face of the die are the same denomination—that is, are both aces, or both kings, etc.

55. Five poker dice (see the preceding problem) are rolled simultaneously. What is the probability of
    a. A full house?
    b. A straight?
    c. One pair?
    d. Two pair?
    e. Three of a kind?

56. A birthday corresponds to a number between 1 and 365. If all birthdays are equally likely, and if there is independence from person to person, the chance that two given people have different birthdays is $365 \times 364/365^2$. What is the chance that three given people have three different birthdays? Four? Five? (The method extends, and it can be shown that in a roomful of 23 people, the chance that they all have different birthdays is just under $\frac{1}{2}$. Most people feel that the probability ought to be much larger than that.)

4.5 Repeated Independent Trials

Consider repeating an experiment several times over and keeping track of the occurrence of some event.

Example 1

To fix ideas, let the experiment be the rolling of a pair of dice, suppose the experiment is repeated three times in a row, and let the event be the rolling of a 7. Let us denote the occurrence of a 7 by S, for *success,* and the occurrence of anything else by F, for *failure* (*success* and *failure,* like *favorable case,* are merely conventional terms). The left-hand column in Table 4.5 lists all the possible histories for the three rolls. The sequence SSS indicates that all three rolls resulted in success (a 7); SSF indicates that the first two

Table 4.5   Independent Trials with $p = \frac{1}{6}$

Sequence	Probability
SSS	$(\frac{1}{6})^3$
SSF	$(\frac{1}{6})^2(\frac{5}{6})$
SFS	$(\frac{1}{6})^2(\frac{5}{6})$
SFF	$(\frac{1}{6})(\frac{5}{6})^2$
FSS	$(\frac{1}{6})^2(\frac{5}{6})$
FSF	$(\frac{1}{6})(\frac{5}{6})^2$
FFS	$(\frac{1}{6})(\frac{5}{6})^2$
FFF	$(\frac{5}{6})^3$

rolls resulted in success and the last roll resulted in failure (non-7); and so on. (Example 7 of Section 4.4 and the rule for words following it show why there must be 2^3, or 8, possibilities.)

If the dice are rolled three times in a row, what is the probability of obtaining some particular sequence, say SSF? If the dice are fair, the probability of success on any one trial is $\frac{1}{6}$, and so the probability of failure is $\frac{5}{6}$. We should further assume, as in Example 12 of Section 4.3, that there is independence from trial to trial—that the result on one trial can in no way influence the result on a different trial. Hence, we multiply probabilities. The chance of getting S on the first roll and then S on the second roll and then F on the third roll is $(\frac{1}{6})(\frac{1}{6})(\frac{5}{6})$. The other probabilities in Table 4.5 are computed the same way.

What is the probability of getting exactly one success (and two failures)? There are three ways this can happen—SFF, FSF, and FFS. Each of these sequences has probability $(\frac{1}{6})(\frac{5}{6})^2$—one factor of $\frac{1}{6}$ for the one S and

SECTION 4.5 REPEATED INDEPENDENT TRIALS

two factors of $\frac{5}{6}$ for the two *F*s. Thus the probability of getting exactly one success is

$$P(1) = 3\left(\frac{1}{6}\right)\left(\frac{5}{6}\right)^2 = \frac{75}{216} = .347$$

The sequences containing two *S*s and one *F* are *SSF*, *SFS*, and *FSS*, and each has probability $(\frac{1}{6})^2(\frac{5}{6})$, so the probability of getting exactly two successes is

$$P(2) = 3\left(\frac{1}{6}\right)^2\left(\frac{5}{6}\right) = \frac{15}{216} = .069$$

Finally, the chance of three successes is the chance of *SSS*, namely,

$$P(3) = \left(\frac{1}{6}\right)^3 = \frac{1}{216} = .005$$

and the chance of no successes is the chance of *FFF*, namely,

$$P(0) = \left(\frac{5}{6}\right)^3 = \frac{125}{216} = .579$$

The general principle is this: We repeat an experiment, singling out an event we call success. We let

p = the probability of success on a single trial,
$1 - p$ = the probability of failure on a single trial,
n = the number of trials,
r = the number of successes.

with replacement.

Binomial probabilities In this circumstance the probability that *S* occurs exactly *r* times (so that *F* must occur $n - r$ times) is

$P(r) = \binom{n}{r}(p)^r(1-p)^{n-r}$

$$P(r) = \binom{n}{r}p^r(1-p)^{n-r} \qquad 4.20$$

For this formula to be valid, *the trials must be independent*. According to the binomial theorem of algebra,

$$(x + y)^n = \sum_{r=0}^{n}\binom{n}{r}x^r y^{n-r}$$

binomial probabilities The quantities $\binom{n}{r}$ are therefore called *binomial coefficients*, which is why the probabilities in Equation (4.20) are called **binomial probabilities.**

The rule (4.20) may be derived in this way. One sequence which contains r Ss and $n - r$ Fs is

$$\underbrace{SS \cdots S}_{r \text{ times}} \underbrace{FF \cdots F}_{n-r \text{ times}}$$

By independence, the probability of r Ss in a row followed by $n - r$ Fs in a row is $p^r(1-p)^{n-r}$. We do not insist that the r Ss occur in the *first* r trials; we insist only that exactly r of the trials produce S and $n - r$ of them produce F. The probability of each such sequence is $p^r(1-p)^{n-r}$. Equation (4.20) follows because the number of such sequences is $\binom{n}{r}$. That is, from the n trials there are $\binom{n}{r}$ ways to choose a combination of r trials in which to put Ss, and the other trials must take Fs. The arguments in Example 1 are special cases of this one, and the computations agree.

Since probabilities for disjoint events can be added, (see (4.4) and (4.5)), the probability that S occurs on x or fewer of the n trials is

$$\sum_{n=0}^{x} P(r) = \sum_{n=0}^{x} \binom{n}{r} p^r (1-p)^{n-r} \qquad 4.21$$

for $x = 0, 1, \ldots, n$. Table 1 in the Appendix gives the values of these probabilities for certain values of n and p.

Example 2

Suppose a marksman has probability .3 of hitting the bull's-eye. What is the probability that in 20 trials he will make 8 or fewer bull's-eyes? Enter Table 1 with $n = 20$, $x = 8$, and $p = .3$; the probability is .887.

What is the probability that he will make *exactly* 8 bull's-eyes in the 20 tries (p still .3)? By the table, the probability of 7 or fewer is .772, and since the probability of 8 or fewer is .887, the probability of exactly 8 must be $.887 - .772$, or $.115$.

Example 3

Suppose the marksman's probability for a bull's-eye is .6. What is the probability that in 15 trials he will make exactly 10 bull's-eyes? Since p of .6 doesn't appear in the table, it is necessary to turn the problem around. If the probability of a hit (a bull's-eye) is .6, the probability of a miss (a non-bull's-eye) is .4. The chance of 10 hits is the same as the chance of 5 misses.

By the table (use $n = 15$, $p = .4$, and $x = 5$ and $x = 4$) the probability is $.403 - .217$, or $.186$.

Example 4

For another example of binomial probabilities, consider the experiment of drawing a card from a deck. Let S be the drawing of a spade and F be the drawing of a nonspade, and suppose we repeat the experiment five times. In this case, since there are 13 spades and 39 nonspades, p is $\frac{13}{52}$, or $.25$, and $1 - p$ is $.75$. It is to be emphasized that after each trial the card is returned to the deck and the deck shuffled. If the card were not returned, there would not be independence from trial to trial. Clearly the occurrence of S on the first trial would decrease the chance of S on the second trial because one spade would be missing on the second drawing. If the card is returned, the binomial formula (4.20) applies with $p = .25$, $1 - p = .75$, and $n = 5$; the probability that exactly two spades are drawn ($r = 2$) is

$$P(2) = \binom{5}{2}(.25)^2(.75)^3 = .2637$$

The problem in this example is translated in the next example into a problem involving polls. The example illustrates important properties of the binomial probabilities and will serve as an introduction to some of the statistical ideas of the following chapters.

Example 5

Suppose that there are 52 voters in a very small voting unit, of which 13 are Republicans and 39 are Democrats. We call this pool of 52 voters the *population* of voters (the general notion of a population is taken up in Section 5.1). Suppose that a political analyst is ignorant of this 13-to-39 split between the two parties and wants to get an idea of what the split in the population is. To do so he draws a voter at random from among the 52; that is, he in effect puts the 52 voters' names on cards, shuffles, and draws one. Then he asks this voter his party affiliation. The probability of drawing a Republican (conventional success, S) is $.25$ and the probability of drawing a Democrat (conventional failure, F) is $.75$.

If voters are drawn successively, we have independent repeated trials, provided the voter drawn is returned to the population or pool of voters after each trial—provided, that is, the voter's card is returned to the deck after each drawing. If the analyst draws five voters ($n = 5$), the chance that he gets two Republicans and three Democrats ($r = 2$) is .2637, exactly as in Example 4. We have merely done Example 4 in a new guise.

This last example raises several questions. If two out of the five voters questioned are Republicans, how should the analyst use this information to estimate or assess the Republican-to-Democrat split in the entire population of 52 voters? This is a central problem of statistics which we set to one side for the present, as it will be studied in detail in the following chapters. Suffice it to say for now that the solution of this statistical problem requires an understanding of binomial probabilities. Also, should the analyst choose his five voters randomly in the first place? He should, but we postpone the discussion of this as well.

Another question is this: If the poller gets Mr. Smith on the first draw and finds he is, say, a Republican, why return his card to the deck representing the 52 voters? To do so makes no sense; indeed, Mr. Smith's card might come up again on the second drawing, in which case no new information regarding party affiliation will be obtained. The sensible thing to do is to leave Mr. Smith's card out of the deck and on the next trial draw a card randomly from among the remaining 51. The second drawing will then be certain to yield new information.

sampling with replacement
sampling without replacement

This is the difference between **sampling with replacement** and **sampling without replacement,** and in practice sampling is done without replacement. But sampling without replacement destroys the independence. If a Republican is drawn on the first trial, there remains a pool of only 12 Republicans together with the 39 Democrats, so the conditional probability of drawing a Republican is $\frac{12}{51}$, which is .2353 instead of .25, the unconditional probability. Thus the formula (4.20) does not apply. Fortunately, in practical situations this effect is negligible because the population of voters is quite large (political analysts are of course generally interested in voting units much larger than 52). Consider a final example.

Example 6

Suppose there are 5200 voters in the population, of whom 1300 are Republicans and 3900 are Democrats. If a voter is drawn at random, the probability he is a Republican is 1300/5200, which comes to .25 as before. Suppose that five voters are drawn in succession and that each voter drawn is returned to the population after each trial. Then there is sampling *with* replacement, the analysis of Examples 4 and 5 still applies, and the probability of getting two Republicans and three Democrats is again .2637.

SECTION 4.5 REPEATED INDEPENDENT TRIALS

Example 6 differs from Example 5 in the following important respect. If in Example 6 the voter is *not* returned to the population after each trial—that is, if there is sampling *without* replacement—the effect on independence is very small. For example, if a Republican is drawn on the first trial and not returned to the pool, there remain 1299 Republicans together with the 3900 Democrats, a total of 5199, so the conditional probability of drawing a Republican on the second trial is $\frac{1299}{5199}$, or .2499, which is very close to the unconditional probability .25. Similarly, the chance of drawing Republicans on the third, fourth, and fifth trials is hardly affected by removing from the population those voters drawn on previous trials.

What is the effect of switching from sampling with replacement to sampling without replacement? If a Republican is drawn on the first trial and not returned to the pool, the probability in Example 5 of a Republican on the second trial changes from .25 to .2352, whereas in Example 6 this probability only changes from .25 to .2499. That the effect of switching from sampling with replacement to sampling without replacement is much less in Example 6 than in Example 5 is also reasonable on intuitive grounds—removing a few voters from a *large* population hardly affects its composition.

This discussion leads to the formulation of two rules. Consider a population (or set, or collection, or pool) of N objects, of which a proportion p are labeled S and a proportion $1 - p$ are labeled F. (In the previous examples the population is the pool of voters, $S =$ Republican, $F =$ Democrat, and $p = .25$. In Example 5, $N = 52$; in Example 6, $N = 5200$.) We draw n objects in succession from the population, the drawing being at random at each trial.

Sampling with replacement If the object drawn is returned to the population after each trial, then the probability that exactly r of the objects drawn are labeled S (and $n - r$ are labeled F) is

$$P(r) = \binom{n}{r} p^r (1 - p)^{n-r} \qquad 4.22$$

Sampling without replacement If the object drawn is not returned to the population after each trial, then (4.22) still gives an accurate approximation to the probability that exactly r of the objects drawn are labeled S, provided n is small compared with N, the population size.

In the examples $n = 5$. In Example 6, n is small in comparison with N, which is 5200, and (4.22) gives a very good approximation in the case of sampling without replacement: (4.22) gives .2637, as we have seen, and the true value is .2638 (to four decimal places). In Example 5, on the other hand, N is only 52, and n is not small enough in relation to N, so here the approximation is not so accurate for sampling without replacement: (4.22) gives .2637, but the true value is .2743. The next section shows how to calculate these true values for sampling without replacement. The section can be omitted because for all the statistical applications in this book (4.22) provides a sufficiently accurate approximation.

Problems

57. A study has shown that 40% of the people using State Route 22 drive faster than the legal speed limit. If a state trooper times five randomly selected vehicles on this highway, what is the probability that <u>three or more</u> will be going over the legal speed limit?

58. A true-false test contains 10 questions. If the passing score is 80%, what is the probability that a student can pass the examination just by guessing? Would the probability change if the test contained 20 questions?

59. A seed company has determined that its seeds have a 90% chance of germinating. If 20 seeds are planted, what is the probability that 18 or more will germinate?

60. If the probability a child is a boy is $\frac{1}{2}$ and sex is independent from child to child, what is the probability in a family of four children that
   a. All will be boys?
   b. At least one will be a boy?
   c. There will be the same number of boys as girls?

61. Do the three parts of Problem 60 for families of six children.

62. A die is rolled seven times. What is the probability that 1 shows
   a. Twice?
   b. Five or more times?
   c. An even number of times?

63. In a triangle test a taster is presented with three food samples, two of which are alike, and is asked to pick out the odd one by tasting. If a taster has no well-developed sense and can pick the odd one only by chance, what is the probability that <u>in five trials he will make</u>
   a. Four or more correct decisions?
   b. No correct decision?
   c. At least one correct decision?

64. If a baseball player with a batting average of .250 comes to bat four times in a game, what is the chance he

SECTION 4.6 FINITE SAMPLING   **141**

   a. Gets one hit?
   b. Gets two hits?
   c. Gets four hits?

65. The World Series terminates when one team wins its fourth game. Suppose the two teams are evenly matched, so each has probability 1/2 of winning any one game. What is the probability that the series will terminate at the end of the fourth game? The fifth game? The sixth game? The seventh game?

66. A manufacturing process is considered to be in control if it produces no more than 10% defectives. The process is stopped and checked if a sample of ten contains more than one defective. What is the probability it will be stopped needlessly if it is producing only 5% defectives?

67. A sharp-shooting basketball player has probability .80 of making a free throw. What is the probability that the player will make fewer than 12 of 15 attempts?

68. Because of atmospheric interference, a signal sent from a space satellite is not always received correctly. If each signal has probability .1 of being received incorrectly, what is the probability that at least one signal will be received incorrectly in 25 transmissions from the satellite?

69. A certain aquatic predator has probability .4 of capturing its prey each time it initiates an attack. How many attacks must the predator initiate to have probability greater than .90 of making at least one capture?

4.6 Finite Sampling

Let us first do Examples 4 and 5 of Section 4.5 (terminology apart, they are the same example) for the case of sampling without replacement.

Example 1

We use the terminology of cards. Let us suppose we draw five cards from a deck, but suppose we do not replace the card drawn after each trial. We ask for the probability of getting exactly two spades and three nonspades. Since the number of spades does not depend on the order in which the cards appear in the drawing, we may ignore order altogether. Suppose then that we choose at random a combination of five of the 52 cards.

   Now the sample space consists of the $\binom{52}{5}$ combinations of the 52 cards taken five at a time. How many favorable cases are there? To make up a hand containing exactly two spades we must make two choices. First we must choose two of the 13 spades; there are $\binom{13}{2}$ ways to do this. Then we must choose three of the 39 nonspades; there are $\binom{39}{3}$ ways to do this. By the multiplication principle there are thus $\binom{13}{2} \times \binom{39}{3}$ favorable cases, so the probability of exactly two spades is

$$\frac{\binom{13}{2} \times \binom{39}{3}}{\binom{52}{5}} = .2743$$

The difference between .2743 and the .2637 of Examples 4 and 5 of the last section reflects the effect of not returning the card to the deck after each trial.

The general rule is this: Suppose we have a population of N objects, of which a are labeled S and b are labeled F; here $a + b$ must equal N. Suppose we take at random a combination of size n and ask for the probability that exactly r of the objects in it are labeled S and $n - r$ are labeled F. Thus, we let

$$a = \text{the number of objects labeled } S,$$
$$b = \text{the number of objects labeled } F,$$
$$a + b = N = \text{the total number of objects,}$$
$$n = \text{the number of objects drawn,}$$
$$r = \text{the number of } S\text{s drawn.}$$

The probability of getting exactly r Ss is

$$\frac{\binom{a}{r}\binom{b}{n-r}}{\binom{a+b}{n}} \qquad 4.23$$

These quantities have the forbidding name *hypergeometric probabilities*.

In Example 1 there were 13 spades (Ss) and 39 nonspades (Fs), so $a = 13$ and $b = 39$. The reasoning underlying (4.23) is the same as in the example. The denominator is the total number of combinations of the $a + b$ objects taken n at a time. Among these combinations the ones that contain exactly r objects labeled S and $n - r$ objects labeled F are those that result from merging some combination of r of the Ss with some combination of $n - r$ of the Fs, and the numerator is the number of pairs of such mergeable combinations.

Example 2

In Example 6 of Section 4.5 the number of objects (voters) labeled S (Republican) is 1300, and the number labeled F (Democrat) is 3900, so $a = 1300$

SECTION 4.6 FINITE SAMPLING   143

and $b = 3900$. In sampling ~~with~~ *without* replacement the probability of two Republicans out of five is then

$$\frac{\binom{1300}{2}\binom{3900}{3}}{\binom{5200}{5}} = .2638$$

The formulas (4.22) and (4.23) give practically identical answers in this case.

Problems

70. A merchant is giving away two dozen ballpoint pens. One dozen are red and one dozen are blue. If the first four customers receiving pens each select one at random, what is the probability that all four select blue?

notes 11 may 83

71. A sample *of two* is taken *with* replacement from a deck of 52 cards. What is the probability that the number of aces in the sample is — *binomial prob*
    a. Zero?
    b. One?
    c. Two?

72. Do Problem 71 for sampling *without* replacement. — *hypergeometric prob.*

73. A sample of two is taken *with* replacement from a population of 520 electronic components, of which 40 are defective and 480 are non-defective. What is the probability that the number of defectives in the sample is
    a. Zero?
    b. One?
    c. Two?

74. Do Problem 73 for sampling *without* replacement. Make a table of the answers to Problems 71, 72, 73, and 74. Explain the agreements and the magnitude of the differences in these answers.

75. From a box containing six chocolates and four hard candies, a child takes a handful of four. What is the chance he gets
    a. No chocolates?
    b. Three chocolates?
    c. At most one chocolate?

76. Since George can't attend Saturday's game, he has decided to give his ticket away to one of three of his friends. He places an ace, king, and queen face down in random order on a table and asks each one of his

friends to select a card. The person who draws the ace gets the football ticket. Is there an advantage in being the first one to draw?

77. A retail store has advertised three new positions. Seven equally qualified people have applied for these positions. Three of the applicants are male and four are female. If three people are selected at random from the seven to fill the positions,
   a. Find the probability that all three are female.
   b. Find the probability that the majority of those hired are male.

78. A physician has 10 antibiotics to choose from to treat bacterial infections. It is known that a particular bacterial strain is resistant to 3 of the 10 antibiotics. If the physician chooses two of the antibiotics at random, what is the probability that at least one of them will be effective in treating this particular bacterial strain?

4.7 Subjective Probability and Bayes' Theorem*

The probabilities discussed in this chapter have all been *objective* in the sense that we have taken the probability of an event as being a property of the event itself. Statistics is sometimes placed in the framework of a different conception of probability—*subjective,* or *personal,* probability. This is the *Bayesian* approach to statistics. Here we can give only a very brief description of the ideas lying behind this approach.

Consider the hypothesis that there is life on Mars; let H denote this hypothesis and let $\bar{H}$ denote the opposite hypothesis, that Mars supports no life. Certainly, people have varying degrees of belief in H and say things like, "It is fairly likely that there is life on Mars" or "It is improbable that there is life on Mars." Adherents of the theory of subjective probability hold that it is possible to assign to H a probability $P(H)$ that represents numerically the degree of a person's belief in H. The idea is that $P(H)$ will be different for different people because they have different information about H and assess it in different ways. This conception of probability does not require ideas of repeated trials and the stabilizing of relative frequencies.

Setting aside the problem of how subjective probabilities are to be arrived at in the first place, we consider here only a method for modifying them in the light of new information or data.

Suppose a space probe equipped with a life-detection device has landed on Mars and has sent back a message saying that there is indeed life

*The contents of this section are in no way required for an understanding of the material in subsequent chapters.

SECTION 4.7 SUBJECTIVE PROBABILITY AND BAYES' THEOREM

there. This new item of information, call it D (for *data*), will certainly alter our attitude towards the hypothesis H. But we know the device is not foolproof. Suppose we know from previous laboratory tests of the life-detection device that if there *is* life on Mars, then the chance that the device will report the existence of life, which we denote $P(D|H)$, has numerical value .8. Suppose we also know that if there is *no* life on Mars, then the chance that the device will report the existence of life, which we denote $P(D|\bar{H})$, has numerical value .1.

We are to compute a *new* probability $P(H|D)$ of the hypothesis given the data from the probe. Since personal probabilities are assumed to obey the rules of Section 4.3, we can proceed as follows. By the definition (4.8) of conditional probability, $P(H|D)$ is equal to $P(H \text{ and } D)/P(D)$, and applying the formula (4.9) to the numerator gives

$$P(H|D) = \frac{P(H) \times P(D|H)}{P(D)} \qquad 4.24$$

Now D happens if "H and D" happens *or* if "$\bar{H}$ and D" happens. Therefore, by the addition rule (4.4)

$$P(D) = P(H \text{ and } D) + P(\bar{H} \text{ and } D)$$

Using the formula (4.9) on each of these last two terms now gives

$$P(D) = P(H)P(D|H) + P(\bar{H})P(D|\bar{H})$$

Substituting this for the denominator in (4.24) gives the answer

$$P(H|D) = \frac{P(H) \times P(D|H)}{P(H) \times P(D|H) + P(\bar{H}) \times P(D|\bar{H})} \qquad 4.25$$

This formula constitutes *Bayes' rule*, or *Bayes' theorem*. Given the *prior* probability $P(H)$ (and $P(\bar{H}) = 1 - P(H)$) and the respective probabilities $P(D|H)$ and $P(D|\bar{H})$ of observing the data D if H and $\bar{H}$ hold, we can use Bayes' rule to compute the *posterior probability* $P(H|D)$—the personal probability for H after the information in D has been taken into account. In our example $P(D|H) = .8$ and $P(D|\bar{H}) = .1$, so the formula gives

$$P(H|D) = \frac{.8 \times P(H)}{.8 \times P(H) + .1 \times P(\bar{H})}$$

If $P(H) = .3$, say, so that $P(\bar{H}) = .7$, then $P(H|D) = .77$. Notice that $P(H|D)$ exceeds $P(H)$ here; observing D increases our personal probability of H because H "explains" D better than $\bar{H}$ does ($P(D|H) > P(D|\bar{H})$).

It is sometimes said that Bayes' theorem is controversial. This is only partly true. If probabilities satisfy the rules of Section 4.3, then they must also satisfy Equation (4.25). The question that is disputed is whether probabilities can sensibly be assigned to hypotheses in the first place. Many people feel quite comfortable with the notion that a scientific hypothesis can be true with a certain probability. Many others find that idea hard to accept, and regard a scientific hypothesis as being simply true or false. They feel that probability has nothing to do with hypotheses as such, and they prefer to formulate all statistical and probabilistic concepts within the framework of the objective, or frequency, theory of probability. That is the course we follow here. For a treatment of statistics from the subjective point of view, see Reference 4.4.

Problems

79. Box I contains three red marbles and two blue marbles. Box II contains one red marble. A marble is drawn at random from box I and placed in box II, and then a marble is drawn at random from box II. Given that the second marble drawn is red, what is the probability that the first marble drawn was red?

80. The chance a patient in a hospital has a certain disease is .05. He is given a test to determine whether he has the disease. If he has it, the chance is .99 that the test will be positive; if he does not have it, the chance is .03 that the test will be positive.
    a. If the test is positive, what is the probability the patient has the disease?
    b. If the test is negative, what is the chance he does not have the disease? Use Bayes' rule.

81. Machines I and II respectively produce 30% and 70% of a factory's output. Machine I produces 3% defectives and machine II produces 4% defectives.
    a. If an item is defective, what is the chance it was produced by machine I?
    b. If an item is not defective, what is the chance it was produced by machine I? Use Bayes' rule.

82. A student takes a multiple-choice test with four answers to each question. Suppose he really knows the answers to 80% of the questions; these he answers correctly. If he does not know the answer, he guesses, so the chance he gets it right is $\frac{1}{4}$.
    a. If he gives the right answer to a question, what is the probability he really knew the answer?
    b. If he gives the wrong answer, what is the chance he guessed?

CHAPTER PROBLEMS

83. Patients receiving emergency treatment at a local hospital were categorized according to the type of emergency and the work shift during which treatment was given. The percentages are given in the table below.

		Type of Emergency			
		Heart Attack	Appendicitis	Auto Accident	Other
Shift	Day	12%	4%	9%	35%
	Night	8%	6%	11%	15%

Suppose that a person receives emergency treatment at this hospital.
  a. What is the probability that the emergency is either a heart attack or appendicitis?
  b. Supposing that treatment is given during the day shift, what is the probability that the emergency is an auto accident?

84. Refer to Problem 83. Show that the occurrence of a heart-attack emergency is independent of the work shift, while the occurrence of an auto-accident emergency is not. What implication does this have for the staffing of the emergency room during the two shifts?

85. Refer to Problem 83. If ten people receive emergency treatment at the hospital during a 24-hour period, what is the probability that at least four have a heart attack?

86. The First Street, Tenth Street, and Baker Street bridges cross the river that passes through town. During rush hour, traffic emergencies will develop independently on these bridges with probabilities .2, .3, and .1, respectively.
  a. What is the probability that no emergency will develop on any of the bridges?
  b. What is the probability that an emergency will develop on at least two of the three bridges?

87. Refer to Problem 86. A traffic officer receives a call that an emergency has occurred on one of the bridges. What is the probability that it has occurred on the Baker street bridge?

Summary and Keywords

The **probability** (p. 101) of an event may be viewed as the fraction of times it would occur in a long series of trials. In the case of equally likely **outcomes** (p. 97), the probability is the number of outcomes in the **event** (p. 99) divided by the number in the **sample space** (p. 98), or

$$\text{Probability} = \frac{\text{Number of favorable cases}}{\text{Total number of cases}}$$

The most important laws probabilities obey are these:

(1) If A and B are **disjoint** (p. 104), then

$$P(A \text{ or } B) = P(A) + P(B)$$

(2) If A and $\bar{A}$ are **complements** (p. 106), then

$$P(\bar{A}) = 1 - P(A)$$

(3) The **conditional probability** (p. 107) of B, given A, is

$$P(B|A) = \frac{P(A \text{ and } B)}{P(A)}$$

This equation is sometimes used in the form

$$P(A \text{ and } B) = P(A)P(B|A)$$

(4) Events A and B are by definition **independent** (p. 110) if

$$P(A \text{ and } B) = P(A)P(B)$$

This condition is the same as

$$P(B|A) = P(B)$$

and the same as

$$P(A|B) = P(A)$$

Many probabilities can be calculated by counting cases. In addition to the basic **multiplication principle** (p. 116), the most important facts about counting are these:

(1) There are a^n **words** (p. 122) of length n from an alphabet of size a.
(2) The number of **permutations** (p. 122) or orderings of n distinct things taken r at a time is $_nP_r = n(n-1) \cdots (n-r+1)$, or, in **factorial notation** (p. 124),

$$_nP_r = \frac{n!}{(n-r)!}$$

(3) The number of permutations of n distinct things is

$$_nP_n = n!$$

(4) The number of **combinations** (p. 125) of n distinct things taken r at a time is

$$\binom{n}{r} = \frac{n!}{r!(n-r)!}$$

If the chance of success at each trial is p, and if the trials are independent, then the chance of exactly r successes in n trials is given by the **binomial probability** (p. 135)

$$P(r) = \binom{n}{r} p^r (1-p)^{n-r}$$

This formula also applies to **sampling with replacement** (p. 138), and it gives an approximation for sampling **without replacement** (p. 138).

REFERENCES

4.1 Feller, William. *An Introduction to Probability Theory and Its Applications*. 3d ed. Volume I. New York: John Wiley & Sons, 1968.

4.2 Mosteller, Frederick; Rourke, Robert E. K.; and Thomas, George B., Jr. *Probability with Statistical Applications*. 2d ed. Reading, Mass.: Addison-Wesley, 1970.

4.3 Parzen, Emanuel. *Modern Probability Theory and Its Applications*. New York: John Wiley & Sons, 1960.

4.4 Savage, I. Richard. *Statistics: Uncertainty and Behavior*. Boston: Houghton Mifflin, 1968.

5

Populations, Samples, and Distributions

5.1 Populations

population

In the broadest meaning of the term a statistical **population** is simply a set or collection; the things of which the population is composed are called its *elements*. Sometimes, the population is specified by a complete list of its elements, as when the voting population of a town is explicitly listed on the voting rolls. More commonly, a population is specified by a definition of some kind, or by the singling out of a characteristic common to its elements. For example, we may speak of

(1) the population of all transistors produced by a given manufacturer in the year 1980;
(2) the population of U.S. males aged 45–55;
(3) the population of U.S. males aged 45–55 and having high blood pressure;
(4) the population of U.S. males aged 45–55 having high blood pressure and undergoing drug treatment for it.

Such examples extend the original notion of the population of a geographic area. Examples 5 and 6 in Section 4.5 involve populations of voters.

The purpose of statistical inference is to establish facts about populations. We want to know the characteristics of the transistors the company produces. What percentage are defective? We want to know the characteristics of the U.S. males, aged 45–55, undergoing drug treatment for high blood pressure. What percentage does the treatment benefit? We want to know the characteristics of a population of voters. What are their political preferences?

finite
infinite

The populations in the preceding examples are all **finite**. Statistics also involves certain hypothetical **infinite** populations. Consider an experiment to measure a physical quantity, such as the speed of light or the percentage of copper in a piece of brass. Carrying through the experiment repeatedly will give a sequence of different answers because of the combined effect of small errors that creep in (this is why experiments are customarily repeated). To analyze the data, the experimenter considers not only the answers already obtained, but the hypothetical, infinite population of all answers that would be obtained if the experiment were repeated infinitely often under identical conditions.

The analysis of a manufacturing process often involves an infinite population, the hypothetical population of all the items (all the transistors, say) that the process would produce if it were to run indefinitely under constant conditions. Sometimes random experiments like those in the examples in Chapter 4 are viewed as based upon infinite populations. There is an infinite population of tosses of a coin or rolls of a pair of dice, and to toss the coin or roll the dice is to observe an element of the population.

Often a finite population is so large as to be effectively infinite for the purposes of statistical analysis. An effective way to acquire an initial understanding of what it means for a population to be infinite is to imagine the population to be finite but so very large that removing a number of elements from it has no discernible effect on the composition of the population. That infinite populations are constructs of the human mind does not make them less important in practical affairs.

Ordinarily, a statistical analysis involves only certain aspects of a population—only certain attributes or characteristics of the elements of the population. In a population of voters we are perhaps concerned only with the voter's party affiliation, and not with his height, blood type, etc. In a population of trees we may be concerned with height but not with age. In a population of transistors we may be concerned only with whether or not a transistor is defective according to certain standards.

Since the primary objective of inductive statistics is to make inferences about populations, it is important that the population concerned be carefully specified. To know how the population of property owners in a town feels on some issue is not necessarily to know how the population as a whole feels. And the population appropriate to one method of measuring a physical quantity may be inappropriate to another method.

5.2 Samples

For inquiring into the nature of a population it would be ideal if we could easily and economically examine each one of its elements, but this is usually out of the question. Sometimes it is impossible because some of the elements are physically inaccessible. In other cases it is uneconomic. Obviously we won't test every item produced if the test destroys the item. Even when all the elements of the population are available for examination, they may be so numerous that a complete census is not justified; often, for most practical purposes, sufficiently accurate results may be more quickly and inexpensively obtained by examining only a small part of the population. And of course, in an infinite population the elements cannot all be examined.

sample

In most situations, then, we must be content with investigating only a part of the whole population. The part investigated is a **sample**. Samples may be collected or selected in a variety of ways. A *systematic sample* is one selected according to some fixed system—taking every hundredth name from the telephone book, for example, or selecting machined parts

random samples

from a tray according to some definite pattern. Most statistical techniques presuppose an element of randomness in the sampling, and here we'll be concerned almost exclusively with **random samples**.

Consider first a population of size N from which we take a sample of size n. In Example 5 of Section 4.5, $N = 52$ and $n = 5$; in Example 6 $N = 5200$ and $n = 5$. Recall that Section 4.5 treated the difference between sampling with replacement and sampling without replacement. In practice, *samples are taken without replacement*. Now in Section 4.5 the elements of the sample were viewed as having been drawn from the population one at a time in sequence. But in sampling without replacement we can instead view the elements of the sample as having all been drawn from the population simultaneously. This is because the order in which the elements come in the sample gives no information about the population itself. Thus we may ignore order and take the sample to be a *combination*, or subset, of size n from the population. The number of such combinations is $\binom{N}{n}$, and *the sampling is defined to be random if each of these* $\binom{N}{n}$ *combinations has the same chance of being drawn as the sample*.

Although the concept of a random sample is fairly easy to grasp, there are situations in which it is not clear how to obtain one. If we can list all the elements of a population and number them, we can get a random sample by drawing numbered tags or tickets from a bowl, or we can use a table of random numbers to decide which elements to include in the sample. But if we cannot enumerate the population, these techniques are useless. How, for instance, would we obtain a random sample of oranges from a tree, fish from a river, or trees from a forest? How can a public opinion poller obtain a random sample of people in a city? We shall not go into such questions of sampling technique here (see References 5.1 and 5.3). We assume we have successfully obtained a random sample from the population in which we are interested, and investigate the inferences that can be based on such a sample.

What is the point of using random samples instead of samples of some other kind? Sampling is done in order to draw inferences about the population sampled, and as we shall see later on, the accuracy of inferences based on random samples can be assessed by means of probability theory.

To understand the notion of sampling from an *infinite* population, return once more to Examples 5 and 6 of Section 4.5. For the population of size 52 the chance of two Republicans and three Democrats in a sample of size five is .2637 for sampling with replacement and .2743 for sampling without replacement (see page 140)—a noticeable difference. But for the larger population of size 5200 this probability is .2637 again for sampling with replacement and .2638 for sampling without replacement—a negligible difference. The larger population is affected less by the removal of the elements sampled. <u>For an infinite population the probability in question</u> would be *exactly the same* for sampling with <u>replacement as for sampling without replacement</u>. In this ideal case removing an element from the population and putting it into the sample (not replacing it) has no effect at all on the composition of the population.

<u>In general, the elements of a sample from an infinite population are independent of one another</u>. Here independence has the meaning given it in the preceding chapter. The event that one element of the sample has a certain characteristic (say, defectiveness) or measurement (say, height) is independent of the event that some other element of the sample has a certain characteristic or measurement.

5.3 Random Variables

[handwritten: R.V. associate a number with an outcome (like a rule)]

Many of the random experiments or observations discussed in Chapter 4 have numbers associated in a natural way with the outcomes. In rolling a pair of dice we can observe the total number of dots; in tossing a coin three times in a row we can count the number of times heads turns up; in a poker hand we can count the number of aces. Such a number, determined by a chance mechanism, is called a **random variable**. We denote random variables by X, Y, etc.

random variable

Before looking further into this idea, consider a physically determined variable that has no element of randomness. To find the area of a square, you measure the length l of one of its sides and then multiply that length by itself, computing l^2. Now l is a variable; it may be 1.5 feet or 8.3 cubits, depending on the square in question. To keep in mind this element of variability, it helps to regard l as a name for the length of the side *before the side is measured*. Measuring the side of a specific square converts l into a specific number. The variable l is useful for making general statements such as, "The area is l^2."

To see the importance of regarding l as a name for the length *before* the side is measured, notice this. "Sometimes l exceeds 2" is a valid general statement. But suppose a specific square is measured and its length turns out to be 1.5 (in feet, say). Now 1.5 is the length *after* the square is measured; if 1.5 is substituted for l in the above statement, there results the absurd assertion that "Sometimes 1.5 exceeds 2."

Now let X be the total on a pair of dice. The dice may show a total of 7 when actually rolled, in which case $X = 7$; or they may show 5, in which case $X = 5$; and so on. Now X is a number whose value depends on the outcome of a random experiment, and the outcome cannot be predicted. To keep in mind this variability and unpredictability, it helps to regard the random variable as a name for the number connected with the outcome of the experiment *before the experiment is performed*. Carrying out the experiment converts the random variable into a specific number.

Selecting a random sample from a population, finite or infinite, is a random experiment, and most random variables we will encounter arise from sampling. The number of defectives in a sample of transistors is a random variable. In a sample of people their individual heights are all random variables, and so is the average of these heights.

discrete

There is a classification of random variables that is based on the set of values the variable can take on. The random variable is **discrete** if there is a definite distance from any possible value of the random variable to the next possible value. The number of defectives in a sample is discrete, the distance between successive possible values being 1. A height measured to the nearest tenth of an inch takes as value one of the numbers . . . , 60.0, 60.1, 60.2, . . . ; it is discrete because the distance from one possibility to the next is 0.1.

continuous

A **continuous** random variable, on the other hand, is assumed able to take any value in an interval. A height measured with complete accuracy can, in principle, take on any positive value, and so is a continuous variable. So are measurements such as weight, temperature, pressure, time, and the like, provided they are measured with complete accuracy. In reality, there is a limit to the precision with which measurements can be made, and continuous random variables are only a useful idealization, just as infinite populations are.

Problems

1. Each of the following experiments results in one measurement (one value of the random variable).
    a. State whether the random variable is discrete or continuous.
    b. Determine, at least in principle, all possible values of the random variable.
       d (i) The number of smokers in a group of 100 persons.
       c (ii) The speed of a passing car.
       d (iii) The number of women in a jury of 12.
       d (iv) The number of leaves on a tree.
       d (v) The number of typing errors on a page.
       c (vi) The amount of rainfall in Denver for the month of September.

SECTION 5.4 DISTRIBUTIONS OF RANDOM VARIABLES   **157**

d (vii) The number of cars being filled at a certain gas pump during one hour.

c (viii) The time required to read the book *How to Lie with Statistics*.

2. The sale of alcoholic beverages at Tampa Stadium created controversy among the city's sports fans. The largest newspaper in the city invited its readers to express their opinions on this issue by filling out and sending in a questionnaire published in the paper.
   a. Would the responses from this survey be a random sample of the opinions of the city's sports fans?
   b. Would the responses from this survey be a random sample of the opinions of the newspaper's sports-minded readership?
   c. What biases may arise in trying to draw inferences about the collective opinion of the city's sports fans from the data of this survey?

5.4 Distributions of Random Variables

The behavior of a discrete random variable can be described by giving the probability with which it takes on each of its values when the experiment is carried out.

Example 1

We know from Chapter 4 that the probability of rolling a 7 with a pair of fair dice is $\frac{6}{36}$. Letting X be the discrete random variable that represents the total for the two dice (X is a name for the total *before the dice are rolled*), we can express this fact by writing $P(X = 7) = \frac{6}{36}$. Similarly, the probability of rolling a 4 is $\frac{3}{36}$, so $P(X = 4) = \frac{3}{36}$. Here is a table of all the probabilities (they can be determined by counting cases in Table 4.2):

$$P(X = 2) = \tfrac{1}{36}, \quad P(X = 3) = \tfrac{2}{36}$$
$$P(X = 4) = \tfrac{3}{36}, \quad P(X = 5) = \tfrac{4}{36}$$
$$P(X = 6) = \tfrac{5}{36}, \quad P(X = 7) = \tfrac{6}{36}$$
$$P(X = 8) = \tfrac{5}{36}, \quad P(X = 9) = \tfrac{4}{36}$$
$$P(X = 10) = \tfrac{3}{36}, \quad P(X = 11) = \tfrac{2}{36}$$
$$P(X = 12) = \tfrac{1}{36}.$$

— note the prob of rolling a 7 is better than any other prob. Maybe this is why 7 is such an important # in dice games (eg) craps.

Figure 5.1a

Distribution of the outcomes when a pair of dice was rolled 100 times

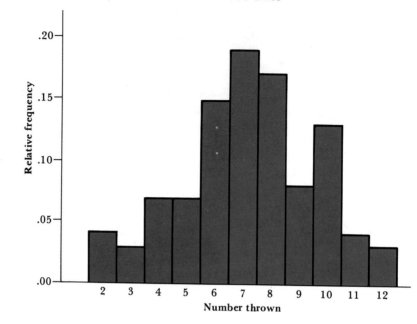

Figure 5.1b

Theoretical distribution of the outcomes when a pair of dice is rolled

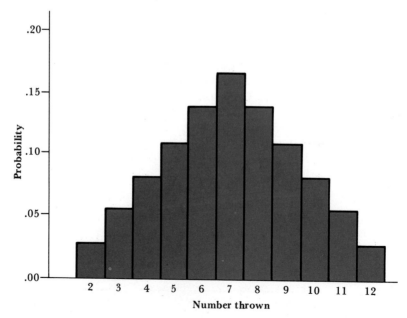

SECTION 5.4 DISTRIBUTIONS OF RANDOM VARIABLES

distribution   This collection of probabilities is called the **distribution** of the discrete random variable X. The distribution can be used to answer any question about X. For instance, the chance of "X is 7 or 11" is

$$P(X = 7 \text{ or } X = 11) = P(X = 7) + P(X = 11) = \tfrac{6}{36} + \tfrac{2}{36} = \tfrac{2}{9}$$

A pair of dice was rolled 100 times and a record kept of the number of times each possible value for the sum of the faces occurred. Figure 5.1a shows the observed relative frequencies; Figure 5.1b shows the corresponding probabilities. The two figures would agree more closely if the dice had been rolled 1000 times, say. The probabilities represent a theoretical limit toward which the observed relative frequencies should tend as the number of rolls increases beyond bound.

Example 2

Suppose a box contains three white balls and two red ones. If we select three balls at random (that is, if we take a random sample of size 3 without replacement), and if we let X be the number of white balls in the sample, then X is a random variable which can take the values 1, 2, and 3. There are $\binom{5}{3}$, or 10, possible samples. The number of samples containing one white ball is 3 (the sample must contain the two reds and one of the whites); the number containing two white balls is 2×3, or 6 (the sample must contain one of the reds and two of the three whites); and the number containing three white balls is 1. Therefore

$$P(X = 1) = .3, \quad P(X = 2) = .6, \quad P(X = 3) = .1$$

These probabilities are shown in Figure 5.2. (They could be computed from the formula (4.23) on page 142 instead of directly.)

     The set of probabilities in Figure 5.2 is the distribution of X. These probabilities sum to 1, and they suffice to answer any question about the probability that X will have a given property. For instance, the probability that X is odd is

$$P(X \text{ is odd}) = P(X = 1) + P(X = 3) = .3 + .1 = .4$$

The probability that X exceeds 1 is

$$P(X > 1) = P(X = 2) + P(X = 3) = .6 + .1 = .7$$

     A discrete random variable is described by its distribution—the list of probabilities for its various possible values. This sort of distribution does not work for a continuous random variable, because such a variable

Figure 5.2

Distribution of number of white balls

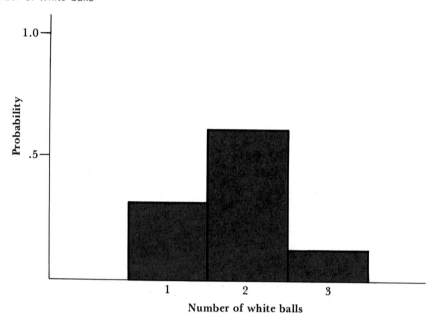

takes on any given value with probability *zero*. If X is continuous, then $P(X = 3.1) = 0$, $P(X = 2.854) = 0$, etc. This seems paradoxical at first, since X must assume *some* value when the experiment is carried out. But think of a line segment; each point in it has length zero, whereas all the points together give a segment with positive length. In the same way, X has probability 0 of taking on any one given number but has probability 1 of taking on *some* number (some number not specified in advance of the experiment). Because of this, the distribution of a continuous X must be given not by a list of probabilities, but by a *continuous curve*.

Example 3

A dart is thrown randomly at a dart board, and the distance X to the center is measured. The behavior of X is described by a curve like that in Figure 5.3. The curve lies entirely above the horizontal axis, and the area under the curve is 1—that is, the area between the curve and the horizontal axis is 1. The probability that, when the experiment is carried out, X will assume a value in a given interval equals the area under the curve and over that interval. The chance that X lies between 1 and 3 is the area of the shaded region R in Figure 5.3; the chance that it lies between 4.5 and 5.8 is the area of the shaded region R'. Notice that since an individual point has probability 0, $P(4.5 < X < 5.8)$ is the same as $P(4.5 \leq X \leq 5.8)$ and the same as $P(4.5 < X \leq 5.8)$; it makes no difference whether or not the endpoints of the interval are included.

All along, we have interpreted probabilities in terms of limiting relative frequencies, and we can give a similar interpretation to the curve in Figure 5.3 by using the histograms of Chapter 2. If we were to throw the dart at the board 100 times and measure the distance to the center with one-inch accuracy, using a class interval of one inch, then the histogram would resemble that in Figure 5.4. If we were to throw the dart 1000 times instead of 100, we could use shorter class intervals, say quarter-inch intervals, and more of them, and the histogram would resemble that in Figure 5.5.

If we continue throwing the dart, using more and more intervals of smaller and smaller length, the histogram will approach the curve in Figure 5.3. In each histogram the area of a bar is proportional to the observed relative frequency in that class, and so an area under the histogram (like the areas of the shaded regions in Figures 5.4 and 5.5) represents an observed relative frequency which converges to a probability (like that represented by the area of the shaded region R in Figure 5.3).

---

frequency curve

The distribution of any continuous random variable is specified by a curve like that in Figure 5.3. The curve is called a **frequency curve,** and the area under the curve between two limits on the horizontal scale is the probability that the random variable will take on a value lying between those two limits. The height of the curve over a point on the horizontal scale is called a _probability density;_ it has no direct probability meaning, the way the areas under the frequency curve do.

If the distribution of a continuous variable is specified by a frequency curve, the question arises how the frequency curve is itself specified. This is only rarely done by an actual curve carefully drawn on graph paper. Sometimes, as in the following example, the distribution is specified by a geometric description.

Example 4

Suppose X always has value between 0 and 2, and suppose its distribution is given by the straight line in Figure 5.6. By the rule for the area of a triangle, the area under the line is 1. To get the probability that X lies between $\frac{1}{2}$ and $\frac{3}{2}$, we compute the area of the shaded trapezoid in the figure.

Figure 5.3

Theoretical distribution of the distance from the center

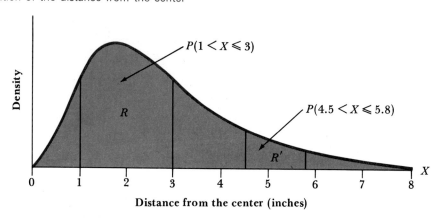

Figure 5.4

Distribution for 100 throws of a dart

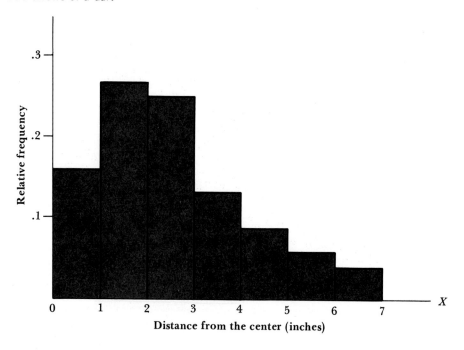

SECTION 5.4 DISTRIBUTIONS OF RANDOM VARIABLES 163

Figure 5.5

Distribution for 1000 throws of a dart

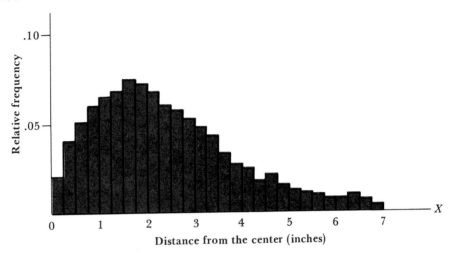

Figure 5.6

A triangular distribution

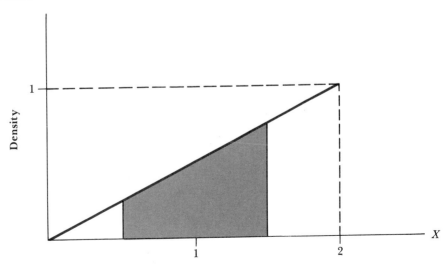

The base of the trapezoid has length 1 and the sides have lengths $\frac{1}{4}$ and $\frac{3}{4}$, so the area rule for trapezoids gives

$$1 \times \frac{\frac{1}{4} + \frac{3}{4}}{2} = \frac{1}{2}$$

and this is the probability sought:

$$P(\tfrac{1}{2} \leq X \leq \tfrac{3}{2}) = \tfrac{1}{2}$$

Usually, distribution curves for continuous random variables are specified by mathematical formulas of greater or lesser complexity (for example, (5.13) on page 178), and the relevant probabilities—that is, the relevant areas—are determined by the methods of integral calculus. Such determinations lie outside the scope of this book. For various important frequency curves often encountered in statistical practice, tables of these areas have been constructed. Several such tables are given in the Appendix; they represent the frequency curves we need in the succeeding chapters.

The distribution of a random variable, whether discrete or continuous, can also be specified by its *cumulative distribution function* (or simply distribution function). This is a function $F(x)$ whose value at x is the probability that

Figure 5.7

Cumulative distribution function for the number of white balls

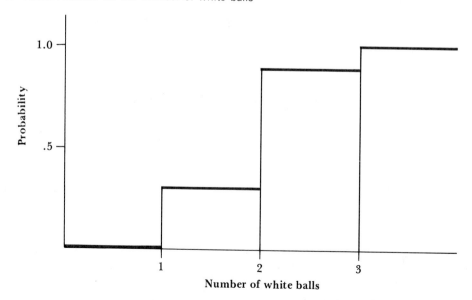

SECTION 5.4 DISTRIBUTIONS OF RANDOM VARIABLES

Figure 5.8

Cumulative distribution function of the distance from the center

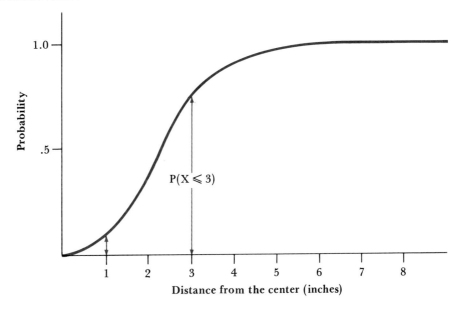

X is at most x; thus $F(x) = P(X \leq x)$. If X is discrete, the cumulative distribution function is a step function. The cumulative distribution function for the random variable in Example 2 is shown in Figure 5.7 (compare Figure 5.2).

If X is continuous, the cumulative distribution function corresponds to a continuously increasing curve. The cumulative distribution function for the distribution in Figure 5.3 is shown in Figure 5.8. Recall that the probability $P(1 < X \leq 3)$ is the area of the shaded region R in Figure 5.3; since $P(1 < X \leq 3) = P(X \leq 3) - P(X \leq 1) = F(3) - F(1)$, this probability can be found by taking the difference of the two heights indicated by arrows in Figure 5.8. In fact, $P(a < X \leq b) = F(b) - F(a)$ always holds for $a < b$, and for this reason the cumulative distribution function is convenient for finding probabilities. The tables in the Appendix are in effect tabulations of these functions.

The cumulative distribution function for a continuous X stands to the frequency curve as the ogive for a set of numbers stands to the histogram.

Problems

3. A marine biologist observes that the number of species of decapod crustaceans on coral heads of a certain size is a random variable having the probability distribution given on page 166.

a. Find the probability that there are 5 or more species on one coral head.
b. Find the probability that there are fewer than 2 species on one coral head.

r	0	1	2	3	4	5	6	7	8	9
p(r)	.05	.09	.12	.19	.25	.13	.08	.04	.03	.02

4. Independently of one another, three consumers are asked to select the more appealing of two television commercials, A or B. Let X denote the number of times commercial A is selected.
   a. List the possible values of X.
   b. If the two commercials have an equal chance of being judged the more appealing each time a choice is made, what is the probability distribution of X?

5. A discrete random variable, X, has possible values $-2, -1, 0, 1, 2, 3$. The cumulative distribution function F is given below.
   a. Find $P(-1 < X \leq 2)$.   .6 } use your fingers
   b. Find $P(0 \leq X \leq 3)$. .7
   c. Find the probability distribution.

x	-2	-1	0	1	2	3
F(x)	0.1	0.3	0.4	0.7	0.9	1.0

6. An integer is selected at random from the integers 1 through 10 and its divisors (including 1) are counted. Let X be the random variable denoting the number of divisors.
   a. Find the probability distribution of X.
   b. Find the cumulative distribution function of X.
   c. Draw the probability distribution.
   d. Draw the cumulative distribution function.
   e. Find the probability of an odd number of divisors.

7. Each of three children—A, B, and C—is asked which of the other two he would prefer as his friend.
   a. Describe explicitly all the possible results of this experiment. How many different possible results are there?
   b. Define X to be the number of times B is chosen. Assuming that each of the above outcomes is equally likely, find the probability distribution of X.
   c. Find the cumulative distribution function of X. step
   d. Draw the probability distribution and the cumulative distribution function.

SECTION 5.4 DISTRIBUTIONS OF RANDOM VARIABLES

8. For the following probability distribution, find the smallest value of b for which $P(X \leq b) > .5$.

X	0	1	2	3	4	5	6
P(X)	.10	.20	.10	.02	.30	.03	.25

9. Given that X is a random variable with the distribution shown below, find the probability that
   a. $.5 \leq X \leq 1$.
   b. $X \geq .75$.
   c. $.25 \leq X \leq .75$.
   d. $X \leq .1$.

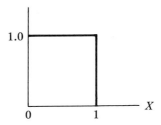

10. If X has the distribution shown below, find
    a. $P(X \geq 1)$.
    b. $P(X < -2)$.
    c. $P(0 < X < 2.5)$.
    d. $P(X > -0.5)$.
    e. $P(-2 \leq X \leq 2)$.
    f. $P(X \geq 3)$.

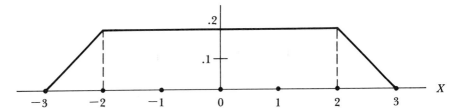

11. To test depth perception, a psychologist has subjects align two distant objects by manipulating rods attached to the objects. Let X denote the number of centimeters by which the objects are misaligned. If X has the distribution shown below, find
    a. a.
    b. $P(X \leq 1)$. .25
    c. $P(X \geq 2)$. .50
    d. $P(1 \leq X \leq 3)$. .50
    e. C such that $P(X \leq C) = .4$. (4×.4) = 1.6

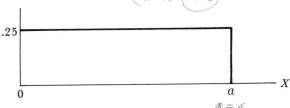

a = 4
(4 ∗ .25 = 1)

12. Sketch the shapes of the probability distributions of the following random variables as you believe they would appear.
    a. The income of an adult wage-earner selected at random from the population of the United States.
    b. The number of inches of rainfall in your city in August.
    c. The amount of time you spend on homework per day.
    d. The speed of an automobile selected at random from a highway in your state.

13. A dart is tossed at a circular dart board with radius 1 foot. Assume that the probability of the dart hitting in a region R is given by

$$P(R) = \frac{\text{Area of } R}{\text{Area of dartboard}}$$

   a. Find the probability that the dart strikes within a quarter of a foot of the center.
   b. Let X denote the distance from the center of the dartboard to the point of impact of the dart. Find the cumulative distribution function of X.

14. The distribution for the number X of times a coin must be tossed to obtain the first head is

$$P(X = r) = (\tfrac{1}{2})^r, \quad r = 1, 2, \ldots$$

   Find the probability that
   a. The first head occurs on the third toss.
   b. Fewer than five tosses are required.
   c. More than three tosses are required.

15. The *geometric distribution* describes the number of trials until a success is encountered. Suppose independent trials are performed, there being probability p of success at each trial. If X is the number of the trial on which success first occurs, then

$$P(X = r) = p(1 - p)^{r-1}, \quad r = 1, 2, 3, \ldots$$

   a. If the probability of hitting a target is .3, find the probability it will be hit for the first time on the third trial.
   b. What is the probability of having to look at 15 one-dollar bills before encountering one (on the sixteenth try) whose serial number ends in 0?

16. A die is rolled until a six appears. Let X be the number of the roll on which this happens for the first time. From the distribution described in Problem 15 find
   a. $P(X = 3)$.   b. $P(X \leq 3)$.   c. $P(X > 3)$.

17. A sportscaster for an Atlanta television station is assigned to interview

people until a hockey fan is found. If 4% of the population in Atlanta are hockey fans, what is the probability that no more than three people will have to be interviewed? (Use Problem 15.)

18. Twenty percent of the children in a school system have an eye defect which can be detected by a screening procedure. What is the probability that a child with a defect is among the first four children tested? (Use Problem 15.)

5.5 Expected Values of Random Variables

expected value

For a discrete random variable X the **expected value** of X, denoted $E(X)$, is the weighted mean of the possible values of X, the weights being the probabilities of these values (see (3.5) on page 53). Thus

$$E(X) = \sum_r rP(X = r) \qquad 5.1$$

where the sum extends over all possible values r of X. For example, suppose that X represents the value of the prize a player can win in a carnival game. Suppose the prizes are worth 1, 2, or 3 dollars, and that the probabilities of winning these prizes are .25, .40, and .35. Then X has the distribution in Table 5.1.

In this example,

$$E(X) = \sum_{r=1}^{3} rP(X = r) = (1 \times .25) + (2 \times .40) + (3 \times .35) = 2.10$$

The words *expected value* here are misleading in a sense, because 2.10 is not to be expected as a value of X at all—only the values 1, 2, and 3 are possible. But $E(X)$ is what is to be expected in an *average* sense. The amount the player expects to win on the average is $2.10, and that is the amount he should be willing to pay as an entry fee to play the game. In any real-life carnival, he will have to pay a lot more than $E(X)$ to play.

Table 5.1

r	1	2	3
$P(X = r)$	.25	.40	.35

Figure 5.9

The expected value as a center of gravity: discrete case

The expected value of X is also called the *mean* of X. This mean—the mean of a random variable—is very closely related to the mean for a set of numerical data, treated in Section 3.4. The two ideas are not identical, though, and the distinction between them will be important further on.

Just as the mean for numerical data can be viewed as a center of gravity (see Figure 3.1 on page 51), so can $E(X)$. For the distribution in Table 5.1 imagine weights positioned on a seesaw at each of the possible values of X, each weight being proportional to the probability of the value it represents, as in Figure 5.9; the expected value 2.10 is where a fulcrum will balance the seesaw.

For Examples 1 and 2 in Section 5.4, the expected values are, respectively,

$$(2 \times \tfrac{1}{36}) + (3 \times \tfrac{2}{36}) + (4 \times \tfrac{3}{36}) + (5 \times \tfrac{4}{36}) + (6 \times \tfrac{5}{36}) + (7 \times \tfrac{6}{36})$$
$$+ (8 \times \tfrac{5}{36}) + (9 \times \tfrac{4}{36}) + (10 \times \tfrac{3}{36}) + (11 \times \tfrac{2}{36}) + (12 \times \tfrac{1}{36}) = 7$$

and

$$(1 \times .3) + (2 \times .6) + (3 \times .1) = 1.8$$

For a continuous X, as for a discrete one, the mean $E(X)$ represents an average. It is the value X can be expected to have on the average, the actual value on a trial being greater or less than $E(X)$ and distributed in such a way as to balance out at $E(X)$. Just as in the case of the discrete variable, $E(X)$ can be viewed as a center of gravity also in the continuous case. If we were to draw the frequency curve on some stiff material of uniform density, such as plywood or metal, cut it out, and balance it on a knife edge arranged perpendicular to the horizontal scale, the value corresponding to the point of balance would be the expected value. Figure 5.10 illustrates this for the curve of Figure 5.3; compare it with the analogous Figure 3.2 on page 51, which relates to the mean for numerical data.

For the frequency curve in Example 4 of the preceding section the expected value works out to $\tfrac{4}{3}$. The computation of expected values in the continuous case, requiring as it does the methods of integral calculus, lies outside the scope of this book. It will be possible nonetheless to use the results of such computations for statistical purposes.

Figure 5.10

The expected value as a center of gravity: continuous case

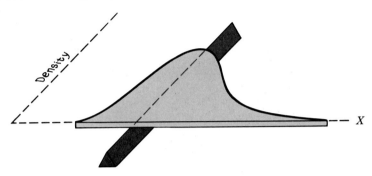

The mean $E(X)$ is often denoted by μ, the Greek letter *mu* (it corresponds to Roman *m* and is pronounced "mew"). If the mean is μ, the **variance** of the random variable is defined as $E(X - \mu)^2$ and the **standard deviation** as the square root of this:

variance
standard
deviation

$$\text{Var}(X) = E(X - \mu)^2$$
$$\text{St Dev}(X) = \sqrt{\text{Var}(X)} = \sqrt{E(X - \mu)^2}$$
5.2

This definition applies in both the discrete and continuous cases. Now $(X - \mu)^2$, which is itself a random variable, is always positive, and the more distant X is from μ, the greater is $(X - \mu)^2$. The expected value of $(X - \mu)^2$ measures the amount by which the distribution is spread around the mean μ, much as does the variance for a set of numerical data as defined on page 67.

In all cases $\text{Var}(X)$ is greater than or equal to 0. In the extreme case in which $\text{Var}(X)$ is equal to 0, the value of X is μ with probability 1, and X is not really random at all.

A discrete random variable X has value r with probability $P(X = r)$, and $(X - \mu)^2$ has value $(r - \mu)^2$ with probability $P(X = r)$. In the discrete case, the variance of X is the weighted mean of these values $(r - \mu)^2$, the weights being the probabilities $P(X = r)$. Thus

$$\text{Var}(X) = \sum_r (r - \mu)^2 P(X = r)$$
5.3

where the sum extends over all possible values r of X. For the value of the carnival prize, with the distribution given in Table 5.1, the possible values of X are 1, 2, and 3, and μ is 2.10. The possible values of $(X - \mu)^2$ are

$(1 - 2.10)^2$, $(2 - 2.10)^2$, and $(3 - 2.10)^2$, and the corresponding probabilities are .25, .40, and .35. According to (5.3), the variance is

$$.25 \times (1 - 2.1)^2 + .40 \times (2 - 2.1)^2 + .35 \times (3 - 2.1)^2 = .59$$

and the standard deviation is $\sqrt{.59}$, or .768.

The mean is often denoted by μ, the variance by σ^2, and the standard deviation by σ; sometimes μ_X, σ_X^2, and σ_X are used instead to indicate that the terms refer to computations involving X. (Here σ is lower-case Greek *sigma*.) Whatever the units of X (feet, degrees, etc.), its mean and standard deviation have these same units.

If X is a random variable, discrete or continuous, and if A and B are fixed numbers, then

$$U = AX + B$$

is another random variable. The expected values of these two random variables are related by the equation

$$E(U) = AE(X) + B \qquad 5.4$$

This is analogous to the change-of-scale formula (3.14) on page 72 and can be understood in the same way (see Figure 3.7 and its accompanying explanation). If X is a random temperature in degrees Celsius, the same temperature in degrees Fahrenheit, U, is $\frac{9}{5}X + 32$; it is only natural that the expected values of U and X should be related by the formula $E(U) = \frac{9}{5}E(X) + 32$.

The variance and standard deviation also obey the rules of Section 3.12:

$$\text{Var}(U) = A^2 \text{Var}(X) \qquad 5.5$$

and

$$\text{St Dev}(U) = A \text{ St Dev}(X) \qquad 5.6$$

for positive A.

Suppose X has mean μ and variance σ^2 (standard deviation σ), and consider the related variable

$$U = \frac{X - \mu}{\sigma} = \frac{1}{\sigma}X - \frac{\mu}{\sigma}$$

According to Equation (5.4),

$$E(U) = \frac{1}{\sigma}E(X) - \frac{\mu}{\sigma} = \frac{1}{\sigma} \cdot \mu - \frac{\mu}{\sigma} = 0$$

SECTION 5.5 EXPECTED VALUES OF RANDOM VARIABLES

and according to Equation (5.6),

$$\text{St Dev}(U) = \frac{1}{\sigma} \text{St Dev}(X) = \frac{1}{\sigma} \cdot \sigma = 1$$

Thus

$$E\left(\frac{X - \mu}{\sigma}\right) = 0 \qquad 5.7$$

and

$$\text{St Dev}\left(\frac{X - \mu}{\sigma}\right) = 1 \qquad 5.8$$

standardized

if X has mean μ and standard deviation σ. Subtracting μ from X centers it at 0, and dividing by σ standardizes the variability; $(X - \mu)/\sigma$ is the random variable X **standardized** to have mean 0 and standard deviation 1.

> For continuous random variables the relations between frequency curves or functions (densities), cumulative distribution functions, means, and variances involve calculus. For those familiar with the subject it may be instructive to set out the relevant formulas. It is to be stressed that these formulas and the calculus concepts they involve will in no way be used in what follows.
> If $F(x) = P(X \leq x)$ is the cumulative distribution function, then the frequency function $f(x)$ is its derivative. Thus $f(x) = F'(x)$; put the other way around, the relation is $F(x) = \int_{-\infty}^{x} f(t)\, dt$. The expected value is $E(X) = \int_{-\infty}^{\infty} x f(x)\, dx$. If $E(X) = \mu$, the variance is $Var(X) = \int_{-\infty}^{\infty} (x - \mu)^2 f(x)\, dx$. If $a < b$, then $P(a < X \leq b) = \int_a^b f(x)\, dx$; hence the above references to the *area under the curve*.

Problems

19. The number of "no-shows" on a scheduled airline flight has the following probability distribution. Find the mean and standard deviation.

r	0	1	2	3	4	5	6	7
$p(r)$	.09	.22	.26	.21	.13	.06	.02	.01

20. The number of photocopiers sold per week by a business-machine

company is a random variable with the probability distribution given below.
a. What is the average number of photocopiers sold per week?
b. What is the probability that the number of sales is greater than 1 standard deviation above the mean?

r	0	1	2	3	4	5
p(r)	.1	.3	.2	.2	.1	.1

21. In a lottery drawing five prizes are awarded as follows: a first prize of $25,000, a second prize of $10,000, and three prizes of $5000 each. What should be the fair cost of a ticket if 100,000 tickets are sold? If a million tickets are sold?

22. A businessman evaluates a proposed venture as follows. He stands to make a profit of $10,000 with probability $\frac{3}{20}$, to make a profit of $5000 with probability $\frac{9}{20}$, to break even with probability $\frac{1}{4}$, and to lose $5000 with probability $\frac{3}{20}$. Find the expected profit.

23. A gambler is offering a bet of $100. If he figures his expected gain to be $50, what is his probability of winning the bet?

24. Given that X has the probability distribution in the following table, find the mean and variance of X.

r	−1	0	1	2
P(X = r)	.1	.2	.3	.4

25. Given that Y has the probability distribution in the following table, find the mean and the variance of Y.

r	0	1	2	3	4
P(Y = r)	$\frac{1}{5}$	0	$\frac{1}{5}$	$\frac{2}{5}$	$\frac{1}{5}$

26. The *hypergeometric distribution* is the appropriate model when samples are selected without replacement from a finite population consisting of two kinds of elements—successes and failures. If the population consists of n successes and m failures, and if we take a sample of

SECTION 5.5 EXPECTED VALUES OF RANDOM VARIABLES

size r, the distribution for the number of successes, X, in the sample is given by (see (4.23))

$$P(X = k) = \frac{\binom{n}{k}\binom{m}{r-k}}{\binom{n+m}{r}}, \quad k = 0, 1, 2, \ldots, r$$

a. Give the distribution for X, the number of red balls in a sample of size 3 taken from an urn which contains three red balls and four white ones.
b. Work out the probabilities for the values that X can take.
c. Find the mean number of red balls in samples of size 3, i.e., the mean of X.

27. At the last stage of a jury selection, 6 jurors are to be selected out of the remaining 15 prospective jurors, 5 of whom are women. Let X be the number of women finally selected.
a. Find the distribution of X.
b. Find the cumulative distribution of X.
c. What is the expected number of women in the jury?
d. Find the variance of X.

28. On his way to work a person passes three stoplights each morning. The stoplights operate independently, and because the distance between them is great, they also appear to operate independently to a person traveling from one to another. The probability of a red light is .4, .8, and .5, respectively, for each of the stoplights.
a. Let Y be the number of red lights the person encounters in a one-way trip. Find the distribution of Y.
b. Compute the expected value and the standard deviation of Y.
c. Assume the waiting time for each red light to be two minutes. What are the expected value and the standard deviation of the waiting time on one trip?

29. The mean of the *geometric distribution* defined in Problem 15 is $1/p$, and the variance is $(1 - p)/p^2$.
a. If a coin is tossed until the first head occurs, what are the mean and standard deviation of the number of tosses?
b. If the variance is 90, what is p?

30. If the probability of making a sale by telephone solicitation is .05 per call, on the average how many calls must be made until the first sale is made, and what is the standard deviation? (See Problem 29.)

31. Random variables U and X are related by the equation $U = 3X + 2$. If $E(U) = 1$ and $Var(U) = 4$, find $E(X)$ and $Var(X)$.

32. A farm cooperative, which accepts implements on consignment and sells them at auction, has a fee of $150 plus 15% of the selling price. If the selling price is a random variable with mean $2500 and standard deviation $1100, find the mean and standard deviation of the fee.

33. A child has an allowance of $1.00 a week, and spends it all. The amount she spends during the first half of the week has mean 65¢ and standard deviation 15¢. What are the mean and standard deviation of the amount she spends during the second half of the week?

5.6 Sets of Random Variables

Random variables often come in pairs. This happens whenever there is a pair of numbers associated with each of the various outcomes of an experiment. If a man is drawn at random from some population, his height X in inches is one random variable and his weight Y in pounds is another. These random variables are associated with each other because they are associated with one experiment (the drawing of the person from the population); X and Y attach to the same man.

Expected values of sums satisfy the formula

$$E(X + Y) = E(X) + E(Y) \qquad 5.9$$

For example, if X is the income of the husband in a family and Y is the income of the wife, $X + Y$ is the family income (other sources excluded). If in a population of families the average income for husbands is $9000 and the average income for wives is $15,000, certainly the average income for families must be $24,000. Equation (5.9) expresses the general form of this fact.

In Chapter 4 we gave a definition of independence of events (Equation (4.10), page 110)—a definition embodying the idea that knowing whether or not one of two events occurred does not in any way help us to guess whether or not the other occurred. There is a similar notion of independence of random variables X and Y; it embodies the idea that knowing the value of X does not in any way help us to guess the value of Y, and vice versa.

For instance, suppose two dice, one red and one green, are rolled. Let X be the number showing on the red die, and let Y be the number showing on the green die. Since there is no interaction between the dice, the value that X takes on when the dice are rolled has no influence on the value Y takes on—X and Y are independent. On the other hand, if X and Y are the height and weight of a man, then X and Y are not independent; if you know that X is very large, then you know that Y is likely to be large also.

The useful fact is that, *if X and Y are independent,* then

$$\text{Var}(X + Y) = \text{Var}(X) + \text{Var}(Y) \qquad 5.10$$

Without giving a full derivation of this fact, we can give a partial argument for it by considering a case of dependence. Suppose that X is some random variable (with positive variance), and suppose that Y is its negative:

$Y = -X$. Certainly, X and Y are dependent—to know X is to know Y exactly. Now for every outcome of the experiment, $X + Y$ is equal to 0, so $X + Y$ has no variability at all and $\text{Var}(X + Y)$ is 0. Thus in Equation (5.10) the right side, certainly positive, exceeds the left side, which vanishes. Here Equation (5.10) fails because X and Y vary in such a way that in $X + Y$ their variability cancels out. In other cases their variability reinforces in such a way that the left side of Equation (5.10) exceeds the right. But if X and Y are independent, they can interact in no way at all, so this sort of canceling and reinforcing is impossible, and the variances exactly add up in accordance with Equation (5.10).

We are often concerned with whole sets of random variables. Suppose we have at hand a random sample of size n from some population (say, a sample of people), and suppose we have measurements $X_1, X_2, \ldots, X_n$ (say, the heights of the people), one measurement for each element of the sample. Choosing a random sample is a random experiment, and each X_i is a number associated with the outcome; thus $X_1, X_2, \ldots, X_n$ are random variables. The procedures of Chapter 3—the computations of means and variances for sets of data—are usually performed on samples.

The rule (5.9) extends to sets of random variables:

$$E\left(\sum_{i=1}^{n} X_i\right) = \sum_{i=1}^{n} E(X_i) \qquad 5.11$$

If $X_1, X_2, \ldots, X_n$ come from a random sample of size n from an infinite population, then they are independent. As explained in Section 5.2, this is the essential feature of samples from infinite populations. The formula (5.10) can be extended to more than two random variables *if they are independent* of one another:

$$\text{Var}\left(\sum_{i=1}^{n} X_i\right) = \sum_{i=1}^{n} \text{Var}(X_i) \qquad 5.12$$

Problems

34. A train arrives at the station at one of the times 12:01, 12:02, 12:03, 12:04, 12:05; the probabilities are $\frac{1}{5}$ each. It leaves the station 45 seconds after arriving. Independently, a commuter arrives at the station at one of the times 12:02, 12:03, 12:04, 12:05, 12:06, 12:07; the probabilities are $\frac{1}{6}$ each. What is the probability that the commuter arrives in time to catch the train?

35. Given the following joint probability distribution of X and Y (where, for example, $P(X = 2 \text{ and } Y = 1) = \frac{2}{8}$), find

a. The probability distribution of X.
b. The probability distribution of Y.
c. $E(X)$ and $E(Y)$.
d. The probability distribution of $X + Y$.
e. $E(X + Y)$. (Check this against $E(X) + E(Y)$.)
f. $\text{Var}(X)$ and $\text{Var}(Y)$.
g. $\text{Var}(X + Y)$.
h. $E(2X - 3Y)$.

X \ Y	1	2	3
1	$\frac{1}{8}$	0	$\frac{1}{8}$
2	$\frac{2}{8}$	$\frac{3}{8}$	0
3	0	0	$\frac{1}{8}$

36. Consider a random variable X that assumes values -1 and $+1$ with probabilities $\frac{1}{3}$ and $\frac{2}{3}$; consider also a random variable Y that assumes values 0, 1, 2 with probabilities $\frac{1}{4}, \frac{1}{2}$, and $\frac{1}{4}$. Find their joint distribution if they are independent. Calculate $E(X - Y)$ and $\text{Var}(X + Y)$.

5.7 Normal Distributions

normal distribution

A **normal distribution,** or normal frequency curve, is given by the formula

$$f(x) = \frac{1}{\sqrt{2\pi\sigma^2}} \exp\left[-\frac{1}{2\sigma^2}(x - \mu)^2\right] \qquad 5.13$$

What is important for us is not this formula, but the fact that the normal distribution is given by the symmetrical, bell-shaped curve of Figure 5.11. We'll see how to find the areas under the curve by a table in the Appendix.

There is a normal distribution for each pair μ and σ^2, where μ is any number and σ^2 is positive. And μ and σ^2 are the mean and variance of the distribution. If we fix μ and let σ^2 vary, we get a family of curves with the same mean but different variances, as shown in Figure 5.12. The lower and more spread out the curve, the greater the variance σ^2. If we fix σ^2 and let μ vary, we get a family of curves with the same shape but different locations along the axis, as shown in Figure 5.13. The further to the right the curve, the greater the mean μ.

Experience has shown that many continuous random variables in diverse fields of application have distributions for which a normal distribution

Figure 5.11

A normal curve

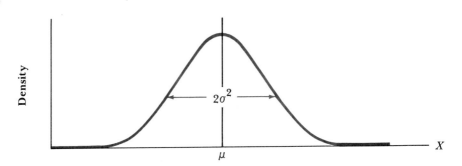

may serve as a mathematical or theoretical model or for which a normal distribution may be used as a good approximation. As we have seen, there is a normal distribution for each pair μ and σ^2. Although it would be impossible to construct a table for each of them, we can select one, tabulate its areas, and use this table with appropriate conversion formulas to find probabilities for any normally distributed variable. The distribution tabulated in Table 2 of the Appendix is the **standard normal distribution**, and it is defined as that normal distribution that has *mean 0 and variance 1*.

standard normal distribution

Figure 5.12

Normal distributions with the same mean but different variances: σ_1, σ_2, σ_3, where $\sigma_1 < \sigma_2 < \sigma_3$

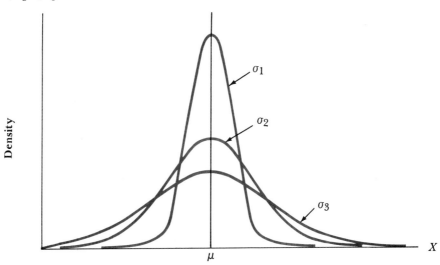

Figure 5.13

Normal distributions with the same variance but different means: μ_1, μ_2, μ_3, where $\mu_1 < \mu_2 < \mu_3$

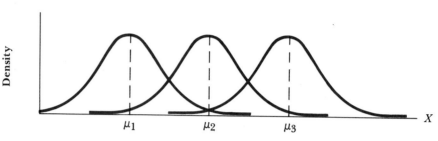

Let Z be a standard normal variable, that is, a normal variable for which $\mu = 0$ and $\sigma^2 = 1$. Table 2 gives the areas under the standard normal curve. Suppose, for example, that we want the area under the curve between 0 and 1.53. Find 1.5 at the left of the table, and single out that row. Find .03 at the top of the table, and single out that column. In the body of the table, where the row for 1.5 and the column for .03 meet, is the entry .4370. This is the area under the standard normal curve between 0 and 1.53, and we use the row for 1.5 and the column for .03 because $1.53 = 1.5 + .03$. To find the area between 0 and 0.86, use row 0.8 and column .06 ($0.86 = 0.8 + .06$); the answer is .3501.

Since the standard normal curve is symmetric about zero, the area between $-Z$ and zero is equal to the area between zero and Z; therefore, only positive values of Z are given in the table. Since area is proportional to probability, the total area under the curve is equal to 1.

We can illustrate the use of the table of normal areas by several examples. A rough sketch of the curve together with the area asked for is a great help in finding probabilities by means of the table.

Example 1

Problem: Find the probability that a single random value of Z will be between .53 and 2.42.

Solution: The shaded area in Figure 5.14 is equal to the required probability. The area given in the table for $Z = 2.42$ is .4922, but this is the area from zero to 2.42. The tabulated area for .53 is .2019. Referring to the sketch, we see that the area between .53 and 2.42 is the difference between their tabular values:

$$P(.53 < Z < 2.42) = .4922 - .2019 = .2903$$

Figure 5.14

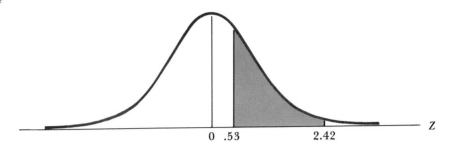

Example 2

Problem: What is the probability that Z will be greater than 1.09?

Solution: The shaded area to the right of $Z = 1.09$ is the probability that Z will be greater than 1.09. In the table the area between zero and 1.09 is .3621. The total area under the curve is equal to 1, so that the area to the right of zero must be .5000; therefore (Figure 5.15),

$$P(Z > 1.09) = .5000 - .3621 = .1379$$

Figure 5.15

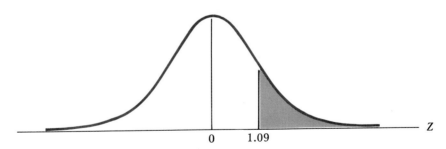

Example 3

The probability that Z will be greater than $-.36$ is the total area to the right of $Z = -.36$ and consists of the area between $-.36$ and zero plus the area under the right-hand half of the curve (Figure 5.16):

$$P(Z > -.36) = .1406 + .5000 = .6406$$

Figure 5.16

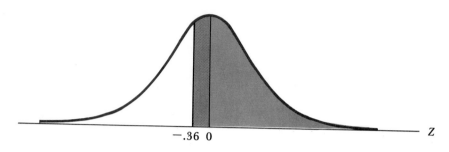

$-.36$ 0 Z

Example 4

The probability that a random value of Z is between -1.00 and 1.96 is found by adding the corresponding areas as indicated by Figure 5.17:

$$P(-1.00 < Z < 1.96) = .3413 + .4750 = .8163$$

Figure 5.17

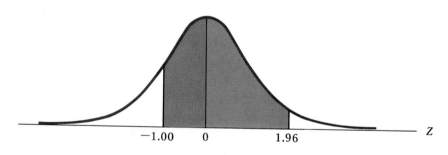

-1.00 0 1.96 Z

As a general rule, if a and b are on the *same side of the origin, we subtract* the corresponding areas. If they are on *opposite sides, we add*.

A slightly different use of the table is required in the following examples.

Example 5

Problem: Find a value for Z such that the probability of a larger value is equal to .0250.

Solution: The probability of a larger value is the area to the right of Z. The area between zero and Z must be $.5000 - .0250$, or $.4750$. We look in the body of Table 2 until we locate .4750 and find the desired value, $Z = 1.96$ (Figure 5.18).

SECTION 5.7 NORMAL DISTRIBUTIONS

Figure 5.18

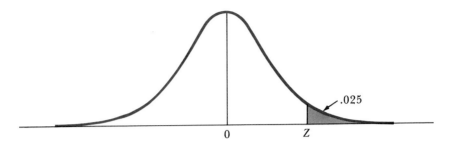

Example 6

Problem: Find the value of Z such that the probability of a larger value is .7881.

Solution: Since the area to the right of Z is greater than .5000, we know that Z must be negative and that the area between Z and zero must be $.7881 - .5000$, or .2881. In the table, we search out the entry .2881 and find that the corresponding Z is $-.80$ (Figure 5.19).

Figure 5.19

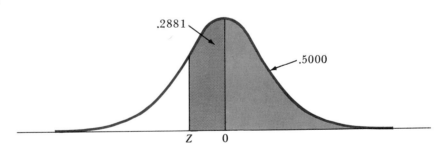

Example 7

Problem: Find the value for b such that $P(-b < Z < b) = .9010$.

Solution: The endpoints $-b$ and b are symmetrically placed; therefore, half the area between them must be between zero and b. In the table we find, opposite the area .4505, the value $Z = 1.65$ (Figure 5.20).

Figure 5.20

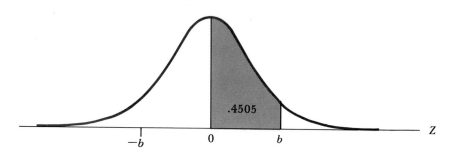

Now that we have seen how the table of normal areas is used to find probabilities for the standard normal variable, we turn our attention to the general case of the normal variable with mean μ and variance σ^2.

The key to using the standard normal areas for finding probabilities for the general normal variable is to pass to the **standardized** variable. If X is a random variable with mean μ and variance σ^2, then, by Equations (5.7) and (5.8) on page 173, the standardized variable

standardized

$$Z = \frac{X - \mu}{\sigma} \qquad 5.14$$

has mean 0 and variance 1. If X is normally distributed, Z is a *standard normal variable*.

The variable Z can be looked on as the *number of standard deviations from the mean*. Suppose, for example, that $Z = 1.5$. By the Equation (5.14), this is the same thing as $(X - \mu)/\sigma = 1.5$, or $X - \mu = 1.5\sigma$. Thus saying that Z is 1.5 is the same thing as saying that the distance from X to the mean μ is 1.5σ, or that the distance from X to the mean μ is 1.5 when the standard deviation σ is taken as the unit of measurement.

Let X be normal with mean μ and variance σ^2, and suppose we want to find the probability that a randomly selected value for X will be between a and b. We use the standardization formula (5.14) and find that when $X = a$,

$$Z = \frac{a - \mu}{\sigma}$$

and that when $X = b$,

$$Z = \frac{b - \mu}{\sigma}$$

SECTION 5.7 NORMAL DISTRIBUTIONS

Therefore, whenever X is between a and b, the standard variable Z will be between $(a - \mu)/\sigma$ and $(b - \mu)/\sigma$. We have, then,

$$P(a < X < b) = P\left(\frac{a - \mu}{\sigma} < Z < \frac{b - \mu}{\sigma}\right)$$

Thus we can turn to the tables and find the probability that Z is between $(a - \mu)/\sigma$ and $(b - \mu)/\sigma$.

Example 8

Suppose that a certain type of wooden beam has a mean breaking strength of 1500 pounds and a standard deviation of 100 pounds and that we want to know the relative frequency of all such beams whose breaking strengths are between 1450 and 1600 pounds. The standardization formula becomes

$$Z = \frac{X - 1500}{100}$$

and

$$P(1450 < X < 1600)$$
$$= P\left(\frac{1450 - 1500}{100} < \frac{X - 1500}{100} < \frac{1600 - 1500}{100}\right)$$
$$= P\left(\frac{1450 - 1500}{100} < Z < \frac{1600 - 1500}{100}\right)$$
$$= P(-.50 < Z < 1.00)$$
$$= .1915 + .3413$$
$$= .5328$$

We may conclude that about 53% of the beams have breaking strengths between 1450 and 1600 pounds.

Example 9

Scholastic Aptitude Test (SAT) scores are normally distributed with mean 500 and standard deviation 100 (nearly, at any rate). What is the probability that a student has an SAT score above 680? That is, what proportion of students have SAT scores above 680? The standardization formula is

$$Z = \frac{X - 500}{100}$$

and

$$P(X > 680) = P\left(\frac{X - 500}{100} > \frac{680 - 500}{100}\right)$$

$$= P(Z > 1.80) = .5000 - .4641 = .0359$$

About 3.6% of students have SAT scores above 680.

Example 10

In the preceding example, find a score S such that 10% of students have SAT scores above S. First we need a value for Z such that the probability of a larger value is .10. Call this value a. We need an a such that $P(Z > a) = .10$, or, what is the same thing, $P(0 \leq Z \leq a) = .40$. In Table 2, the nearest entry to .40 is .3997, and this corresponds to a Z value of 1.28. Thus $a = 1.28$, and $P(Z > 1.28) = .10$ (nearly). But 1.28 must be converted to the scale for the SAT score X: we must find a score S such that $P(X > S) = .10$. Since the standardized variable is $Z = (X - 500)/100$,

$$P(X > S) = P\left(\frac{X - 500}{100} > \frac{S - 500}{100}\right) = P\left(Z > \frac{S - 500}{100}\right)$$

Therefore, choose S so that $(S - 500)/100$ equals the value 1.28 already found:

$$\frac{S - 500}{100} = 1.28$$

The solution is $S = 500 + 1.28 \times 100 = 628$. About 10% of students have SAT scores above 628. This can be checked by working backwards:

$$P(X > 628) = P\left(\frac{X - 500}{100} > \frac{628 - 500}{100}\right)$$

$$= P(Z > 1.28) = .5000 - .3997 = .1003$$

Thus 10.03% of students have SAT scores above 628. The discrepancy between 10.03% and 10% is due to the fact that the nearest entry to .40 in the table was .3997—not an exact match.

Problems

37. Given that Z is the standard normal variable, use the table of normal areas to find

SECTION 5.7 NORMAL DISTRIBUTIONS

    a. $P(Z \geq 2)$.
    b. $P(Z < -3)$.
    c. $P(0 \leq Z \leq 1.5)$.
    d. $P(-1 < Z < 2)$.

38. Given that Z is the standard normal variable, find
    a. $P(Z \leq .5)$.
    b. $P(Z \geq -.25)$.
    c. $P(-1.75 \leq Z < 0)$.
    d. $P(-.40 < Z < .60)$.

39. Given that Z is the standard normal variable, find
    a. $P(Z \leq .28)$.
    b. $P(Z > -1.23)$.
    c. $P(Z < -2.38)$.
    d. $P(Z > 2.31)$.
    e. $P(-1.20 < Z \leq -.35)$.
    f. $P(-1.45 \leq Z < 2.10)$.

40. Given that Z is the standard normal variable, find C such that
    a. $P(Z \leq C) = .1587$.
    b. $P(Z \geq C) = .975$.
    c. $P(-C \leq Z \leq C) = .2358$.

41. Given that Z is the standard normal variable, find C such that
    a. $P(Z \geq C) = .0764$.
    b. $P(Z \geq C) = .7881$.
    c. $P(Z \leq C) = .8508$.
    d. $P(Z \leq C) = .2743$.
    e. $P(-C \leq Z \leq C) = .9108$.
    f. $P(-1 \leq Z \leq C) = .7528$.

42. Given that X is a normal variable with $\mu = 50$ and $\sigma = 10$, find
    a. $P(X \leq 70)$.
    b. $P(X > 35)$.
    c. $P(36 \leq X < 48)$.
    d. $P(40 < X < 62)$.

43. Given that X is a normal variable with $\mu = 25$ and $\sigma = 5$, find
    a. $P(X > 20)$.
    b. $P(X \leq 40)$.
    c. $P(21 \leq X \leq 30)$.
    d. $P(18 < X < 23)$.

44. Given that X is normal with $\mu = .05$ and $\sigma = .024$, find
    a. $P(X \geq .074)$.
    b. $P(.071 \leq X \leq .077)$.

45. Given that X is normal with $\mu = .130$ and $\sigma^2 = .000625$, find
    a. $P(X \geq .124)$.
    b. $P(.100 \leq X \leq .175)$.

46. Given that X is normally distributed, find the value of C such that

$$P(\mu - C\sigma \leq X \leq \mu + C\sigma) = .95$$

47. If X is normal with $\mu = 50$ and $\sigma^2 = 100$, find
    a. The value of C such that $P(X \leq C) = .4207$.
    b. Two numbers, A and B, that are equidistant from μ and are such that $P(A \leq X \leq B) = .966$.

48. Given that X is normal, $\sigma = 10$, and $P(X \geq 57.5) = .2266$, find μ.

49. Given that X is normal, $\mu = 20$, and $P(X \leq 30) = .9773$, find σ.

50. The length of time needed to complete a certain test is normally distributed with $\mu = 60$ minutes and $\sigma = 10$ minutes. How long a time should be allowed to permit 95% of the people taking the test to complete it?

51. The mean annual precipitation in Illinois is 33.18 inches, and the standard deviation is 4.23 inches. What is the probability of 45 or

more inches of precipitation next year? Assume a normal distribution for rainfall.

52. If the resistances of carbon resistors of 1200 ohms nominal value are normally distributed with $\mu = 1200$ ohms and $\sigma = 120$ ohms,
   a. What proportion of these resistors have resistances greater than 1000 ohms?
   b. What proportion have resistances that do not differ from the mean by more than 1% of the mean?
   c. What proportion do not differ from the mean by more than 5% of the mean?

53. The mean income per household in a certain city is $13,500, with a standard deviation of $1800. Assuming a normal distribution, find
   a. The proportion of households with incomes under $12,000.
   b. The probability a randomly chosen household will have an income exceeding $17,000.

54. Given that X is normally distributed, that $P(X \geq 6.08) = .9750$, and that $P(X \geq 13.52) = .0392$, find μ and σ.

55. The service life of a certain product is normally distributed. Suppose 92.5% of the items have lives exceeding 2160 hours and 3.92% have lives exceeding 17,040 hours. Find the mean and standard deviation of the service life.

56. An environmental study has shown that the daily average noise level on a busy street is a normal random variable with mean 37 decibels and standard deviation 6 decibels.
   a. What is the probability that the noise level exceeds 46 decibels?
   b. What decibel range contains the middle 49.72% of the probability?

57. Back-radiation from the ocean surface at a certain location is a normal random variable with mean -44 watts/m² and a standard deviation of 5 watts/m².
   a. What is the probability that back-radiation is less than -47 watts/m²?
   b. What is the probability that the back-radiation is between -47 and -40 watts/m²?

58. If the weekly average market price of wheat is a normal random variable with mean $4.25 per bushel and standard deviation $.40 per bushel, what is the probability that the price will exceed $4.87 per bushel?

59. The number of miles each automobile in a company's fleet is driven in a year's time is a normal random variable with mean 17,800 miles and standard deviation 1,600 miles. Those that have mileage in the upper 15% are sold at auction at the end of the year; the rest are traded. What is the minimum number of miles an auto in the fleet must be driven for it to be sold at auction?

5.8 The Binomial Distribution

In Section 4.5 we discussed repeated independent trials. There is a sequence of trials, each trial results either in success or in failure, and the trials are independent—the outcome on one trial in no way influences the outcome on another trial. Let

$n =$ the number of trials,
$p =$ the probability of success on a single trial,
$1 - p =$ the probability of failure on a single trial,
$X =$ the number of successes in the n trials.

We saw in Section 4.5 that the probability of exactly r successes is given by

$$P(r) = P(X = r) = \binom{n}{r} p^r (1 - p)^{n-r}, \quad r = 0, 1, \ldots, n \quad 5.15$$

The integers $0, 1, 2, \ldots, n$ are the possible values of the random variable X, and Equation (5.15) gives its distribution. By substituting the proper value for r in Equation (5.15), we find the probability of this value. We call X a *binomial random variable,* and we call its distribution (5.15) a **binomial distribution**.

binomial distribution

Suppose we toss a fair coin five times. The probability p of a head (success) is $\frac{1}{2}$, and the probability $1 - p$ of a tail (failure) is $\frac{1}{2}$. Hence, the probability of getting exactly r heads in the five tosses is

$$P(r) = P(X = r) = \binom{5}{r} (\tfrac{1}{2})^r (\tfrac{1}{2})^{5-r}, \quad r = 0, 1, 2, 3, 4, 5$$

The probability of exactly three heads is

$$P(3) = \binom{5}{3} (\tfrac{1}{2})^3 (\tfrac{1}{2})^2 = 10(\tfrac{1}{2})^5 = \tfrac{10}{32}$$

and the probability of no heads is

$$P(0) = \binom{5}{0} (\tfrac{1}{2})^0 (\tfrac{1}{2})^5 = 1 \times 1 \times (\tfrac{1}{2})^5 = \tfrac{1}{32}$$

Calculating the remaining probabilities the same way gives the distribution in Table 5.2. Note that these probabilities add to 1.

In practical problems we are usually interested not in the probability that X assumes some individual value, but instead in the probability that X will not exceed a specified value or that X will lie in a given interval. These may be found by adding terms in the distribution. For example, the probability of two or more heads in five tosses is

Table 5.2 Binomial Probability Distribution, $p = \frac{1}{2}$, $n = 5$

r	0	1	2	3	4	5
$P(r) = P(X = r)$	$\frac{1}{32}$	$\frac{5}{32}$	$\frac{10}{32}$	$\frac{10}{32}$	$\frac{5}{32}$	$\frac{1}{32}$

$$P(X \geq 2) = \sum_{r=2}^{5} P(r) = P(2) + P(3) + P(4) + P(5) = \tfrac{26}{32}$$

The probability that the number of heads will be greater than two but less than or equal to four is

$$P(2 < X \leq 4) = \sum_{r=3}^{4} P(r) = P(3) + P(4) = \tfrac{15}{32}$$

The probability that there will be at least one head is

$$P(1 \leq X \leq 5) = \sum_{r=1}^{5} P(r) = \tfrac{31}{32}$$

Note that this last probability may more easily be found by using the rule for complementary events ((4.7), page 106); that is, the probability of at least one head is

$$P(X \geq 1) = 1 - P(X = 0) = 1 - \tfrac{1}{32} = \tfrac{31}{32}$$

> In Section 4.5 it was shown (on page 136) how to use Table 1 in the Appendix to answer questions about binomial probabilities. For example, in 25 tosses of a fair coin, the chance of 13 or fewer heads ($n = 25$, $x = 13$, $p = .5$) is .655. The chance of 8 or fewer heads ($n = 25$, $x = 8$, $p = .5$) is .054, and the chance of between 9 and 13 heads is $.655 - .054 = .601$.

Notice that for writing probabilities associated with discrete variables it is important to distinguish between *greater than or equal to* ($\geq$) and *greater than* ($>$) and also between *less than or equal to* ($\leq$) and *less than* ($<$). These distinctions are unnecessary in the case of a continuous variable, since there the probability of a single specified value is 0.

A *binomial population* is one in which the elements may be split into two classes—male and female, defective and nondefective, inoculated and not inoculated, and the like. The classes may be conventionally labelled *success* and *failure*. Let p represent the proportion of successes in the population and $1 - p$ the proportion of failures. Suppose we take a random sample of size n from the population. The probability that a particular element in the sample turns out to be a success is p, since that is the proportion of successes in the population. If the population is infinite, the sample elements are independent and so the rule (5.15) applies. The probability of exactly r successes in the random sample of n is given by the rule (5.15). Moreover, this rule will be sufficiently accurate for all practical purposes if the population is large compared with the sample size n. In this case the elements in the sample are virtually independent, because transferring an element from the population to the sample hardly affects the composition of the population. See the discussion in Section 4.5, in particular Examples 5 and 6. (If the sample and population were of comparable sizes, the hypergeometric probabilities of Section 4.6 would be called for.)

As an example, suppose an advertising agency claims that of the college students who smoke, 25% smoke brand A. If we obtain a random sample of four students who smoke, and if the claim is true, what is the probability that at least one of them will be found to smoke brand A? Since the population of smoking students is large, the number X of students in the sample who smoke follows the binomial distribution to all intents and purposes. Under the assumption that the claim is true, p is $\frac{1}{4}$, and so

$$P(r) = P(X = r) = \binom{4}{r} (\tfrac{1}{4})^r (\tfrac{3}{4})^{4-r}, \qquad r = 0, 1, 2, 3, 4$$

The probability of our finding at least one smoker of brand A is

$$P(X \geq 1) = 1 - P(0) = 1 - \binom{4}{0}(\tfrac{1}{4})^0(\tfrac{3}{4})^4 = 1 - \tfrac{81}{256} = \tfrac{175}{256}$$

or

$$P(X \geq 1) = P(1) + P(2) + P(3) + P(4) + P(5)$$
$$= \tfrac{108}{256} + \tfrac{54}{256} + \tfrac{12}{256} + \tfrac{1}{256} = \tfrac{175}{256}$$

There is a separate binomial distribution for each integer n and each p between 0 and 1. The mean and variance of the binomial distribution can be shown to be

$$\mu = np \qquad \qquad 5.16$$

and

$$\sigma^2 = np(1 - p) \qquad \qquad 5.17$$

Figure 5.21a

The binomial distribution with $p = \frac{1}{2}$, $n = 6$.

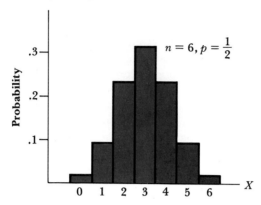

Figure 5.21b

The binomial distribution with $p = \frac{1}{4}$, $n = 6$.

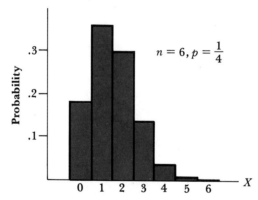

Figure 5.21c

The binomial distribution with $p = \frac{1}{16}$, $n = 6$.

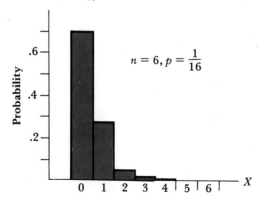

SECTION 5.8 THE BINOMIAL DISTRIBUTION

The binomial distribution is symmetric if $p = .5$; otherwise, it is skew, or asymmetric, the degree of skewness increasing as p moves toward 0 or 1. The probability distributions for $n = 6$ and $p = \frac{1}{2}, \frac{1}{4}$, and $\frac{1}{16}$ are shown graphically in Figures 5.21a, 5.21b, and 5.21c respectively.

Problems

60. For the binomial distribution

$$P(X = r) = \binom{5}{r} \left(\frac{1}{3}\right)^r \left(\frac{2}{3}\right)^{5-r}, \quad r = 0, 1, \ldots, 5$$

find
  a. $P(X = 2)$. b. $P(X \leq 3)$. c. $P(X = 5)$. d. $P(X \leq 4)$.

61. In Portland, rain falls on the average one day out of every three days during which the sky is overcast. Give the distribution of the number of days with rainfall among the next three overcast days, assuming complete independence. Find the mean and the variance of the number of rainy days.

62. The probability of conviction by jury is $\frac{9}{10}$. Let X be the number of acquittals on the next six trials.
  a. Give the distribution of X.
  b. Find the probability of at least two acquittals.
  c. Compute $E(X)$ and $\text{Var}(X)$.

63. In a large orchard 10% of the apples are wormy. If four apples are selected at random, what is the probability that
  a. Exactly one will be wormy?
  b. None will be wormy?
  c. At least one will be wormy?

64. The in-flight probability of failure of each of the four engines of a plane is .00005. Assuming the four engines operate independently, find the probability that on a single flight
  a. No engine failure occurs.
  b. No more than one engine failure occurs.
  c. Exactly two engine failures occur.

65. In a test to detect learning disabilities, a child is given 10 questions each of which has possible answers labeled 1, 2, or 3. Children with a disability of Type A almost always answer 1 or 2 on every question, while children with a disability of Type B almost always answer 3 on every question. Normal children have an equal chance of answering 1, 2, or 3 for each question.

a. What is the probability that a normal child's answers will be all 1s and 2s, thereby incorrectly indicating a Type A disability?

b. A child is tested further for a Type B disability if he answers 3 five or more times. What is the probability that a normal child will be tested further for a Type B disability?

66. To win a radio quiz game, a contestant must answer at least two out of three questions correctly. The probability of a correct answer is $\frac{1}{2}$ for each question. If the quiz game is played 10 times a day, what is the probability distribution of the number of winners?

5.9 The Normal Approximation to the Binomial

Since, as the sample size increases, the number of individual binomial probabilities increases and each one involves higher powers of p and $1 - p$, calculating these probabilities becomes tedious for moderate values of n and practically impossible for larger values. Binomial tables have been tabulated for sample sizes up to 100, but if tables are not available or if n exceeds 100, we need an approximation. The most important approximation uses the table areas for the standard normal distribution. The approximation is important not just for computational purposes, but also for a general understanding of the binomial distribution.

Section 6.5 deals with a very general case of approximation by the normal distribution. The **approximation of the binomial distribution** is an important special instance of it.

approximation of the binomial distribution

It can be shown that as the sample size (or number of trials) n increases, the binomial distribution becomes very close to a normal distribution with mean np and variance $np(1 - p)$. (As remarked in the preceding section, np and $np(1 - p)$ are the mean and variance of the number of successes.) Whenever our sample is sufficiently large, the number X of successes in the sample is approximately normally distributed, and we can use the normal tables to approximate the probabilities in which we are interested.

If X is nearly normal with mean np and variance $np(1 - p)$—that is, with standard deviation $\sqrt{np(1 - p)}$—then the *standardized* binomial variable

$$Z = \frac{X - np}{\sqrt{np(1 - p)}} \qquad 5.18$$

is approximately a standard normal variable (see Equations (5.7) and (5.8), and also Equation (5.14)). Therefore, the probability that X will lie between

SECTION 5.9 THE NORMAL APPROXIMATION TO THE BINOMIAL

a and b will be approximately the probability that a standard normal variable Z will lie between

$$a' = \frac{a - np}{\sqrt{np(1-p)}} \quad \text{and} \quad b' = \frac{b - np}{\sqrt{np(1-p)}}$$

Thus,

$$P(a \leq X \leq b) \cong P(a' \leq Z \leq b')$$

where the symbol $\cong$ means *is approximately equal to*.

Example 1

Suppose that from the voting population of a city we take a sample of 100. Suppose that 36% of the voters in the population favor candidate A over candidate B in a coming election. We ask for the probability that the number X of voters in the sample who favor candidate A will be between 24 and 42, inclusive. Identify success with a voter who favors A (this is of course just a convention). Then we have a sample of 100 from a binomial population in which the proportion of successes is .36, and we ask for the probability that in the sample the number X of successes will be between 24 and 42, inclusive.

Here the mean, variance, and standard deviation of X are

$$\mu = np = 100 \times .36 = 36$$
$$\sigma^2 = np(1-p) = 100 \times .36 \times .64 = 23.04$$
$$\sigma = \sqrt{np(1-p)} = \sqrt{23.04} = 4.8$$

The standardization formula is

$$Z = \frac{X - 36}{4.8}$$

Hence (here Z represents a random variable with exactly a standard normal distribution),

$$P(24 \leq X \leq 42) = P\left(\frac{24 - 36}{4.8} \leq \frac{X - 36}{4.8} \leq \frac{42 - 36}{4.8}\right)$$

$$\cong P\left(\frac{24 - 36}{4.8} \leq Z \leq \frac{42 - 36}{4.8}\right)$$

$$= P\left(-\frac{12}{4.8} \leq Z \leq \frac{6}{4.8}\right) = P(-2.5 \leq Z \leq 1.25)$$

From the table for normal areas we get

$$P(-2.5 \leq Z \leq 1.25) = .4938 + .3944 = .8882$$

The exact probability (taken from binomial tables) is .907386. Our approximation, .8882, is accurate to about 2%. Rounding off .8882 to .89, we see that the normal approximation gives

$$P(24 \leq X \leq 42) \cong .89$$

If we take a sample of 100 voters from a population in which 36% favor candidate A, there is probability about .89 that the sample will contain between 24 and 42 voters who favor A.

Example 2

Suppose again that we take a sample of 100 voters, but this time suppose that 40% of the voters in the population favor A. Again we ask for the probability that the sample will contain between 24 and 42 voters who favor A. That is, again we take a binomial sample of 100 and again we ask for the probability that the number X of successes will be between 24 and 42. This time, however, we suppose the proportion p of successes in the population is .40.

Here

$$\mu = np = 100 \times .40 = 40$$
$$\sigma^2 = np(1-p) = 100 \times .40 \times .60 = 24$$
$$\sigma = \sqrt{np(1-p)} = \sqrt{24} = 4.9$$

The standardization is

$$Z = \frac{X - 40}{4.9}$$

so

$$P(24 \leq X \leq 42) = P\left(\frac{24-40}{4.9} \leq \frac{X-40}{4.9} \leq \frac{42-40}{4.9}\right)$$
$$\cong P\left(-\frac{16}{4.9} \leq Z \leq \frac{2}{4.9}\right) = P(-3.27 \leq Z \leq .41)$$

By the normal table,

$$P(-3.27 \leq Z \leq .41) = .4994 + .1591 = .6585$$

Rounding off gives

$$P(24 \leq X \leq 42) \cong .66$$

If we take a sample of 100 voters from a population in which 40% favor candidate A, there is probability about .66 that the sample will contain between 24 and 42 voters who favor A.

SECTION 5.9 THE NORMAL APPROXIMATION TO THE BINOMIAL

Example 3

It is also possible to approximate probabilities defined by a single inequality. Suppose we ask for the probability that the number X of successes in a sample of 100 is 45 or more. If p is .36, we standardize by the mean and standard deviation of Example 1:

$$P(X \geq 45) = P\left(\frac{X - 36}{4.8} \geq \frac{45 - 36}{4.8}\right)$$

$$\cong P\left(Z \geq \frac{45 - 36}{4.8}\right) = P(Z \geq 1.88)$$

$$= .5000 - .4699 = .0301$$

If p is .40, we use the standardization of Example 2:

$$P(X \geq 45) = P\left(\frac{X - 40}{4.9} \geq \frac{45 - 40}{4.9}\right)$$

$$\cong P\left(Z \geq \frac{45 - 40}{4.9}\right) = P(Z \geq 1.02)$$

$$= .5000 - .3461 = .1539$$

Rounding off to two places in each result, we have

$$P(X \geq 45) \cong .03 \quad \text{if } p = .36$$

and

$$P(X \geq 45) \cong .15 \quad \text{if } p = .40$$

Of course, the larger p is, the more likely we are to get 45 or more successes.

To put it in terms of the voting populations of Examples 1 and 2, if 36% of the voters in the population favor A, the probability is .03 that 45 or more voters in the sample of 100 will favor A; if 40% of the voters in the population favor A, the probability is .15 that 45 or more voters in the sample of 100 will favor A.

It is sometimes convenient to work not with the number X of successes in n trials, but instead with the fraction $f = X/n$ of successes. Dividing both the numerator and denominator in (5.18) by n gives

$$Z = \frac{f - p}{\sqrt{p(1 - p)/n}} \qquad 5.19$$

This ratio thus has approximately a standard normal distribution for large n; it is, in fact, f standardized, because f has mean p and variance $p(1-p)/n$.

Example 4

Suppose n is 100 and p is .36. What is the probability that the fraction f of successes will lie between .24 and .42? Here the mean, variance, and standard deviation of f are

$$E(f) = p = .36$$

$$\text{Var}(f) = p(1-p)/n = .36 \times .64/100 = .002304$$

$$\text{St Dev}(f) = \sqrt{p(1-p)/n} = .048$$

Hence by (5.19) the standardized f is

$$Z = \frac{f - .36}{.048}$$

and

$P(.24 \le f \le .42)$

$$= P\left(\frac{.24 - .36}{.048} \le \frac{f - .36}{.048} \le \frac{.42 - .36}{.048}\right)$$

$$\cong P\left(\frac{.24 - .36}{.048} \le Z \le \frac{.42 - .36}{.048}\right) = P(-2.5 \le Z \le 1.25)$$

$$= .8882 \cong .89$$

Note that the answer arrived at in Example 4 is the same as the answer in Example 1, because n and p are the same in the two examples and because $.24 \le f \le .42$ is the same thing as $24 \le X \le 42$. Any given problem can be solved in terms of X or in terms of f; the answer will be the same in either case.

We noted in Example 1 that the normal approximation gave an answer about 2% off from the true answer. This accuracy suffices for many practical purposes. Further accuracy can be achieved by using the _continuity correction_. The exact value of the probability sought in Example 1 is

$$P(24 \le X \le 42) = \sum_{r=24}^{42} \binom{100}{r}(.36)^r(.64)^{100-r}$$

SECTION 5.9 THE NORMAL APPROXIMATION TO THE BINOMIAL **199**

Figure 5.22

Normal approximation to a binomial distribution

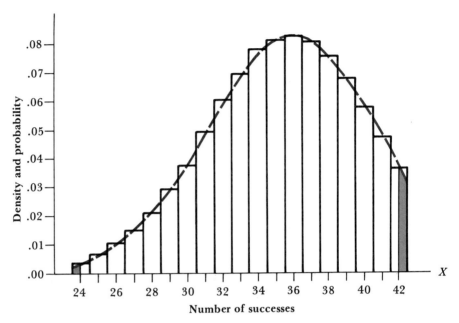

It is represented by the combined areas of the bars in Figure 5.22. In Example 1 we approximated this probability by the area under the normal curve (the one for $\mu = 36$ and $\sigma = 4.8$) between 24 and 42. Examination of the figure shows that we should obtain a better approximation if we find the area under the normal curve between 23.5 and 42.5, because this will include the areas of the two shaded regions, omitted in the previous approximation. With this procedure we get

$$P(24 \leq X \leq 42) = P(23.5 \leq X \leq 42.5)$$
$$= P\left(\frac{23.5 - 36}{4.8} \leq \frac{X - 36}{4.8} \leq \frac{42.5 - 36}{4.8}\right)$$
$$\cong P(-2.60 \leq Z \leq 1.35)$$
$$= .4953 + .4115 = .9068$$

The error in this second approximation is less than 0.1%.

In general, in applying the continuity correction, we subtract $\frac{1}{2}$ from the lower value on the X-scale and add $\frac{1}{2}$ to the upper value, and then proceed as before. Thus in the preceding calculation we first passed from $P(24 \leq X \leq 42)$ to $P(23.5 \leq X \leq 42.5)$. Suppose, however, that we have a strict inequality, as in $P(31 < X \leq 82)$. This probability is the same thing as $P(32 \leq X \leq 82)$, and we first pass to $P(31.5 \leq X \leq 82.5)$ and then proceed as before. The point is, it would be wrong to subtract $\frac{1}{2}$ from 31 (instead of 32), passing to $P(30.5 \leq X \leq 82.5)$. The rule is this: Convert the probability to the form $P(a \leq X \leq b)$, where a and b are integers (and the inequalities are of the form $\leq$ and not of the form $<$), pass to $P(a - \frac{1}{2} \leq X \leq b + \frac{1}{2})$, and then proceed as in the examples above.

The continuity correction enables us to approximate the probability that X will take on a single specified value, as shown in the following example.

Example 5

Suppose that a novice at archery has probability .16 of hitting the target on each shot, and suppose he shoots 80 times during an afternoon. What is the probability that he hits the target exactly 20 times? Here n is 80, p is .16, and we ask for the chance of exactly 20 successes.

In this case

$$\mu = np = 80 \times .16 = 12.8$$
$$\sigma = \sqrt{np(1-p)} = \sqrt{80 \times .16 \times .84} = 3.279$$

Using the continuity correction gives

$$P(X = 20) = P(19.5 \leq X \leq 20.5)$$
$$= P\left(\frac{19.5 - 12.8}{3.279} \leq \frac{X - 12.8}{3.279} \leq \frac{20.5 - 12.8}{3.279}\right)$$
$$\cong P(2.04 \leq Z \leq 2.35) = .4906 - .4793 = .0113$$

The exact probability is .012234.

As pointed out in Section 4.5 and again in the preceding section, in sampling without replacement from a population divided into two categories, success and failure, the number of successes in the sample is a random variable which has approximately the binomial distribution, provided the population size is much greater than the sample size. But if the sample size is large, and the population size is larger still, the normal approximation applies.

Problems

67. For a binomial distribution with $n = 25$ and $p = .4$, consider the three probabilities

SECTION 5.9 THE NORMAL APPROXIMATION TO THE BINOMIAL

      a. $P(8 \leq X \leq 11)$.   **b.** $P(X \leq 7)$.   **c.** $P(X \geq 12)$.

      (i) Calculate these three probabilities by the normal approximation without the continuity correction. Add them.

      (ii) Calculate them again with the continuity correction. Add them.

      (iii) Compare the two sums in (i) and (ii). What does the comparison tell you about the continuity correction?

68. For a binomial experiment with $n = 400$ and $p = .1$, consider the three probabilities

      a. $P(X \geq 50)$.   **b.** $P(X < 55)$.   **c.** $P(31 < X < 52)$.

      (i) Calculate these three probabilities by the normal approximation without the continuity correction.

      (ii) Calculate them with the continuity correction.

69. Let X be a binomial random variable with $p = .1$ and $n = 10$. Use Table 1 to find $P(X = 1)$, and compare it with the value obtained by the normal approximation.

70. Let X be a binomial random variable with $n = 25$ and $p = .2$. Use Table 1 to find $P(X \leq 3)$, and compare it with the values obtained by the normal approximation with and without the continuity correction.

71. If a fair die is rolled 200 times,

    a. What is the probability that a one will occur between 30 and 50 times, inclusive?

    b. What is the probability a one will occur fewer than 20 times?

72. The unemployment rate in a certain city is 8.5%. A sample of 100 people from the labor force is drawn. Find the probability that the sample contains

    a. At least ten unemployed people.

    b. No more than five unemployed people.

    c. Exactly eight unemployed people.

73. If one crosses garden peas having the gene pair (red, white) with peas also having the gene pair (red, white), one-fourth of the progeny are expected to have white flowers. If 64 plants from such a cross are examined, what is the probability that there will be exactly 16 with white flowers?

74. If 50% of the population favors a given political candidate, what is the probability that a sample of size 300 will contain more than 52% who favor this candidate?

75. Of the customers who do business at a rental car agency, 25% prefer a large automobile. If there are 45 large automobiles on hand, what is the probability that the agency will not be able to meet the demand for these automobiles in a randomly selected group of 160 customers?

76. A company has determined that an average of 7% of its manufactured items will be damaged during shipment. What is the probability that 10% or more of these items will be damaged in a shipment of 150 items?

5.10 The Poisson Distribution

Poisson distribution

The **Poisson distribution** is discrete. A random variable X has the Poisson distribution if its possible values are the nonnegative integers (0, 1, 2, 3, . . .) and if

$$P(X = r) = e^{-\mu}\frac{\mu^r}{r!}, \qquad r = 0, 1, 2, 3, \ldots \qquad 5.20$$

Here μ is the mean of the distribution, and e is a fundamental mathematical constant which approximately equals 2.718. Table 3 in the Appendix lists the values of $e^{-\mu}$ for certain values of μ.

The Poisson distribution is used to describe the number of events of a given kind which occur in a fixed interval of time. In many cases it gives an accurate description if the events are rather rare.

Example 1

Consider incorrect connections (wrong numbers) at a telephone switchboard. Let X be the number of these that occur during a period of one day. Suppose the mean of X is 2.3: there are on the average 2.3 incorrect connections made per day. The distribution of X is well described by (5.20) with $\mu = 2.3$. The chance of three incorrect connections during the day is by Table 1

$$P(X = 3) = e^{-2.3}\frac{(2.3)^3}{3!} = .100 \times \frac{12.17}{6} = .203$$

The chance of no incorrect connections is (recall that $0! = 1$ and $\mu^0 = 1$)

$$P(X = 0) = e^{-2.3}\frac{(2.3)^0}{0!} = e^{-2.3} = .100$$

Example 2

Suppose that on a particular stream a fisherman can catch one trout per hour on the average. Under the Poisson model, the chance that he catches 5 trout in a single hour is (here $\mu = 1$)

$$P(X = 5) = e^{-1}\frac{1^5}{5!} = .368 \times \frac{1}{120} = .0031$$

SECTION 5.10 THE POISSON DISTRIBUTION

The Poisson distribution can also apply to counting the number of events that occur in a given portion of space instead of a given interval of time:

Example 3

If a drop of water is viewed through a microscope, the number of bacteria visible will closely follow the Poisson distribution. If $\mu = 6.0$, then the probability of seeing at most two bacteria is

$$P(X \leq 2) = P(X = 0) + P(X = 1) + P(X = 2)$$
$$= e^{-6.0}\frac{(6.0)^0}{0!} + e^{-6.0}\frac{(6.0)^1}{1!} + e^{-6.0}\frac{(6.0)^2}{2!}$$
$$= .002 + .002 \times 6.0 + .002 \times \frac{36.00}{2} = .050$$

The probability of seeing at least one is

$$P(X \geq 1) = 1 - P(X = 0) = 1 - e^{-6.0} = 1 - .002 = .998$$

The Poisson distribution is useful for describing the frequency of accidents of various kinds.

Example 4

Suppose that a driver experiences a flat tire every 6,000 miles on the average. Suppose that she sets out on a 15,000 mile trip. She can then expect 15,000/6,000 flat tires on the trip, which comes to 2.5. The Poisson distribution applies with $\mu = 2.5$. The probability of more than one flat tire during the trip is

$$P(X > 1) = 1 - P(X = 0) - P(X = 1)$$
$$= 1 - e^{-2.5}\frac{(2.5)^0}{0!} - e^{-2.5}\frac{(2.5)^1}{1!}$$
$$= 1 - .082 - .082 \times 2.5 = .713$$

Problems

77. Find the following probabilities for the Poisson random variable X:
  a. $P(X \leq 2)$, $\mu = 2$.
  b. $P(X = 1)$, $\mu = 5.5$.
  c. $P(X \geq 3)$, $\mu = 1.5$.

78. The number of particles emitted per second by a certain radioactive substance is a Poisson random variable with mean $\mu = 3$. What is the probability that 5 or more particles are emitted in a second?

79. The number of emergency appendectomies per day performed at Highpoint Hospital is a Poisson random variable with a mean of 1.
   a. What is the probability that no appendectomies are performed on two consecutive days?
   b. Would it be unusual for 4 or more appendectomies to be performed on a given day?

80. If the number of telephone calls passing through a given switchboard has a Poisson distribution with mean θ equal to $3t$, where t is the time in minutes, find the probability of
   a. Two calls in any one minute.
   b. Five calls in two minutes.
   c. At least one call in one minute.

81. At a particular location on a river the number of fish caught per man-hour of fishing has a Poisson distribution with θ equal to 1.2. If a man fishes there for one hour, what is the probability he will catch exactly two fish? At least one fish?

82. The Poisson distribution has the property that the variance is equal to the mean. For the Poisson random variable X, find $P(X \geq \mu + 2\sigma)$ when $\mu = 4$ and when $\mu = 2$.

CHAPTER PROBLEMS

83. The purpose of this problem is to examine the data for the active group in Chapter 2 to determine whether or not it is appropriate to use the normal distribution to describe the change in blood pressure of this group. (See Table 2.9 and Figure 2.7.) Using the values of $\bar{X}$ and s from these data for the values of μ and σ, respectively, find the probability that the normal random variable is less than -30, -20, -10, 0, 10, 20, and 30. Compare each probability to its corresponding observed proportion (use Table 2.9). Does it appear that the change in blood pressure of the active group is a normal random variable? See Section 10.4.

84. Given that Z is a standard normal variable, find
   a. $P(Z^2 \leq 3)$.
   b. $P(2 \leq Z^2 \leq 3)$.

85. For a binomial random variable X with a large enough value of n ($n \geq 20$) and a small enough value of p ($p \leq .05$), the distribution of X can

be approximated rather well by the Poisson distribution with $\mu = np$. Find the following probabilities using the binomial distribution and this Poisson approximation.

a. $P(X \leq 2)$, $n = 20$, $p = .05$.
b. $P(X = 1)$, $n = 30$, $p = .02$.

86. A certain genetic trait occurs in one out of every 100 people on the average. Find the probability that the trait occurs in fewer than 3 of 200 randomly selected people. (See Problem 85.)

87. When a sample is taken without replacement from a finite population, and the sample size is small relative to the size of the population, the binomial distribution with $p = n/(n + m)$ can be used to approximate the hypergeometric probabilities. (See Problem 26 above and the discussion in Section 4.6.) Make use of this approximation to calculate the probability of finding at least 1 red ball when sampling 5 balls without replacement out of an urn that contains 3 red balls and 17 black ones. Evaluate the error in the approximation by calculating the exact probability.

88. A manufacturer of heavy equipment recorded the number of times 100 earth-movers needed to be repaired in the first 12 months of operation. The frequency distribution is given below. Compute the Poisson probabilities of 0, 1, etc., using $\mu = \bar{X}$ and compare to the observed proportions of 0, 1, etc. Comment on the appropriateness of the Poisson distribution for these data.

Number of repairs	0	1	2	3	4	5	6	7
Frequency	14	19	30	20	10	3	2	2

89. The score on an aptitude examination a company gives to prospective draftsmen is a normal random variable with mean 75 and standard deviation 5. To be considered for employment, the applicant must score 79 or more. What is the probability that at least 3 out of the next 10 applicants will be considered for employment?

Summary and Keywords

The purpose of statistical inference is to establish facts about **populations** (p. 152). In addition to **finite** populations (p. 152), some problems involve hypothetical **infinite** populations (p. 152). Statistical inferences are made on the basis of **samples** (p. 153), usually **random** samples (p. 154). The elements of a sample are **random variables** (p. 155), either **discrete** (p. 156) or **continuous** (p. 156), and a random variable is described by its

distribution (p. 159) or its **frequency curve** (p. 161); its average is measured by the **expected value** (p. 169) and its dispersion by the **variance** and **standard deviation** (p. 171). The **normal distribution** (p. 178) describes many random variables encountered in practice; probabilities relating to a normal random variable are found by passing to the **standardized variable** (p. 173 and p. 184) and using the tables of the **standard normal distribution** (p. 179). The normal distribution is also important as an **approximation** (p. 194) to the **binomial distribution** (p. 189). The **Poisson distribution** (p. 202) is an appropriate model in many situations involving counts.

REFERENCES

5.1 Cochran, W. G. *Sampling Techniques*. 2d ed. New York: John Wiley & Sons, 1963.

5.2 Hogg, Robert V., and Craig, Allen T. *Introduction to Mathematical Statistics*. 3d ed. New York: Macmillan, 1970.

5.3 Kish, L. *Survey Sampling*. New York: John Wiley & Sons, 1965.

5.4 Parzen, Emanuel. *Modern Probability Theory and Its Applications*. New York: John Wiley & Sons, 1960.

6

Sampling Distributions

6.1 Sampling and Inference

sampling
inference

It is in order to discover facts about populations that we draw samples from them. Several times in the previous chapters we have touched on the ideas underlying opinion polling. A more detailed discussion of these ideas will make clear the connection between **sampling** from populations and **inference** about populations.

Consider the voting population of a city split between those who intend to vote for candidate A in a coming election and those who intend to vote for candidate B. An opinion poller trying to predict the outcome of the election draws a random sample of size 100 from the population of voters. If in the population a proportion .36 favor candidate A (and a proportion .64 favor B), what is the chance that the number of voters in the sample who favor A will be 45 or more? In Example 3 of Section 5.9 we approximated this probability by using the normal distribution (favoring A was there called success). If X is the number of voters in the sample who favor A, then

$$P(X \geq 45) \cong .03 \qquad \text{if } p = .36$$

If a proportion .36 of the voting population favor A, then the chance that $X \geq 45$ is about .03.

It is natural to challenge the relevance of this sort of calculation to the problem of making inferences about the population of voters. Indeed, if the opinion poller actually knew that the proportion in the population favoring A was .36—the assumption underlying the computation—then he would not have gone to the trouble of drawing a sample in the first place. He drew the sample, after all, in order to get some idea of what proportion of all the voters do in fact favor A. This proportion is a number p which is unknown to the poller, so why assume that p is .36?

The poller makes guesses about p on the basis of X, the number in the sample of 100 who favor A; for his sample X will have some specific value like 72 or 18. Even without the benefit of statistical theory, we do know that if X is 72, then the poller should guess that p is somewhere around .72; that p is .36 would be a bad guess. If X is 18, the poller should guess that p is something like .18; that p is .95 would be a bad guess. We know also that the poller could make more accurate guesses about p if the sample size were 1000 instead of 100. But we want to go beyond these initial ideas to a more detailed understanding of how the poller should guess at p and what kind of precision he can hope for.

In the case where X is 72 we regard .36 as a very bad guess for p because if p really *is* only .36, then it is very strange indeed that X should be so large as 72. And this is the relevance of probability computations

SECTION 6.1 SAMPLING AND INFERENCE

based on the assumption that p is .36. Our computation showed that if p were .36, then the chance would be only about .03 that X is even as great as 45. A detailed understanding of how the poller ought to guess requires a detailed knowledge of the distribution of X for the case where p is .36. But of course there is nothing special about .36, and we must also know the distribution of X in the case where p is .40, say. In Example 3 of Section 5.9 we treated this case as well:

$$P(X \geq 45) \cong .15 \qquad \text{if } p = .40$$

The advantage of a mathematical treatment is that by letting p be general, we can in principle consider all values of p at the same time; that is, we can give a description of the distribution of X for the general p between 0 and 1. As explained in Section 5.8 (see page 189), if a large population is divided into two classes, there labeled *success* and *failure*, and if a random sample is drawn from the population, then the number of elements X in the sample that are in the *success* category has the binomial distribution (5.15) (see also the related discussion on page 135 in Section 4.5). Here the classes are "favor A" and "favor B" instead of "success" and "failure," but the situation is otherwise the same. The number X who favor A in the random sample of size 100 has the distribution (5.15), where n is 100 and p is the proportion in the voting population who favor A. Thus

$$P(X = r) = \binom{100}{r} p^r (1-p)^{100-r}, \qquad r = 0, 1, \ldots, 100$$

For purposes of computation we use the normal approximation, as described in the first four examples in Section 5.9. We require the binomial formula above in general form—that is, for general p—precisely because the poller does not know the actual value of p in advance.

parameters

statistics

We draw samples to discover facts about a population, and these facts are usually expressed in terms of numbers called **parameters**. In the preceding discussion the parameter was p, the proportion in the voting population who favor A. In general, a parameter is a number describing some aspect of a population. In making inferences about a parameter on the basis of a sample, we usually deal with **statistics**, which are numbers that can be computed from the sample. In the preceding discussion the statistic was X, the number of people in the sample of 100 who favor candidate A. The poller is ignorant of the value of the parameter p; once he has drawn his sample, he knows the value of the statistic X and can use it to make inferences (guesses) about p.

The example above is typical. Generally, we are concerned with one or more parameters that help describe a population. We do not know the values of the parameters (and usually never will). We draw a sample and compute one or more statistics on the basis of the sample. We have actual numerical values for the statistics. And we use the statistics to make inferences about the parameters. In order to know how to make these inferences, we must know about the distributions of the statistics.

For a further example, consider the average weight μ (in pounds) of the population of trout in Crater Lake. We do not know the value of the parameter μ and we never will. To get an idea of what μ is, we take a sample of trout and compute the mean $\bar{X}$ of their weights (as in Chapter 3); the statistic $\bar{X}$ is something we find the actual numerical value of. On the basis of $\bar{X}$ we make inferences about μ. Without statistical theory, we know that $\bar{X}$ somehow estimates μ, but to know just how to make the inferences and how exact they will be, we need to know the distribution of the random variable $\bar{X}$—we need to know its **sampling distribution**.

sampling distribution

6.2 Expected Values and Variances

The mean and variance of a population of numbers are defined as the mean and variance of a single random observation from that population, that is, as $E(X_1)$ and $Var(X_1)$ for a random sample X_1 of size 1. The population is said to be discrete or continuous depending on whether a single observation X_1 has discrete or continuous distribution. A discrete population may be finite or infinite; a continuous population must be infinite, though it may be approximated by a large but finite population.

There are certain formulas connected with sampling—(6.1) through (6.6) below—that will be used over and over again in the following chapters. To understand the meaning and significance of these formulas consider a random sample of size 2 from a population of size 5. Suppose the population consists of the elements 2, 4, 6, 8, and 10. This example is entirely artificial (actual populations number in the hundreds, at least), but it is sufficiently simple to make clear the ideas lying behind the formulas.

The population mean μ is

$$\mu = \tfrac{1}{5}(2 + 4 + 6 + 8 + 10) = 6$$

Now suppose that we take from the population a random sample of size 2 without replacement. There are $\binom{5}{2}$, or 10, different outcomes—different samples of size 2—and each of them has probability $\tfrac{1}{10}$ of being the sample actually drawn. Recall that sampling from a finite population is random if each possible sample has the same probability. Each of these 10 samples has associated with it a sample mean $\bar{X}$:

SECTION 6.2 EXPECTED VALUES AND VARIANCES

Sample	Sample Mean $\bar{X}$
2, 4	3
2, 6	4
2, 8	5
2, 10	6
4, 6	5
4, 8	6
4, 10	7
6, 8	7
6, 10	8
8, 10	9

The sample being random, each outcome has probability $\frac{1}{10}$. The probability that $\bar{X}$ has the value 5 is the probability of getting the sample 2,8 or the sample 4,6, so $P(\bar{X} = 5) = \frac{2}{10}$. This and analogous computations give the sampling distribution of $\bar{X}$:

r	3	4	5	6	7	8	9
$P(\bar{X} = r)$	.1	.1	.2	.2	.2	.1	.1

By the definition (5.1), page 169, of the expected value for a discrete random variable,

$$E(\bar{X}) = (3 \times .1) + (4 \times .1) + (5 \times .2) + (6 \times .2) + (7 \times .2) \\ + (8 \times .1) + (9 \times .1) = 6$$

The point is that $E(\bar{X})$ coincides with the population mean μ.

This is always so. Let $X_1, X_2, \ldots, X_n$ be a random sample of size n from a population, finite or infinite, discrete or continuous. The sample mean is, as in Chapter 3,

$$\bar{X} = \frac{1}{n} \sum_{i=1}^{n} X_i \qquad 6.1$$

and it is always true that

$$E(\bar{X}) = \mu \qquad 6.2$$

The formulas (6.1) through (6.6) will be used constantly in the rest of the book. Although we give derivations of them at the end of this section,

the mean of the mean is the mean

the derivations are not needed in the sequel; an understanding of the *meaning* of the formulas will be enough.

Equation (6.2) is sometimes expressed as: **The mean of the mean is the mean.** Each *mean* here has a different meaning, and the statement is short for: The expected value E (first *mean*) of the sample mean $\bar{X}$ (second *mean*) is the population mean μ (third *mean*).

If the population is infinite, so that the elements $X_1, X_2, \ldots, X_n$ in the random sample are *independent*, then we have for the variance of $\bar{X}$ the formula

for infinite populations

$$\text{Var}(\bar{X}) = \frac{\sigma^2}{n} \qquad \sigma_{\bar{X}}^2 \qquad 6.3$$

where σ^2 is the population variance. The standard deviation is of $\bar{X}$

$$\text{St Dev}(\bar{X}) = \frac{\sigma}{\sqrt{n}} \qquad \sigma_{\bar{X}} \qquad 6.4$$

The formulas (6.3) and (6.4) actually only apply if the population is infinite, but they give a good approximation if the population is large.

In addition to the mean and variance of $\bar{X}$, we can find the mean of the sample variance s^2. Recall from Chapter 3 that

$$s^2 = \frac{1}{n-1} \sum_{i=1}^{n} (X_i - \bar{X})^2 \qquad 6.5$$

For a sample from an *infinite* population, we have the formula

$$E(s^2) = \sigma^2 \qquad 6.6$$

Equation (6.6) concerns the mean of the variance, whereas (6.3) concerns the variance of the mean. To keep these straight, the words must be "spelled out"—(6.6) concerns the expected value of the sample variance, and (6.3) concerns the variance of the sample mean.

Equation (6.2) holds because of the rules for manipulating expected values. Applying to (6.1) the rule (5.4) and then (5.11), we have

$$E(\bar{X}) = \frac{1}{n} E\left(\sum_{i=1}^{n} X_i\right) = \frac{1}{n} \sum_{i=1}^{n} E(X_i)$$

Since each individual X_i is a single observation from the population, $E(X_i)$ has the same value as μ. Equation (6.2) follows by Rule 3 for summations, page 46.

SECTION 6.2 EXPECTED VALUES AND VARIANCES

To see why (6.3) holds for an infinite population, use (5.5) and then (5.12); the latter formula is applicable because the X_i are independent:

$$\text{Var}(\bar{X}) = \text{Var}\left(\frac{1}{n}\sum_{i=1}^{n} X_i\right) = \frac{1}{n^2}\text{Var}\left(\sum_{i=1}^{n} X_i\right) = \frac{1}{n^2}\sum_{i=1}^{n}\text{Var}(X_i)$$

But each $\text{Var}(X_i)$ is σ^2, since X_i is a single observation from the population, and so Rule 3 for summations (page 46) yields (6.3).

For the sake of simplicity, we derive (6.6) only for the case where μ is 0. By (3.11),

$$\Sigma(X_i - \bar{X})^2 = \Sigma X_i^2 - \frac{1}{n}(\Sigma X_i)^2 = \Sigma X_i^2 - n\bar{X}^2$$

If $\mu = 0$, so that $E(X_i) = 0$ and $E(\bar{X}) = 0$, then $E(X_i^2) = \sigma^2$ and $E(\bar{X}^2) = \text{Var}(\bar{X}) = \sigma^2/n$; therefore

$$E\left[\sum_{i=1}^{n}(X_i - \bar{X})^2\right] = \sum_{i=1}^{n}\sigma^2 - n\frac{\sigma^2}{n} = (n - 1)\sigma^2$$

which gives (6.6).

The formulas (6.3) and (6.4) apply only when *the population is infinite*. If the population is finite, (6.3) must be replaced by the formula

$$\text{Var}(\bar{X}) = \frac{N - n}{N - 1} \cdot \frac{\sigma^2}{n} = \frac{\sigma^2}{n}\left(1 - \frac{n - 1}{N - 1}\right)$$

where N is the size of the population. The factor $(N - n)/(N - 1)$ is the *finite-population correction factor*. In the example above, the population variance is

$$\sigma^2 = E(X_1 - 6)^2 = \tfrac{1}{5}(16 + 4 + 0 + 4 + 16) = 8$$

The variance of $\bar{X}$ is

$$E(\bar{X} - 6)^2 = (9 \times .1) + (4 \times .1) + (1 \times .2)$$
$$+ (1 \times .2) + (4 \times .1) + (9 \times .1) = 3$$

which agrees with the above formula for $N = 5$ and $n = 2$:

$$3 = \frac{5 - 2}{5 - 1} \cdot \frac{8}{2}$$

If the population size N is large in comparison with the sample size n, then the correction factor is nearly 1, so that the formula for $\text{Var}(\bar{X})$ for a finite population is practically the same thing as (6.3) in this case. For example, if the population size N is 10,000, and if the sample size n is 200, then the finite-population correction factor, $(N - n)/(N - 1)$, is $(10,000 - 200)/(10,000 - 1) = 9,800/9,999 = .98$, which is near 1. This is analogous to the fact (Section 4.5) that sampling without replacement is practically the same thing as sampling with replacement if the population is large compared with the sample.

Problems

1. Various samples are drawn from an infinite population with $\mu = 25$ and $\sigma^2 = 25$. If the population is normally distributed, find the mean, the variance, and the standard deviation of the sample mean for each of the following sample sizes.
   a. $n = 10$.   b. $n = 100$.   c. $n = 1000$.
2. Answer Problem 1 under the assumption that the population has an unknown distribution.
3. A sample of size 50 is drawn from a population with $\sigma = 10$. Find the variance of the sample mean if
   a. The population is infinite.
   b. The population size is $N = 100{,}000$.
   c. The population size is $N = 1000$.
   d. The population size is $N = 100$.
4. Let the numbers 1, 2, 3, 4, 5, and 6 constitute a population. If samples of size three are drawn without replacement, there are 20 different possible samples. List these 20 samples and find the mean and median for each of them. Find the mean and variance of the 20 sample means. Do the same for the medians. Which measure, mean or median, shows the smaller variation from sample to sample?
5. An infinite population consists of the values 1, 2, 3, 4, 5 in the proportion $\frac{1}{5}$ each.
   a. List the possible samples of size 2, and give the probability of obtaining each sample.
   b. List the possible values of $\bar{X}$ for $n = 2$, and give the probability of obtaining each value.
   c. Verify that $E(\bar{X}) = \mu$ and $\text{Var}(\bar{X}) = \sigma^2/n$.
6. Find the probability distribution of s^2 for samples of size $n = 2$ where the population is that of Problem 5. Verify that $E(s^2) = \sigma^2$.
7. A finite population consists of the five values 1, 2, 3, 4, 5.
   a. List the samples of size 2, where sampling is done without replacement, and give the probability of obtaining each sample.
   b. List the possible values of $\bar{X}$ and give the probability of obtaining each value.
   c. Verify that $E(\bar{X}) = \mu$ and $\text{Var}(\bar{X}) = (N-n)\sigma^2/(N-1)n$.

6.3 Sampling from Normal Populations

Let $\bar{X}$ be the sample mean for a sample of size n of independent observations. As we saw in the last section, if the population mean and variance are μ and σ^2, then $\bar{X}$ has mean μ and variance σ^2/n, and this is true whatever the form of the parent population may be. Suppose now that the population is *normal*. In this case $\bar{X}$ is *normally distributed with mean μ and variance σ^2/n*. This is a fundamental fact about normal populations. That $\bar{X}$ has mean μ and variance σ^2/n is true whatever the population; what is new here is that if the population is normal, then $\bar{X}$ is normally distributed.

This characteristic is plainly seen in the results of a sampling exercise. Each of 158 statistics students took two random samples of five values each from a population approximately normal with mean 40 and variance 100.

Table 6.1 Distributions of the Sample Means in Samples of 5 and 10

Mean	Frequency ($n = 5$)	Frequency ($n = 10$)
25–27	1	1
27–29	1	1
29–31	4	2
31–33	18	0
33–35	18	5
35–37	37	14
37–39	47	36
39–41	60	43
41–43	41	26
43–45	45	19
45–47	27	9
47–49	12	1
49–51	1	1
51–53	4	0
Totals	316	158

Each computed the two means and also the mean of the ten values combined into one sample. Table 6.1 shows the resulting distributions of sample means, and Figures 6.1 and 6.2 show the histograms.

Figure 6.1

Histogram for the means of samples of size 5

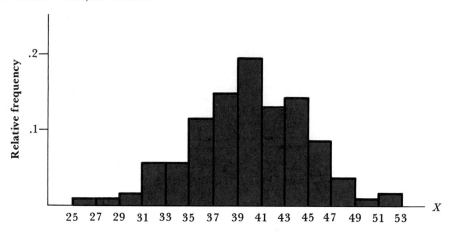

The means for these empirical distributions are 40.06 for samples of size 5 and 39.98 for samples of size 10. The variances are 20.95 and 12.64. Even with fairly small numbers of means, the histograms show a tendency toward the symmetrical, bell-shaped normal curve; the means and variances agree well with the theoretical values $\mu = 40$, $\sigma^2/5 = 20$, and $\sigma^2/10 = 10$.

Figure 6.2

Histogram for the means of samples of size 10

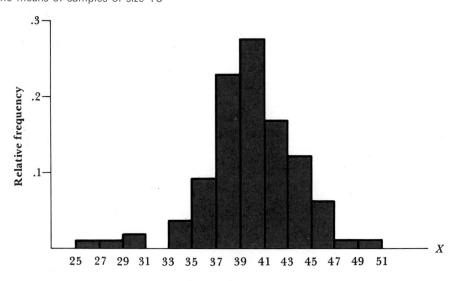

6.4 The Standardized Sample Mean

Since $\bar{X}$ has mean μ and standard deviation $\sigma/\sqrt{n}$, the standardized variable

$$Z = \frac{\bar{X} - \mu}{\sigma/\sqrt{n}} \qquad 6.7$$

has mean 0 and variance 1; see Equations (5.7) and (5.8) on page 173.
 As stated in the preceding section, if the population is normal, then Z, defined by Equation (6.7), is a standard normal variable and we may use the table of normal areas to calculate probabilities for the sample mean.

Example 1

Problem: Given a normal distribution with mean 50 and variance 100, find the probability that the mean of a sample of 25 observations will differ from the population mean by less than four units.

Solution: We want

$$P(-4 < \bar{X} - \mu < 4)$$

Now $\sigma/\sqrt{n} = 2$; if we divide the middle member of this last inequality by $\sigma/\sqrt{n}$ and the two outer members by 2, we obtain

$$P(-4 < \bar{X} - \mu < 4) = P\left(\frac{-4}{2} < \frac{\bar{X} - \mu}{\sigma/\sqrt{n}} < \frac{4}{2}\right)$$

$$= P(-2 < Z < 2) = .9544$$

Example 2

Problem: For the normal distribution of the preceding example find two values equidistant from the mean such that 90% of the means of all samples of size 400 will be contained between them.

Solution: From the table of normal areas we have

$$.90 = P(-1.64 \leq Z \leq 1.64)$$

From Equation (6.7),

$$.90 = P\left(-1.64 \leq \frac{\bar{X} - \mu}{\sigma/\sqrt{n}} \leq 1.64\right)$$

This time, $\sigma/\sqrt{n} = 10/\sqrt{400} = 10/20 = .5$; therefore

$$.90 = P(-1.64 \leq \frac{\bar{X} - 50}{.5} \leq 1.64)$$
$$= P(50 - .5 \times 1.64 \leq \bar{X} \leq 50 + .5 \times 1.64)$$
$$= P(49.18 \leq \bar{X} \leq 50.82)$$

So the numbers asked for are 49.18 and 50.82.

Problems

8. The annual precipitation in Illinois is a normal random variable with mean 33.18 inches and standard deviation 4.23 inches.
   a. What is the probability that the average rainfall for the next 10 years will be less than 30 inches?
   b. What is the probability that the total rainfall for the next 10 years will be greater than 350 inches?

9. A sample of 100 items is taken from the population described in Problem 55 of Chapter 5. Find the approximate probability that the mean life of the items in the sample will
   a. Exceed 10,000 hours.
   b. Be less than 8000 hours.
   c. Fall between 8000 and 10,000 hours.

10. If the population of times measured by three-minute egg timers is normally distributed with μ equal to 3 minutes and σ equal to .2 minutes, and if we test samples of 25 timers, find the time that would be exceeded by 95% of the sample means.

11. What minimum sample size should be taken from a normal population with $\mu = 10$ and $\sigma = 20$ in order for the probability of the sample mean exceeding twice the population mean to be under .025?

12. A wholesale distributor has found that the amount of a customer's order is a normal random variable with a mean of $200 and a standard deviation of $50. If 20 orders are received, what is the probability that the total amount is greater than $4500?

13. In Problem 12, what is the minimum number of orders that must be received so that there is probability .975 of having a total amount greater than $5000?

6.5 The Central Limit Theorem

central limit theorem

The **central limit theorem** concerns the approximate normality of means of random samples or of sums of random variables; we accordingly state it in two forms.

Suppose $X_1, X_2, \ldots, X_n$ is a sample from an infinite population with mean μ and variance σ^2; the X_i are independent random variables.

First form If n is large, then

$$\frac{\bar{X} - \mu}{\sigma/\sqrt{n}} \qquad 6.8$$

has approximately a standard normal distribution, or (what is the same) $\bar{X}$ has approximately a normal distribution with mean μ and standard deviation $\sigma/\sqrt{n}$.

We know that whatever the parent population may be, the standardized variable (6.8) has mean 0 and standard deviation 1; and we know that if the parent population is normal, then the variable (6.8) has exactly a standard normal distribution. The remarkable fact is that even if the parent population is not normal, the standardized mean is approximately normal if n is large. The importance of the theorem lies in the fact that it permits us to use normal theory for inferences about the population mean regardless of the form of the population, provided only that the sample size is large enough.

To prove the central limit theorem requires the full apparatus of mathematical probability. But we can illustrate it by starting with a specific nonnormal distribution—a J-shaped exponential distribution with mean μ of 2 and variance σ^2 of 4. Figure 6.3a shows this distribution (solid line) and the normal distribution (dashed line) with the same mean and variance. The two distributions are very dissimilar.

For samples with size n equal to 4 from the exponential population of Figure 6.3a, the sample mean $\bar{X}$ has mean 2 and variance 1. Figure 6.3b shows the exact distribution (solid line) of $\bar{X}$, together with the normal distribution (dashed line) with mean 2 and variance 1. The two curves are rather similar.

Figure 6.3c shows the same pair of curves for a sample size n of 12 (the mean and variance are now 2 and $\frac{1}{3}$), and here the agreement is really quite close. These graphs show insufficient detail for the tails of the distribution; the normal approximation is usually better for values near the mean than for

Figure 6.3a

Exponential and normal distributions with $\mu = 2$ and $\sigma^2 = 4$

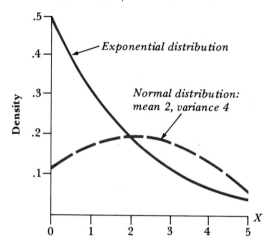

values far removed. As appears from this example, however, the sample need not be excessively large before we can feel reasonably safe in using the central limit theorem.

A second version of the central limit theorem concerns the sum ΣX_i of a

Figure 6.3b

Distribution of $\bar{X}$ for samples of 4 and the normal distribution with the same mean and variance

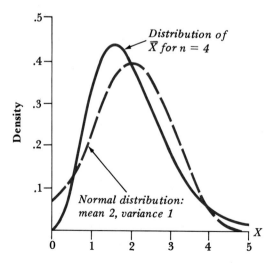

Figure 6.3c

Distribution of $\bar{X}$ for samples of 12 and the normal distribution with the same mean and variance

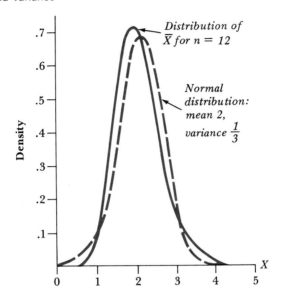

set $X_1, X_2, \ldots, X_n$ of independent random variables all having the same distribution with mean μ and variance σ^2.

Second form If n is large, then

$$\frac{\sum_{i=1}^{n} X_i - n\mu}{\sigma \sqrt{n}} \qquad 6.9$$

has approximately a standard normal distribution, or (what is the same) ΣX_i has approximately a normal distribution with mean $n\mu$ and standard deviation $\sigma \sqrt{n}$.

This theorem is really the same as the first one because ΣX_i is $n\bar{X}$, so the variable (6.9) is just $(n\bar{X} - n\mu)/\sigma \sqrt{n}$, and algebra reduces this to the variable (6.8).

This theorem (together with more general versions of it) is one of the reasons why the normal distribution often arises in nature. If one performs a complicated physical measurement, the measurement error is the sum ΣX_i of many small independent random errors X_i. The height of a plant is the sum ΣX_i of many small independent increments X_i. In such cases the normal distribution at least roughly approximates the distribution of the sum.

> The normal approximation to the binomial, as treated in Section 5.9, really comes under the central limit theorem. In an infinite (or finite but large) population split into two categories, success and failure, label each element in the success category with a 1 and each element in the failure category with a 0. Now a random sample of size n from the population becomes converted into a set $X_1, X_2, \ldots, X_n$ of independent random variables, each having value either 0 or 1. The number of successes is the number of 1s, and this is ΣX_i, so the central limit theorem applies. In this case (see (5.16) and (5.17)), $n\mu = np$ and $\sigma\sqrt{n} = \sqrt{np(1-p)}$. The variable (6.9) is the same thing as the standardized number of successes (see (5.18)); it has approximately a standard normal distribution.

Problems

14. A member of the company bowling team has over the years averaged 480 points per match with a standard deviation of 60 points. What is the probability that he will bowl an average greater than 490 in a 20 match season?

15. The average vitamin B-2 content of a certain brand of vitamins is 30 mg with a standard deviation of 2 mg. A quality-control inspector selects 25 pills for testing. What is the probability that the average vitamin B-2 content of these 25 pills is between 29 and 31 mg?

16. The cost of individual long-distance phone calls for a company is a random variable with mean $\mu = \$3.20$ and standard deviation $\$.80$. Find an interval that has probability .95 of containing the total cost of 100 phone calls.

17. When a two-decimal number is rounded to one decimal, the difference between the number and its rounded value will be one of the following: $-.05, -.04, -.03, -.02, -.01, .00, .01, .02, .03, .04$. Suppose that this difference is a random variable with values having probability $\frac{1}{10}$ each. What is the probability that the difference between the average of 16 randomly chosen two-decimal numbers and the average of their rounded values is more than .01?

18. Consider the person in Problem 28 of Chapter 5. During one year he takes 250 trips to work. Let $\bar{Y}$ be the mean of the number of red lights encountered in each of the trips.

CHAPTER PROBLEMS

a. Find the mean and the standard deviation of $\bar{Y}$.
b. Compute $P(\bar{Y} \geq 1.5)$ by using the central limit theorem.

19. In a certain country $\frac{1}{3}$ of the families have no cars, $\frac{1}{3}$ have one car, $\frac{1}{6}$ have two cars, $\frac{1}{12}$ have three cars, and $\frac{1}{12}$ have four cars. Let X be the random variable representing the number of cars per family.
    a. Find $E(X)$ and $Var(X)$.
    b. For a random sample of 100 families, find $E(\bar{X})$ and $Var(\bar{X})$.
    c. If each car has five tires, what are the mean and standard deviation of the number of tires to a family?
    d. With $\bar{X}$ as in part b, find $P(\bar{X} < 1)$, approximately.

20. The number of service calls per day made by a computer repair company is a Poisson random variable with a mean of 2 calls per day. Use the central limit theorem to find the probability that more than 45 service calls are made in an 18-day period. Note that $\sigma^2 = \mu$ for the Poisson distribution.

21. For sampling from a normal distribution with variance σ^2, the distribution of s^2 has mean σ^2 and variance $2\sigma^4/(n-1)$. As the sample size increases, the distribution of s^2 approaches a normal distribution. If we take a sample of size 73 from a normal population with $\sigma^2 = 80$, what is the approximate probability that s^2 will be greater than 100? Less than 40? Between 50 and 110?

CHAPTER PROBLEMS

Problems 22, 23, and 24 are based on a sampling experiment. Let X denote the number of times a coin must be tossed to obtain a head. The random variable X has a geometric distribution with mean $\mu = 2$ and variance $\sigma^2 = 2$. (See Problems 14, 15, and 29 of Chapter 5.) Each student in class is to obtain 10 values of X and compute the value of $\bar{X}$ for these data. For Problems 23 and 24, each student will also compute s^2. The instructor will arrange to make all the values of $\bar{X}$ and s^2 available to the entire class. If the class size is small, it may be necessary for students to perform this experiment more than once. It is recommended that at least 30 values of $\bar{X}$ and s^2 be obtained.

22. This problem uses the values of $\bar{X}$ from the sampling experiment to illustrate the central limit theorem.
    a. Make a histogram of the sample means. Does the histogram suggest an approximate normal distribution for $\bar{X}$?
    b. Find the average of the sample means and compare with $E(\bar{X}) = 2$.

c. Find the variance of the sample means; that is, find $s_{\bar{X}}^2$. Compare with $\text{Var}(\bar{X}) = .2$.

23. a. Use the central limit theorem to find $P(2 - \sqrt{.2} \leq \bar{X} \leq 2 + \sqrt{.2})$, where $\bar{X}$ is the average of 10 observations from a population whose mean and variance are 2. Compare this probability with the fraction of the sample means from the sampling experiment that fall in the interval $2 - \sqrt{.2}$ to $2 + \sqrt{.2}$.
    b. Make similar computations and comparisons for other intervals to determine how well the outcome of the sampling experiment is predicted by the central limit theorem.

24. This problem uses the values of s^2 from the sampling experiment to illustrate properties of the sampling distribution of s^2.
    a. Make a histogram of the values of s^2 from the sampling experiment. Does the histogram have the appearance of a normal distribution?
    b. Find the average of the values of s^2 and compare with $E(s^2) = 2$.

Summary and Keywords

The science of statistics rests on the connection between **sampling** and **inference** (p. 210). A **statistic** (p. 211) is a number that can be computed from a sample—that can be calculated by means of a formula from the actual data at hand. On the basis of statistics, we make inferences about **parameters** (p. 211), which are numbers describing aspects of the population under investigation. In order to understand how to use statistics to make inferences about parameters, it is necessary to understand something about **sampling distributions** (p. 212). The most important fact (and it requires interpretation) is: **The mean of the mean is the mean** (p. 214). The **central limit theorem** (p. 221) describes the approximate sampling distribution of the sample mean for large samples.

REFERENCES

6.1 Hoel, Paul G. *Elementary Statistics*. 3d ed. New York: John Wiley & Sons, 1971. Chapter 6.

6.2 Hogg, Robert V., and Craig, Allen T. *Introduction to Mathematical Statistics*. 3d ed. New York: Macmillan, 1970. Chapter 4.

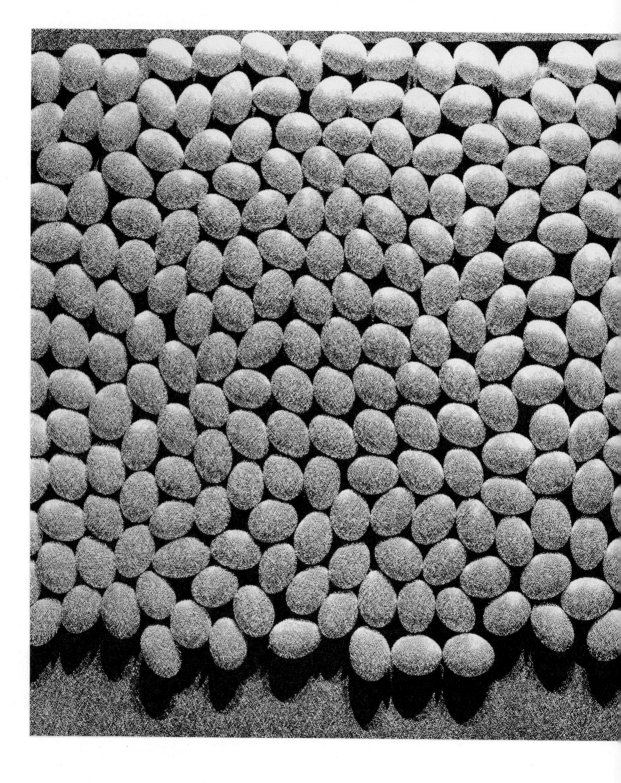

7

Estimation

7.1 Introduction

A statistician collects data by experiment, sample, or sample survey in the hope of drawing conclusions about the phenomenon under investigation. From his experimental results or sample values he wants to pass to *inferences* about the underlying population. He may use his data for the *estimation* of the values of unknown parameters or for *tests of hypotheses* concerning these values.

In Section 6.1 we discussed an opinion poller who takes a sample from a voting population divided into those who favor candidate A and those who favor B, the unknown parameter being the proportion p who favor A. The poller may estimate p (try to guess its value), or he may test a hypothesis about p—test the hypothesis, say, that $p > \frac{1}{2}$ (try to guess whether A will win the election). In either case he infers something about p.

This chapter deals with methods of estimation and the principles underlying them. Testing is taken up in Chapter 8.

problem of estimation

The Public Health Service hypertension study also illustrates the **problem of estimation.** According to Table 1.2 on page 3, 37 of the 193 patients (mild hypertensives) on the blood-pressure-reducing drug experienced hypertensive events—sustained damage associated with high blood pressure (such as enlargement of the heart). The fraction was 37/193, which is 19%. If similar patients are given the same drug in the future, it seems reasonable to predict that again something like 19% of them will sustain high-blood-pressure damage. The job of the theory of statistical estimation is to make the phrase "something like" more precise—to assess the accuracy of the figure 19% as a predictor.

Chapters 2 and 3 dealt partly with the change in blood pressure among the patients on the blood-pressure-reducing drug (Table 2.1, page 13). The average change was -9.85 (page 50); that is, on the average the drug reduced blood pressure (among mild hypertensives) by 9.85 points. Again, if similar patients are given the same drug in the future, it seems reasonable to predict that their blood pressures will be reduced by something like 9 or 10 points on the average. And again, the theory of statistical estimation will provide a way of assessing the accuracy of the prediction.

7.2 Estimation

We have at hand a sample from a population involving an unknown parameter, such as the mean; the problem is to construct a sample quantity that

SECTION 7.2 ESTIMATION

estimator
estimate

will serve to estimate the unknown parameter. Such a sample quantity we call an **estimator**; the actual numerical value obtained by evaluating an estimator in a given instance is the **estimate**. For example, the sample mean $\bar{X}$ is an estimator for the population mean μ; if for a specific sample the sample mean is 9.85, we say 9.85 is our estimate for μ. Notice that an estimator must be a statistic; it must depend only on the sample and not on the parameter to be estimated.

Now the sample mean $\bar{X}$ is an estimator for the population mean μ, but then so is the quantity $\bar{X} + 1000$. We will all agree that, as an estimator for μ, $\bar{X}$ is more reasonable than $\bar{X} + 1000$, but why? Because $\bar{X}$, unlike $\bar{X} + 1000$, is equal to μ *on the average:*

$$E(\bar{X}) = \mu \qquad 7.1$$

for all μ (see Equation (6.2) on page 213). This is a statement typical of statistical theory. Putting ourselves in the position of the experimenter, we do not know μ (and doubtless never will); given the sample, we do know $\bar{X}$; and *whatever* the unknown μ may be, $\bar{X}$ balances out at μ in the sense that

unbiased

$E(\bar{X}) = \mu$. An estimator with this property is called **unbiased**:

> **Unbiased estimator** An estimator $\hat{\theta}$ of a parameter θ is said to be unbiased if it has expected value θ, whatever θ may be, that is, if
>
> $$E(\hat{\theta}) = \theta \qquad 7.2$$
>
> for all θ.

We have seen that $\bar{X}$ has this desirable property. So has the sample variance. If $X_1, X_2, \ldots, X_n$ is an independent sample, then the sample variance

$$s^2 = \frac{1}{n-1} \sum_{i=1}^{n} (X_i - \bar{X})^2 \qquad 7.3$$

is an unbiased estimator of σ^2:

$$E(s^2) = \sigma^2 \qquad 7.4$$

(See Equation (6.6), page 214.) Notice that s^2 is a statistic; it does not depend on the unknown μ and σ^2. The point of Equation (7.4), the condition for unbiasedness, is not that we can somehow check it for the true values of μ and σ^2—we do not know what these true values are. The point is that Equation (7.4) holds *whatever* values μ and σ^2 may happen to have. The same remark applies to any unbiased estimator.

It was exactly in order to make the sample variance unbiased that in the original definition (3.9) on page 67 we divided by the then-mysterious $n - 1$; division by n would have introduced bias. As an estimator of σ, the sample standard deviation

$$s = \sqrt{\frac{1}{n-1} \sum_{i=1}^{n} (X_i - \bar{X})^2} \qquad 7.5$$

is ordinarily used, even though it is somewhat biased.

The precision of an unbiased estimator $\hat{\theta}$ of a parameter θ is customarily measured by its variance or by its standard deviation:

$$\text{St Dev}(\hat{\theta}) = \sqrt{E(\theta - \hat{\theta})^2}$$

For the mean $\bar{X}$ this quantity is $\sigma/\sqrt{n}$:

$$\text{St Dev}(\bar{X}) = \frac{\sigma}{\sqrt{n}} \qquad 7.6$$

(see (6.4), page 214). If the population parameter σ is unknown, an idea of the precision of X as an estimator of μ can be got by passing from $\sigma/\sqrt{n}$ to $s/\sqrt{n}$. The quantity $s/\sqrt{n}$, which can be calculated on the basis of the sample alone, is called the **standard error:**

standard error

$$\text{Standard error} = \frac{s}{\sqrt{n}} \qquad 7.7$$

In summarizing the information concerning μ contained in the sample, it is customary to give the standard error as well as the sample mean.

A *minimum-variance unbiased estimator* is an estimator that in the first place is unbiased, and in the second place has smaller variance than *any other* unbiased estimator.

Returning to the mean, consider a sample $X_1, X_2, \ldots, X_{100}$ of size 100, say. Here

$$\bar{X} = \frac{1}{100} \sum_{i=1}^{100} X_i$$

is an unbiased estimator of the population mean μ. But if we throw out the last 50 observations, the remaining ones—$X_1, X_2, \ldots, X_{50}$—form a sample of size 50 by themselves, and *their* mean

$$\bar{Y} = \frac{1}{50} \sum_{i=1}^{50} X_i$$

SECTION 7.2 ESTIMATION

→ Smaller sample will have larger variance.

is another unbiased estimator of μ. Given the full sample of size 100, we will all agree that $\bar{X}$ is a better estimator of μ than the mean $\bar{Y}$ of the partial sample is, but why? Each of the two estimators is unbiased, but one feels that $\bar{X}$ is better than $\bar{Y}$ because there is somehow more information in it. This idea can be made precise by using variances. Now $\bar{X}$ has variance $\sigma^2/100$ (as follows by squaring (7.6); here σ^2 is the population variance, also unknown to us). On the other hand, $\bar{Y}$ has the larger variance $\sigma^2/50$. Since variance measures the spread, of two unbiased estimators we naturally prefer the one with smaller variance.

Problems

1. For the following results from samples from normal populations, give the best estimates for the mean μ, the variance σ^2, and the variance of the mean $\sigma_{\bar{x}}^2$. Also give estimates for the standard deviation of the population σ and for the standard deviation of the mean (see (7.7)). $\sigma_{\bar{x}}$
   a. $n = 16$, $\Sigma X_i = 80$, $\Sigma (X_i - \bar{X})^2 = 240$.
   b. $n = 10$, $\Sigma X_i = 250$, $\Sigma (X_i - \bar{X})^2 = 1089$.
   c. $n = 25$, $\Sigma X_i = 100$, $\Sigma X_i^2 = 1000$.

2. A random sample of 49 children visiting a state fair produced the following data on their ages: $\Sigma X_i = 343$, $\Sigma (X_i - \bar{X})^2 = 4800$.
   a. Find unbiased estimates for the mean and variance of the ages of children visiting the fair.
   b. Find an unbiased estimate for the variance of the sample mean. Find the standard error.

3. One hundred students randomly selected from the seniors in a school system were given two different but similar mathematics examinations. Data on the differences X_i between the two exam scores show the following results: $\Sigma X_i = 325$, $\Sigma X_i^2 = 2046.25$.
   a. Find unbiased estimates for the mean and variance of the population of differences of scores.
   b. Find an estimate for the standard deviation of the sample mean— that is, find the standard error.

4. For each of the following samples from normal populations, find best estimates for μ, σ^2, and $\text{Var}(\bar{X})$. Also give estimates for the standard deviation of the population and standard deviation of the mean.
   a. 5, 6, 4, 2, 0, 5, 1, 3.   b. 1, 2, 5, 5, 7, 6, 4, 4, 2.
   c. 0, 4, 6, 0, -4, 6.   d. -1, 5, 13, 17, 21.

5. Nine randomly selected college athletes were timed twice in a 200-meter dash. The differences between the two times (in seconds) are: -1.5, -1, -1, 0, 0, .5, .5, 1, 1.5.

a. Find unbiased estimates for the mean and variance of the population of differences of scores.
b. Find an unbiased estimate for the variance of the sample mean.
6. The numbers of meals served by a large cafeteria on five randomly selected days are 1289, 1361, 1469, 1347, 1290. Find estimates for the standard deviation of the population and the standard deviation of the sample mean.

7.3 Confidence Intervals for Normal Means: Known Variance

We have seen that the sample mean $\bar{X}$ is an unbiased estimator of the population mean μ. This section deals with the problem of estimating the mean μ of a normal population when the variance σ^2 is known. This is a rather artificial circumstance—anyone ignorant of the mean is usually ignorant of the variance as well. Assuming σ^2 known, however, serves to simplify the reasoning and make clear the principles underlying estimation; it also serves to introduce the more complicated and more realistic case, treated in the next two sections, where σ^2 is unknown.

Example 1

Ten patients were given two soporific drugs, A and B. In each case the patient slept longer under the effect of B than of A, and Table 7.1 shows the amount of the increase in each case.

Let us assume we know from experience that the increase is normally distributed with some mean μ and that the variance σ^2 is 1.664. The estimator $\bar{X}$ then has a variance $\sigma^2/10$, or .1664, and a standard deviation (see (7.6)) $\sigma/\sqrt{10}$, or .408. Knowing this and the fact that $\bar{X}$ is normally distributed, we can calculate the probability that $\bar{X}$ is within .8, say, of the population mean μ. We do not know μ, but whatever value it may have, the standardized variable $(\bar{X} - \mu)/.408$ has a standard normal distribution and hence (see Figure 7.1)

$$P(\mu - .8 \leq \bar{X} \leq \mu + .8) = P\left(-\frac{.8}{.408} \leq \frac{\bar{X} - \mu}{.408} \leq \frac{.8}{.408}\right)$$

$$= P\left(-1.96 \leq \frac{\bar{X} - \mu}{.408} \leq 1.96\right) = .95$$

Thus the chance is 95% that $\bar{X}$ is off by .8 or less. Since $\mu - .8 \leq \bar{X} \leq$

SECTION 7.3 CONFIDENCE INTERVALS FOR NORMAL MEANS: KNOWN VARIANCE

Table 7.1 Additional Hours Sleep Gained by Using Drug B Instead of Drug A

Patient	Increase
1	1.2
2	2.4
3	1.3
4	1.3
5	0.0
6	1.0
7	1.8
8	0.8
9	4.6
10	1.4

$\bar{X} = 1.58$ hours

$\mu + .8$ is the same as $\bar{X} - .8 \leq \mu \leq \bar{X} + .8$ (since each is the same as $|\bar{X} - \mu| \leq .8$), we can say that

$$P(\bar{X} - .8 \leq \mu \leq \bar{X} + .8) = .95 \qquad 7.8$$

whatever μ may be. Now $\bar{X}$ for our data in Table 7.1 has the value 1.58, and it is tempting to replace the $\bar{X}$ in Equation (7.8) by 1.58:

$$P(1.58 - .8 \leq \mu \leq 1.58 + .8) = .95 \qquad 7.9$$

Figure 7.1

Standard normal curve

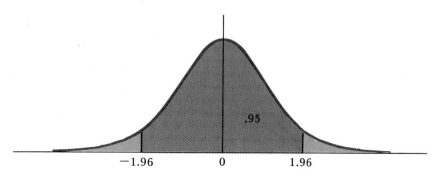

That is, it is tempting to conclude that there is probability .95 that the unknown μ lies between .78 (that is, 1.58 − .8) and 2.38 (that is, 1.58 + .8). But this would be incorrect. Although μ is unknown to us, it is not a random variable, but a fixed number. Thus $.78 \leq \mu \leq 2.38$ is simply either true or false; the probability on the left in Equation (7.9) is accordingly either 1 or 0—and we do not know which, since we do not know μ.

confidence intervals

Although Equation (7.9) is wrong as it stands, the idea lying behind it can be made sense of by means of **confidence intervals.** The first thing to understand is the source of the error in passing from (7.8) to (7.9). Consider a related but simpler case. Let Y be the total obtained in rolling a pair of fair dice. As we know from Chapter 4, the chance of rolling a 7 is $\frac{1}{6}$: $P(Y = 7) = \frac{1}{6}$. Suppose we actually roll the dice and obtain a total of 3; if in the equation $P(Y = 7) = \frac{1}{6}$ we replace Y by 3, we get $P(3 = 7) = \frac{1}{6}$, which is nonsense (the probability of $3 = 7$ is 0, not $\frac{1}{6}$). Or suppose we happen to roll a 7; if in the equation $P(Y = 7) = \frac{1}{6}$ we replace Y by 7, we get $P(7 = 7) = \frac{1}{6}$, again nonsense (the probability of $7 = 7$ is 1, not $\frac{1}{6}$).* In passing from Equation (7.8), which is true, to Equation (7.9), which is false, we have made the same error—that of replacing inside a probability statement the random variable $\bar{X}$ by the specific value 1.58 that it happened to assume when the experiment was carried out.

For Example 1 above,

$$L = \bar{X} - .8 \quad \text{and} \quad R = \bar{X} + .8$$

are 95% *confidence limits* for μ, and the interval bounded by L and R is a 95% *confidence interval* for μ. Now L and R are random variables, and the interval they determine is a random interval. What is the chance that the confidence limits surround μ, or that the confidence interval includes μ?

Now $L \leq \mu \leq R$ if $\bar{X} - .8 \leq \mu \leq \bar{X} + .8$, and this is the same as $|\bar{X} - \mu| \leq .8$ or $-.8 \leq \bar{X} - \mu \leq .8$. Since $(\bar{X} - \mu)/.408$ has a standard normal distribution, whatever μ may be,

$$P(L \leq \mu \leq R) = P(-.8 \leq \bar{X} - \mu \leq .8)$$

$$= P\left(-\frac{.8}{.408} \leq \frac{\bar{X} - \mu}{.408} \leq \frac{.8}{.408}\right)$$

$$= P\left(-1.96 \leq \frac{\bar{X} - \mu}{.408} \leq 1.96\right) = .95$$

---

*It is perhaps illuminating to observe also that in the expected value $E(Y)$ the random variable Y cannot be replaced by a numerical value. In fact, this expected value is 7: $E(Y) = 7$. If we roll a 3, substitution yields the equation $E(3) = 7$, nonsense once more. The illegitimacy of these substitutions becomes clearer if we keep in mind that Y is a name for the total number of dots

SECTION 7.3 CONFIDENCE INTERVALS FOR NORMAL MEANS: KNOWN VARIANCE

Figure 7.2

Standard normal curve

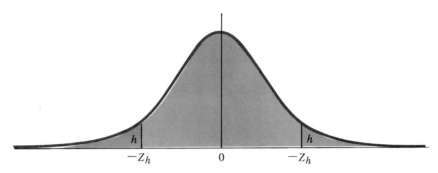

Thus the variables $L = \bar{X} - .8$ and $R = \bar{X} + .8$ have the property that the probability of $L \leq \mu \leq R$ is *always* .95 no matter *what* value the unknown μ may happen to have.

Since the limits L and R have a 95% chance of enclosing the true μ between them, whatever μ may be, they are called 95% confidence limits; 95% is called the **confidence level**. For the data in Table 7.1, L is .78 and R is 2.38, and we say we are 95% confident that $.78 \leq \mu \leq 2.38$. This is not to be interpreted as saying μ is random and has a 95% chance of lying between .78 and 2.38; it is really only a rephrasing of Equation (7.8). The confidence we have in the limits .78 and 2.38 really derives from our confidence in the *statistical procedure* that gave rise to them.* The procedure gives random variables L and R that have a 95% chance of enclosing the true μ; whether their specific values .78 and 2.38 actually enclose μ we have no way of knowing.

confidence level

To construct confidence intervals for the general case, we need some auxiliary concepts. If Z is a standard normal variable, the quantity Z_h, defined by the relationship

$$P(Z > Z_h) = h$$

upper percentage point

(see Figure 7.2), is the **upper percentage point** of the standard normal distribution corresponding to a probability of h. It is the point on the Z-scale such that the probability of a *larger* value is equal to h. The upper 2.5% point for the standard normal distribution is denoted by $Z_{.025}$ and is the

---

"before the dice are rolled." Recall a nonrandom analogue, the length l of the side of a square, discussed on page 155; substitution of a specific measured value of l (say, $l = 1.5$) can make nonsense of a valid general statement like "Sometimes l exceeds 2."

*This accords with at least one everyday use of the word *confidence*. What confidence a layman has in a physician's advice very likely derives from what confidence he has in the physician.

value on the Z-scale such that the area to the *right* of it is .025. By the table of normal areas (Table 2) we find that $Z_{.025}$ is equal to 1.96.

Because of the symmetry of the standard normal distribution, its lower percentage points are equal in absolute value to the corresponding upper ones but are negative. Thus $-Z_h$ is the point on the Z-scale such that the probability of a *smaller* value is h—the area to the *left* of $-Z_h$ is h. There is probability .025 that a standard normal variable is less than $-Z_{.025}$, or -1.96. In terms of percentage points, we have by definition,

$$P(-Z_{\alpha/2} \leq Z \leq Z_{\alpha/2}) = 1 - \alpha \qquad 7.10$$

(Here α is *alpha*, the Greek letter corresponding to *a*.) Note that we use the $\frac{1}{2}\alpha$ percentage points in order to have $100(1 - \alpha)\%$ of the area between them; each tail has area $\frac{1}{2}\alpha$, so the two together have area α.

Now consider a normal population with unknown mean μ and known variance σ^2. Since the standardized variable $(\bar{X} - \mu)/(\sigma/\sqrt{n})$ has the standard normal distribution, it follows that

$$P\left(-Z_{\alpha/2} \leq \frac{\bar{X} - \mu}{\sigma/\sqrt{n}} \leq Z_{\alpha/2}\right) = 1 - \alpha \qquad 7.11$$

Since multiplication by a positive number leaves inequalities undisturbed, the probability in Equation (7.11) is the probability that

$$-Z_{\alpha/2}\frac{\sigma}{\sqrt{n}} \leq \bar{X} - \mu \leq Z_{\alpha/2}\frac{\sigma}{\sqrt{n}}$$

Now these two inequalities are together the same thing as

$$|\bar{X} - \mu| \leq Z_{\alpha/2}\frac{\sigma}{\sqrt{n}}$$

and hence they are the same thing as

$$-Z_{\alpha/2}\frac{\sigma}{\sqrt{n}} \leq \mu - \bar{X} \leq Z_{\alpha/2}\frac{\sigma}{\sqrt{n}}$$

Adding $\bar{X}$ to all three members of this last expression gives

$$P\left(\bar{X} - Z_{\alpha/2}\frac{\sigma}{\sqrt{n}} \leq \mu \leq \bar{X} + Z_{\alpha/2}\frac{\sigma}{\sqrt{n}}\right) = 1 - \alpha \qquad 7.12$$

The equations (7.11) and (7.12) are the same thing because of the algebraic rules for manipulating inequalities. The $100(1 - \alpha)\%$ confidence interval is the interval bounded by the confidence limits

$$L = \bar{X} - Z_{\alpha/2}\frac{\sigma}{\sqrt{n}} \quad \text{and} \quad R = \bar{X} + Z_{\alpha/2}\frac{\sigma}{\sqrt{n}} \qquad 7.13$$

SECTION 7.3 CONFIDENCE INTERVALS FOR NORMAL MEANS: KNOWN VARIANCE

By Equation (7.12), there is probability $1 - \alpha$ that the confidence interval will contain the unknown μ, whatever μ may be.

In Example 1

$$1 - \alpha = .95, \quad \alpha = .05, \quad \tfrac{1}{2}\alpha = .025 \quad Z_{.025} = 1.96$$

Since $\sigma = \sqrt{1.664}$ and $n = 10$, and since the data give $\bar{X} = 1.58$,

$$L = \bar{X} - 1.96 \frac{\sqrt{1.664}}{\sqrt{10}} = 1.58 - .8 = .78$$

and

$$R = \bar{X} + 1.96 \frac{\sqrt{1.664}}{\sqrt{10}} = 1.58 + .8 = 2.38$$

Example 2

Problem: Lifetimes (in hours) of a certain kind of battery are known to be normally distributed, and it is known that the standard deviation σ of the lifetime is 10. In a random sample of size 25, the sample mean $\bar{X}$ was found to be 75. We are to calculate 90% confidence limits.

Solution: Here

$$1 - \alpha = .90, \quad \alpha = .10, \quad \tfrac{1}{2}\alpha = .05, \quad Z_{.05} = 1.645$$

By computation,

$$L = 75 - 1.645 \frac{10}{\sqrt{25}} = 75 - 3.29 = 71.71$$

$$R = 75 + 1.645 \frac{10}{\sqrt{25}} = 75 + 3.29 = 78.29$$

We are 90% confident that μ lies between 71.71 and 78.29, in that the procedure itself will give limits that enclose μ 90% of the time.

Example 3

Problem: For the data of Example 2 we are to calculate 95% confidence limits.

Solution: Here $Z_{\alpha/2}$ is 1.96, as in Example 1, and so

$$L = 75 - 1.96 \frac{10}{\sqrt{25}} = 75 - 3.92 = 71.08$$

$$R = 75 + 1.96 \frac{10}{\sqrt{25}} = 75 + 3.92 = 78.92$$

The length of the 95% confidence interval is 2 × 3.92, or 7.84; the length of the 90% confidence interval (Example 2) is 2 × 3.29, or 6.58. The 90% confidence interval is better in that it is shorter; it apparently gives greater precision. But of course we have less confidence in it (90% as opposed to 95%). Confidence has been traded for precision, and neither interval is really better than the other.

If the parent population is *normal*, it is possible to show that $\bar{X}$ is a minimum-variance unbiased estimator of μ. In this sense we may say that for a given sample size, $\bar{X}$ is the *best* estimator of μ.

To prove that $\bar{X}$ is best or optimal in the sense of having the smallest possible variance lies outside the purpose of this book (see Reference 7.2 for a proof). It is simple and natural to estimate μ by $\bar{X}$, and we do know that $\bar{X}$ is unbiased and has variance σ^2/n—even if we have not proved this variance to be minimal.

Problems

7. Assuming random samples from normal populations with known variance, find confidence intervals for the means that have the specified degree of confidence.
   a. $n = 9$, $\bar{X} = 4$, $\sigma^2 = 16$, confidence = 90%.
   b. $n = 100$, $\bar{X} = 29$, $\sigma^2 = 49$, confidence = 95%.
   c. $n = 64$, $\bar{X} = 105$, $\sigma^2 = 100$, confidence = 99%.

8. Assuming samples from normal populations with known variance find
   a. The degree of confidence used if $n = 16$, $\sigma = 8$, and the total width of a confidence interval for the mean is 3.29 units.
   b. The sample size when $\sigma^2 = 100$ and the 95% confidence interval for the mean is from 17.2 to 22.8.
   c. The known variance when $n = 100$ and the 98% confidence interval for the mean is 23.26 units in width.

9. What happens to the width of the confidence interval for the mean of a normal population when the sample size is doubled? Quadrupled?

10. The length of a bolt made by a machine parts company is a normal

SECTION 7.4 THE t-DISTRIBUTION

random variable with a standard deviation σ of .01 mm. The lengths of four randomly selected bolts are: 20.01, 19.98, 20.00, 19.99.
  a. Make a 95% confidence interval for the mean.
  b. Specifications require a mean length μ of 20.00 mm for the population of bolts. Do the data indicate that the bolts meet the specification?

11. The weight of trout grown and sold by a commercial hatchery is a normal random variable with a standard deviation .25 lb. A sample of 10 such trout produces a sample mean $\bar{X}$ of 2.10 lb. Make a 95% confidence interval for the mean weight of the population of trout.

12. A machine to measure the "bounce" of a ball is used on 45 randomly selected tennis balls. Experience has shown that the standard deviation σ of the "bounce" is .30. If $\bar{X} = 1.70$, make a 90% confidence interval for the mean "bounce" of the population of such tennis balls.

7.4 The t-Distribution

The interval estimates of the last section, where σ^2 was assumed known, were based on the fact that in normal sampling the standardized mean

$$\frac{\bar{X} - \mu}{\sigma/\sqrt{n}} \qquad 7.14$$

has a distribution that does not depend on μ and σ^2, namely, the standard normal distribution. In trying to construct a confidence interval for the case of unknown variance, we are led to ask what happens if in the expression (7.14) we merely replace σ by its estimator s, the sample standard deviation. We then have the statistic

$$t = \frac{\bar{X} - \mu}{s/\sqrt{n}} \qquad 7.15$$

and happily it turns out that this statistic has a distribution that does not depend on μ and σ^2 (although it does depend on n). The denominator in (7.14) is the standard deviation of $\bar{X}$; the denominator in (7.15) is the standard error.

t-statistics
t-distributions
degrees of freedom

The expression (7.15) is the first of a number of **t-statistics** we will encounter. The **t-distributions** form a family of distributions dependent on a parameter called the (number of) **degrees of freedom**. (The origin of this phrase needn't concern us.) For the *t*-variable (7.15) the degrees of freedom are $n - 1$, where n is the sample size.

Figure 7.3

The *t*-distribution with 5 degrees of freedom and the standard normal distribution

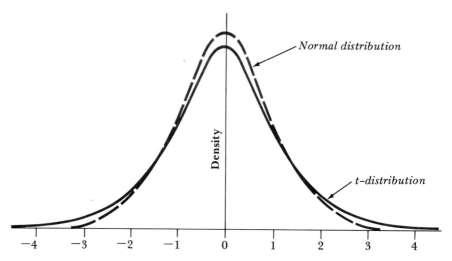

The *t*-distribution is a symmetric distribution with mean zero. Its graph is similar to that of the standard normal distribution, as Figure 7.3 shows. There is more area in the tails of the *t*-distribution, and the standard normal distribution is higher in the middle. The larger the number of degrees of freedom, the more closely the *t*-distribution resembles the standard normal distribution. As the number of degrees of freedom increases without limit, the *t*-distribution approaches the standard normal distribution, and it is convenient to regard the standard normal distribution as a *t*-distribution with an infinite number of degrees of freedom.

Table 4 of the Appendix gives percentage points of the *t*-distribution—points on the *t*-scale such that the probability of a larger *t* is equal to a specified value. The percentage point t_h is defined as that point at which

$$P(t > t_h) = h$$

Since the distribution is symmetric about zero, only positive *t*-values are tabulated. The lower *h* percentage point is $-t_h$, because

$$P(t < -t_h) = P(t > t_h) = h$$

In general, we denote a percentage point for *t* by $t_{h,d}$, where *h* is the probability level and *d* is the degrees of freedom.

To find a percentage point in Table 4, we locate the row of the table that corresponds to the given degrees of freedom *d* and then take the value in that row that is also in the column headed by the given probability level *h*. For example, $t_{.01, 15}$ is the .01 percentage point of *t* with 15 degrees of

SECTION 7.5 CONFIDENCE INTERVALS FOR NORMAL MEANS: UNKNOWN VARIANCE

freedom. Using Table 4 we find that it is equal to 2.602. If 15 is the degrees of freedom for t, then $P(t > 2.602) = .01$. The bottom line of the table corresponds to an infinite number of degrees of freedom, that is, to the standard normal distribution. Thus, the .05 percentage point for the standard normal, $Z_{.05}$, is $t_{.05, \infty} = 1.645$, and the .025 percentage point, $Z_{.025}$, is $t_{.025, \infty} = 1.96$.

Problems

13. Find the value of t for which
    a. The probability of a larger value is .025 when d (the degrees of freedom) is 20.
    b. The probability of a smaller value is .995 when d is 25.
    c. The probability of a larger value, sign ignored, is .10 when $d = 60$.
14. For random samples from normal populations the quantity
$$t = \frac{\bar{X} - \mu}{s/\sqrt{n}}$$
has a t-distribution with $n - 1$ degrees of freedom. In each of the following find the indicated quantity.
    a. $n = 25$, $\bar{X} - \mu = 3$, $s = 2$. Find t.
    b. $t = 2$, $n = 16$, $\bar{X} - \mu = 8$. Find s^2.
    c. $n = 25$, $s = 20$, probability of a larger t is .05. Find $\bar{X} - \mu$.
    d. $n = 16$, $\bar{X} - \mu = 10$, probability of a smaller t is .90. Find s.
15. Given that t_d has a t-distribution with d degrees of freedom, find a C such that
    a. $P(t_{13} \geq C) = .025$.
    b. $P(t_{22} < C) = .90$.
    c. $P(-C < t_{30} < C) = .99$.
    d. $P(0 < t_{15} < C) = .495$.

7.5 Confidence Intervals for Normal Means: Unknown Variance

By the definition of percentage points,
$$P(-t_{\alpha/2, d} \leq t \leq t_{\alpha/2, d}) = 1 - \alpha$$
if t has a t-distribution with d degrees of freedom. The fact underlying the construction of confidence intervals in the case where σ^2 is unknown is that the statistic (7.15) has a t-distribution with $n - 1$ degrees of freedom,

where n is the sample size and the population is assumed normal. Therefore,

$$P\left(-t_{\alpha/2,\,n-1} \le \frac{\bar{X}-\mu}{s/\sqrt{n}} \le t_{\alpha/2,\,n-1}\right) = 1 - \alpha \qquad 7.16$$

If we multiply each member of the expression within the parentheses by $s/\sqrt{n}$, subtract $\bar{X}$ from each, and then multiply through by -1, we arrive at

$$P\left(\bar{X} - t_{\alpha/2,\,n-1}\frac{s}{\sqrt{n}} \le \mu \le \bar{X} + t_{\alpha/2,\,n-1}\frac{s}{\sqrt{n}}\right) = 1 - \alpha \qquad 7.17$$

On the basis of this equation, we define the $100(1-\alpha)\%$ confidence interval as the interval bounded by the limits

$$L = \bar{X} - t_{\alpha/2,\,n-1}\frac{s}{\sqrt{n}} \quad \text{and} \quad R = \bar{X} + t_{\alpha/2,\,n-1}\frac{s}{\sqrt{n}} \qquad 7.18$$

The point is that whereas the confidence limits (7.13) involve the population standard deviation σ, assumed known in Section 7.3 but not here, the confidence limits (7.18) instead involve the sample standard deviation s, an estimate of σ. Thus these limits can be computed from the sample alone; we need not know σ. By Equation (7.17) the probability that the confidence interval includes μ—the probability that the confidence limits L and R surround μ—is just $1 - \alpha$. The interpretation of these limits is just as for those in Section 7.3.

Example 1

Let us return to the data of Example 1 in Section 7.3, but this time without the unrealistic assumption that we know the population variance. The sample variance s^2 for the data in Table 7.1 is 1.513, so the sample standard deviation s is $\sqrt{1.513}$, or 1.230, and $s/\sqrt{n}$ is $s/\sqrt{10}$, or .389. The degrees of freedom $n - 1$ is 9. If we are to compute 95% confidence limits ($\alpha = .05$, $\alpha/2 = .025$), the appropriate percentage point is $t_{.025,\,9}$, or 2.262. Since $\bar{X}$ is equal to 1.58, the 95% confidence limits are

$$L = \bar{X} - 2.262\frac{s}{\sqrt{10}}$$
$$= 1.58 - 2.262 \times .389$$
$$= 1.58 - .88 = .70$$

and

$$R = \bar{X} + 2.262\frac{s}{\sqrt{10}}$$
$$= 1.58 + 2.262 \times .389$$
$$= 1.58 + .88 = 2.46$$

SECTION 7.5 CONFIDENCE INTERVALS FOR NORMAL MEANS: UNKNOWN VARIANCE

We are 95% confident that μ lies between .70 and 2.46, in the sense that the random variables L and R (which for our data take values .70 and 2.46) have probability .95 of enclosing μ between them, whatever μ may be.*

The sample mean $\bar{X}$ has standard deviation $\sigma/\sqrt{n}$, and the estimate for this is the standard error $s/\sqrt{n}$ (see p. 232). Note that the standard error appears in the formulas (7.18). It is a common mistake to use s instead of $s/\sqrt{n}$—that is, to calculate $\bar{X} \pm t_{\alpha/2, n-1}s$ instead of $\bar{X} \pm t_{\alpha/2, n-1}s/\sqrt{n}$. It should be remembered that the standard error is the basic unit of measurement.

Example 2

Return to the Public Health Service hypertension study discussed in Chapters 1 and 2. For 171 patients on the active blood-pressure-reducing drug, Table 2.1 (page 13) records the actual change in blood pressure (blood pressure after 12 months on the drug minus baseline blood pressure), a negative entry representing a decrease. For these 171 numbers, the mean $\bar{X}$ is -9.85 and the sample standard deviation s is 11.55. The standard error, $s/\sqrt{n}$, is $11.55/\sqrt{171}$, which is .88. In this case the degrees of freedom, $n - 1$, is 170. This puts us down near the last line of Table 4—the line labeled ∞. In other words, with a sample size n as large as this one, $(\bar{X} - \mu)/(s/\sqrt{n})$ is almost normally distributed, and we can use the percentage points for the normal distribution, which are given in the last line of Table 4. Since we are to find the 95% confidence limits, $\alpha = .05$, $\alpha/2 = .025$, and $Z_{\alpha/2} = 1.96$. Therefore the confidence limits are (since $s/\sqrt{n} = .88$)

$$L = -9.85 - 1.96 \times .88 = -9.85 - 1.72 = -11.57$$

and

$$R = -9.85 + 1.96 \times .88 = -9.85 + 1.72 = -8.13$$

If in the future all mild hypertensives similar to the ones in the PHS study were put on the blood-pressure-reducing drug, the mean change in their blood pressure would be μ, the parameter being estimated. We are 95% confident that μ lies between -11.57 and -8.13. To put it the other way around, we are 95% confident that in this population the mean *reduction* in blood pressure would be between 8.13 and 11.57.

---

*The t-distribution was discovered in 1908 by W. S. Gosset, who wrote under the name *Student*. The data in the example are his; see Reference 7.1.

Example 3

Now let us turn to Table 2.2 (page 14), which records the changes in blood pressure for the 176 patients given the placebo instead of the active drug. For these data $\bar{X}$ is .24 and s is 11.13. The degrees of freedom are $n - 1 = 175$, which is again so large that we use the percentage points for the normal distribution (the last line of Table 4). The standard error is $s/\sqrt{n} = 11.13/\sqrt{176} = .84$, and the 95% confidence limits are

$$L = .24 - 1.96 \times .84 = .24 - 1.65 = -1.41$$

and

$$R = .24 + 1.96 \times .84 = .24 + 1.65 = +1.89$$

If in the future all mild hypertensives similar to the ones in the PHS study were given a placebo, we are 95% confident that the mean change in their blood pressure would be between -1.41 and $+1.89$. In other words, there would not be much change at all.

In Examples 2 and 3 the sample means are -9.85 and .24, which is good evidence that the drug is effective. But this can be evaluated more precisely only by a consideration of the two standard errors, .88 and .84. If the standard errors had been around 20, for example, then the contrast in the sample means (-9.85 versus .24) would *not* be impressive. This shows the importance of standard error in assessing the evidence in a sample mean. The question of variability must not be overlooked.

The data in Examples 2 and 3 are good evidence that the drug is effective because the two 95% confidence intervals—-11.57 to -8.13 in Example 2, and -1.41 to $+1.89$ in Example 3—are far removed from one another. In Chapter 9 we will study in more detail how to compare the means of two different samples.

Problems

16. Suppose the values of σ^2 given in Problem 7 are actually values of s^2. Find the confidence intervals using the *t*-distribution, and compare them with the results in Problem 7.
17. Assuming unknown variances, find 95% confidence intervals for the means based on the sample results given in Problem 1.
18. Assuming unknown variances, and using the data given in Problem 4, find

SECTION 7.5 CONFIDENCE INTERVALS FOR NORMAL MEANS: UNKNOWN VARIANCE **247**

    a. 95% confidence intervals for the means for parts a and b.
    b. 99% confidence interval for the mean for part c.
    c. 98% confidence interval for the mean for part d.

19. The mean and standard deviation for the life of a random sample of 25 light bulbs were found to be 1980 and 150 hours, respectively. Find a 99% confidence interval for the lifetime of such bulbs.

20. A sample of nine plots had a mean yield of 100 grams, and the estimated population standard deviation was 15 grams. Find a 98% confidence interval for the mean yield.

21. In a sentencing study of 25 prisoners convicted of the same crime, the mean sentence was found to be 132.3 months with a standard deviation of 34.7 months. Find a 95% confidence interval for the sentence length for this offense.

22. A car owner wants to know the mean weekly mileage he puts on his car. He records his mileage for 52 weeks and finds a mean of 176 miles per week, with a standard deviation of 69. Construct a 95% confidence interval for the mean weekly mileage.

23. Nine samples of a solution were analyzed for copper concentration in grams per liter, giving $\bar{X} = 12.37$ and $s^2 = .0256$. Find a 95% confidence interval for the true unknown concentration.

24. An important physical constant is the ratio e/m of the charge of the electron to its mass. The following experimental values for e/m were obtained:

1.7604×10^7	1.7638×10^7	1.7609×10^7
1.7563×10^7	1.7556×10^7	1.7582×10^7
1.7525×10^7	1.7663×10^7	1.7624×10^7
1.7620×10^7	1.7605×10^7	1.7621×10^7

    a. From these data what inferences can we make in terms of point estimates?
    b. Find the 99% confidence interval for e/m.

25. A stimulus was tested for its effect on blood pressure. Twenty men had their blood pressure measured before and after the stimulus. The results are given below. Find a 99% confidence interval for the mean change in blood pressure.

8,	7,	1,	9,	−8,	−3,	1,	2,	−8,	2,
3,	8,	1,	7,	−5,	−4,	0,	7,	−1,	5

26. A product comes in cans labeled 38 ounces. A sample of six cans had these weights:

    34.60, 39.65, 34.75, 40.00, 39.50, 34.25

    Give a 98% confidence interval for the mean weight per can.

27. A random sample of 30 scores on an achievement test given to high school students has a mean of 423 and a standard deviation of 68. Find a 95% confidence interval for the population mean of the test.

28. In five consecutive time trials a skier is found to have a mean time of 69.23 seconds and a standard deviation of 1.38 seconds. Give a 99% confidence interval for the mean time of the skier on this run.

29. In order to estimate the amounts owed, a city clerk takes a random sample of 16 files from the file of all delinquent accounts and finds the mean amount owed to be $230.28 with a standard deviation of $40.23. Find a 99% confidence interval for the mean delinquent account.

30. Use the data of Problem 34, Chapter 3, to find a 99% confidence interval for the mean spring constant of this type of wood beam.

31. A random sample of 60 patients of the Douglass Allergy Clinic gave the following frequency distribution of the ages of the patients. Make a 95% confidence interval for the mean age of the population of such patients.

Class	Relative Frequency
10 but less than 20	.10
20 but less than 30	.30
30 but less than 40	.40
40 but less than 50	.20

32. A random sample of 24 college-age males produced the following data on their weights. Make a 95% confidence interval for the mean weight of the population of college-age males.

155	140	155	153	150	180
155	145	163	150	160	140
170	180	175	150	180	160
145	190	228	150	165	145

SECTION 7.6 SAMPLE SIZE

33. The following distribution is that for the lengths of mayfly larvae. Find a 95% confidence interval for the mean length.

Length X (mm)	Number of Larvae
$11 \leq X < 15$	2
$15 \leq X < 17$	4
$17 \leq X < 19$	7
$19 \leq X < 21$	9
$21 \leq X < 23$	6
$23 \leq X < 25$	15
$25 \leq X < 27$	12
$27 \leq X < 29$	5
$29 \leq X < 33$	4

34. A random sample of 60 stores showed that the mean price of milk was 77.3 cents, with a standard deviation of 4.2 cents. Find the 95% confidence interval for the mean price.

35. For a sample of 100 tires the mean lifetime was 43,500 miles, with a standard deviation of 5,500 miles. Find the 99% confidence interval for the mean lifetime of tires of this type.

36. Because of variability in the components of candies, the number of calories in a one-pound box cannot be kept within narrow limits. A sample of 500 boxes gave a mean of 473 calories per box, with a standard deviation of 84. Find the 99.9% confidence interval for the mean number of calories per box.

7.6 Sample Size

For the case of a known population variance σ^2, the confidence interval given by the limits (7.13) has width

$$w = 2Z_{\alpha/2} \frac{\sigma}{\sqrt{n}} \qquad 7.19$$

If σ^2 is unknown, so that we must use the limits (7.18), the confidence interval has width

$$w = 2t_{\alpha/2,\, n-1} \frac{s}{\sqrt{n}} \qquad 7.20$$

Commonly we would like to know, in advance of sampling, how large a sample is required to estimate the mean with specified precision, as measured by the confidence level together with the width of the confidence interval.

Example 1

Problem: Suppose, as in Example 2 in Section 7.3, that we are investigating the lifetimes of batteries of a certain design, and that we know that σ is 10. Suppose further that we want a 95% confidence interval of width 5; in other words, we want a 95% chance that our estimate $\bar{X}$ will be within 2.5 of the true μ. What sample size is required?

Solution: We have

$$w = 5, \quad \sigma = 10, \quad Z_{.025} = 1.96$$

and so by Equation (7.19)

$$5 = 2 \times 1.96 \frac{10}{\sqrt{n}}$$

Thus $\sqrt{n}$ must be 7.8, so n must be 60.8—or actually 61.

Example 2

If we have σ equal to 10 as before, and still want a w of 5, but require a 99% confidence interval, then

$$w = 5, \quad \sigma = 10, \quad Z_{.005} = 2.576$$

so Equation (7.19) gives

$$5 = 2 \times 2.576 \frac{10}{\sqrt{n}}$$

This time $\sqrt{n}$ is 10.3, so n must be 106. *For a fixed width, greater confidence requires larger sample size.*

Example 3

Problem: Suppose σ is 10 and we want a 95% confidence interval, as in Example 1, but suppose we require a w of 3.

Solution: Here

$$w = 3, \quad \sigma = 10, \quad Z_{.025} = 1.96$$

so

$$3 = 2 \times 1.96 \frac{10}{\sqrt{n}}$$

SECTION 7.6 SAMPLE SIZE

which implies that $\sqrt{n}$ is 13.1; n must be 172. *For a fixed level of confidence, attaining a smaller width w requires a larger sample*.

If the variance is not known, Equation (7.20) applies. In this case the width w depends on the sample itself because of the factor s, and the situation is more complicated. To find the n needed for a given precision requires some advance notion of the size of σ, either from past experience or from a preliminary sample. Using Equation (7.19) in place of (7.20) will give a very rough idea of what n must be. See Reference 7.4.

Problems

37. If a normal population has a known standard deviation σ of .75, how large a sample must be taken in order that the total width of the 95% confidence interval for the population mean will not be greater than .8?

38. A statistician wants to determine the mean hourly earnings (in cents) for employees in a given occupation. She wants a 95% confidence interval of length 10. Because of other similar studies she is willing to assume a normal distribution with a variance of 650. What sample size should she plan to use?

39. If a normal population is known to have σ equal to 5, how large a sample should we take in order to be 95% confident that the sample mean will not differ from the population mean by more than one unit?

40. Suppose the diameters of ball bearings made by a given process are known to be normally distributed with a known standard deviation σ of .5. How large a sample must we take to be 95% confident that our estimate for the mean will not differ from the true mean diameter by more than .01?

41. A consumer agency wants to estimate the mean weight of a product whose package is marked 8 oz. Although the mean and variance are unknown, it is guessed that 95% of all such packages actually weigh between $6\frac{1}{2}$ and 8 oz. How large a sample must be taken to estimate the mean weight to within $\frac{1}{8}$ oz. with probability .99? Assume a normal distribution for the actual package weight.

42. An advertising firm in Peoria is going to conduct a survey to estimate the average amount of time an adult in that city watches television during the weekend. If viewing time is a normal random variable with

a standard deviation of 2 hours, how large a sample must be taken to estimate the mean to within 15 minutes with probability .99?

7.7 Estimating Binomial p

In Section 5.8 we defined a binomial population as a population whose elements are classified as belonging to one of two classes conventionally labelled success and failure. We represent the proportion of the population that belongs to the first class by p, and the proportion that belongs to the second class by $1 - p$. In general, p is unknown and the problem is to estimate it.

Suppose that we have at hand a random sample of size n; let X be the number of successes in the sample, and let f be the **proportion of successes** (or fraction of successes), so that $f = X/n$. Now f is the natural estimator of the unknown p. (For this reason, the proportion of successes is sometimes denoted by $\hat{p}$ instead of by f.) As stated in Section 5.8, X has mean np and variance $np(1 - p)$, and f has mean, variance, and standard deviation as follows:

$$E(f) = p$$
$$\text{Var}(f) = p(1 - p)/n$$
$$\text{St Dev}(f) = \sqrt{p(1 - p)/n}$$

proportion of successes

The first equation shows that f is an unbiased estimator of p. The second gives the variance of this estimator. Its value, of course, cannot be found without knowledge of the value of the parameter p to be estimated. However, this variance is largest (for fixed n) when p is $\frac{1}{2}$, so that

$$\text{Var}(f) \leq \frac{1}{4n}$$

This gives a conservative idea of the variance. Or we can get an idea of the variance by replacing p with its estimator f—the variance is approximated by $f(1 - f)/n$.

We now turn our attention to the problem of finding interval estimates for the proportion p. The normal approximation to the binomial provides a method of finding approximate confidence limits for large sample sizes. A knowledge of the method is of practical importance and also deepens our understanding of binomial estimation.

As stated in Section 5.9, the standardized fraction of successes

$$\frac{f - p}{\sqrt{p(1 - p)/n}}$$

7.21

SECTION 7.7 ESTIMATING BINOMIAL p

has for large n approximately a standard normal distribution; see Equation (5.19), page 197, and the example following it. If in the expression (7.21) we replace each p in the denominator by its estimator f, we get a ratio

$$\frac{f - p}{\sqrt{f(1 - f)/n}} \qquad 7.22$$

which turns out also to have approximately a normal distribution for large n.

By the definition of the percentage points $Z_{\alpha/2}$ for the standard normal distribution, we therefore have the approximation

$$P\left(-Z_{\alpha/2} \leq \frac{f - p}{\sqrt{f(1 - f)/n}} \leq Z_{\alpha/2}\right) \cong 1 - \alpha \qquad 7.23$$

If we operate on the inequalities here just as we did in constructing the confidence intervals in Sections 7.3 and 7.5, we arrive at

$$P\left(f - Z_{\alpha/2}\sqrt{\frac{f(1 - f)}{n}} \leq p \leq f + Z_{\alpha/2}\sqrt{\frac{f(1 - f)}{n}}\right) \cong 1 - \alpha$$

Therefore, the random variables

$$L = f - Z_{\alpha/2}\sqrt{\frac{f(1 - f)}{n}} \quad \text{and} \quad R = f + Z_{\alpha/2}\sqrt{\frac{f(1 - f)}{n}}$$

have approximate probability $1 - \alpha$ of containing between them the true value of p, whatever it may be; they therefore can be used as approximate $100(1 - \alpha)\%$ confidence limits if n is larger than 20 or so.

Example 1

Problem: In a town a sample of 100 voters contained 64 persons who favored a bond issue. Between what limits can we be 95% confident that the proportion of voters in the community who favor the issue is contained?

Solution: Here

$$n = 100, \quad X = 64, \quad f = .64, \quad Z_{.025} = 1.96$$

Hence

$$\sqrt{\frac{f(1 - f)}{n}} = \sqrt{\frac{.64 \times .36}{100}} = .048$$

and

$$L = .64 - 1.96 \times .048 = .64 - .094 = .546$$
$$R = .64 + 1.96 \times .048 = .64 + .094 = .734$$

We are 95% confident the true proportion is covered by the interval from .55 to .73.

There is a common misconception about sampling that can be understood in terms of this example. The sample is of size 100. If the voting population of the town is 10,000, then the sample constitutes 1% of the population. If the voting population of the town is 100,000, then the sample constitutes only 0.1% of the population. In each case the confidence limits are .546 and .734, and the width of the confidence interval is .188 (that is, .734 − .546).

This seems paradoxical. In one case the sample is 1% of the population, in the other case only 0.1%. How can there be the same precision (as measured by the width of the 95% confidence interval) in the two cases? The fact is, the precision depends on the variance of f, which in turn depends on the *sample* size. People sometimes doubt the possibility of making accurate estimates on the basis of a tiny fraction of a population. But the estimates can be accurate if the tiny fraction constitutes a *random sample* and if its *absolute size* (not its size relative to that of the population) is large enough.

Example 2

According to Table 1.2 (page 3), there were 193 mildly hypertensive patients on the active blood-pressure-reducing drug in the PHS study, and of these, 37 suffered hypertensive events—that is, damage related to high blood pressure (enlargement of the heart, etc.). We want 95% confidence limits for the proportion p of future drug-treated mild hypertensives who will experience hypertensive events. Here

$$n = 193, \quad X = 37, \quad f = .19, \quad Z_{.025} = 1.96$$

and

$$\sqrt{\frac{f(1-f)}{n}} = \sqrt{\frac{.19 \times .81}{193}} = .028$$

The 95% confidence limits are

$$L = .19 - 1.96 \times .028 = .19 - .05 = .14$$

and

$$R = .19 + 1.96 \times .028 = .19 + .05 = .24.$$

If p is the probability that a mild hypertensive will suffer this damage if he is put on the drug, then we are 95% confident that p lies between .14 and .24.

SECTION 7.7 ESTIMATING BINOMIAL p

Example 3

Table 1.2 also shows that among the 196 mild hypertensives put on the placebo, 89 experienced hypertensive events. If we set the confidence level at 95%, then

$$n = 196, \quad X = 89, \quad f = .45, \quad Z_{.025} = 1.96$$

and

$$\sqrt{\frac{f(1-f)}{n}} = \sqrt{\frac{.45 \times .55}{196}} = 0.36$$

The 95% confidence limits are

$$L = .45 - 1.96 \times .036 = .45 - .07 = .38$$

and

$$R = .45 + 1.96 \times .036 = .45 + .07 = .52$$

If p is the probability that a mild hypertensive on placebo has a hypertensive event, then we are 95% confident that p lies between .38 and .52.

A comparison of the confidence intervals in Examples 2 and 3 shows that the drug is indeed effective in preventing damage related to high blood pressure. In Chapter 9 we study in more detail how to compare the observed proportions in two binomial samples.

OMIT

The calculation of "exact" confidence intervals based on the binomial probabilities is complex; therefore, tables and graphs of these intervals have been constructed for various confidence probabilities. Table 5 of the Appendix gives the 95% confidence intervals for p for a number of sample sizes from 10 to 1000. If our sample size is one of those tabulated and is less than 100, we find the number observed in the leftmost column and then move horizontally into the column corresponding to our sample size. The two numbers we find there are the endpoints of the interval and may be expressed as either percentages or relative frequencies. For example, suppose I take a sample of 30 students and find that eight of them do not smoke. In the column for $n = 30$ I find the figures 12 and 46 opposite the observed number 8. I may then state that, on the basis of this sample, I am 95% confident that the percentage of students who do not smoke is between 12% and 46%. Alternatively, the confidence interval for the relative frequency of nonsmoking students is .12 to .46.

For sample sizes greater than or equal to 100, we use the column for observed relative frequency rather than that for the observed number of successes. If $n = 250$ and $X = 75$, then the observed fraction is 75/250, or .3. We locate .3 in the "Fraction Observed" column, and in the column for values of $n = 250$ we find the corresponding pair of values to be 24 and 36. This means our 95% confidence interval for p is from .24 to .36.

If the observed fraction f is greater than .50, we use $1 - f$ and subtract each limit from 100. For example, out of 1000 heads of families questioned, it was found that 930 owned one or more television receivers. The observed fraction, .93, is greater than .50. We subtract .93 from 1.00 to get .07. The corresponding figures in the column for $n = 1000$ are 6 and 9. We subtract these from 100 to get 94 and 91, respectively. The 95% confidence interval for the percentage of families that own TV sets is 91% to 94%.

Interpolation may be used in this table and is straightforward for the three largest sample sizes. For example, if $n = 500$ and $X = 150$, then the observed fraction f is $\frac{150}{500}$, or .30. For $n = 250$ the confidence limits would be .24 and .36. For $n = 1000$ the limits would be .27 and .33. The proportion we want is

$$\frac{500 - 250}{1000 - 250} = \frac{250}{750} = \frac{1}{3}$$

Therefore, the lower limit is

$$.24 + \tfrac{1}{3}(.27 - .24) = .25$$

and the upper limit is

$$.36 + \tfrac{1}{3}(.33 - .36) = .35$$

For values of n from 10 to 50, inclusive, the interpolation is not quite as simple, for we must calculate for each sample size the observed number that corresponds to a given observed fraction. Let $n = 25$ and $X = 15$. Then $f = 15/25$, or .60. For $n = 20$ the observed number corresponding to $f = .60$ is $20(.60)$, or 12, and the confidence limits are .36 and .81. For $n = 30$ the value of X used to enter the table is $30(.60)$, or 18, and the confidence limits are .40 and .77. Since our sample size, 25, is midway between $n = 20$ and $n = 30$, our confidence limits are

$$.36 + \tfrac{1}{2}(.40 - .36) = .38$$

and

$$.81 + \tfrac{1}{2}(.77 - .81) = .79$$

In Example 1, the approximate 95% confidence limits (from the normal approximation) were .55 and .73. The 95% confidence limits from Table 5 are .54 and .73, which shows that the normal approximation works very well.

Problems

43. Given the following results from random samples from binomial populations, find the best estimates for p, the variance and standard devia-

SECTION 7.7 ESTIMATING BINOMIAL p   **257**

tion of the number X of successes, and the variance and standard deviation of the observed fraction, $f = X/n$.
  a. $n = 25$, $X = 5$.   b. $n = 48$, $X = 12$.
  c. $n = 100$, $X = 80$.   d. $n = 300$, $X = 160$.

44. In a sample of 250 cars tested, 25 were rejected by a pollution-control test. What are the best estimates for
  a. The proportion of polluting cars?
  b. The standard deviation of the estimator?

45. Use the table of binomial confidence intervals to find 95% confidence intervals for p when
  a. $n = 20$, $X = 8$.   b. $n = 15$, $X = 0$.
  c. $n = 100$, $X = 76$.   d. $n = 250$, $X = 50$.
  e. $n = 1000$, $X = 500$.   f. $n = 1000$, $X = 700$.

46. Use the table of binomial confidence intervals to find 95% confidence intervals for p when
  a. $n = 15$, $X = 12$.   b. $n = 30$, $X = 30$.
  c. $n = 100$, $X = 87$.   d. $n = 250$, $X = 125$.
  e. $n = 1000$, $X = 120$.   f. $n = 250$, $X = 200$.

47. Use interpolation in the table of binomial confidence intervals to find 95% confidence intervals for p in each of the following situations:
  a. $n = 40$, $X = 20$.   b. $n = 25$, $X = 5$.
  c. $n = 150$, $X = 50$.   d. $n = 500$, $X = 100$.

48. Use interpolation in the table of binomial confidence intervals to find 95% confidence intervals for p when
  a. $n = 500$, $X = 125$.   b. $n = 150$, $X = 115$.
  c. $n = 12$, $X = 6$.   d. $n = 25$, $X = 15$.

49. In a random sample of 100 people in the labor force, 10 were found to be unemployed. Give a 95% confidence interval for the percentage of unemployment.

50. In a lake a sample of 1000 fish was obtained by use of nets. It was found that 290 were members of the bass family. Give a 95% confidence interval for the percentage of bass among fish in the lake.

51. In a random sample of 1000 adults in the general population, 60 reported never having had any legal problems. Find a 95% confidence interval for the percentage of adults who have never had a legal problem by using
  a. Table 5.
  b. The normal approximation.

52. If 5.5% of a sample of 600 items were found to be defective, what is the 99% confidence interval for the proportion defective in the lot from which the sample was taken? Use the normal approximation to the binomial.

53. One-fourth of 300 persons interviewed were opposed to a certain program. Calculate a 99% confidence interval for the fraction of the population who are in opposition to the program. Use the normal approximation.

54. In a random sample of 450 families it was found that in 87 cases at least one of the family members watched a certain TV program. Give a 95% confidence interval for the proportion of homes in which that program was seen. Use the normal approximation.

55. A poll of 1500 voters showed that 52% favor candidate A over candidate B in the primary elections. Give a 99% confidence interval for the proportion of voters that prefer candidate A.

56. A test to detect a certain type of cancer has been developed by a medical investigator. It is of interest to know the probability of a false negative indication which occurs when the test fails to show the presence of a cancer that is actually there. The test is given to 250 patients known to have cancer, and 5 tests fail to show its presence. Make a 95% confidence interval for the probability of a false negative indication using both Table 5 and the normal approximation.

57. A cosmetic company found that 180 of 1000 randomly selected women in New York City have seen the company's latest newspaper advertisement. Make a 95% confidence interval for the percentage of women in New York City who have seen this advertisement. Use Table 5 and also the normal approximation.

CHAPTER PROBLEMS

58. The upper and lower confidence bounds in (7.18) may be used to obtain an *approximate* $100(1 - \alpha)\%$ confidence interval for the mean of a nonnormal population. In practice, this confidence interval may be applied if the sample size is large enough for the central limit theorem to be effectively applied. Obtain 16 values of the geometric random variable with $p = \frac{1}{2}$ by tossing a coin until heads is observed, recording the number of tosses required, and repeating this experiment 16 times. (See Problem 22 in Chapter 6.) Make an approximate 95% confidence interval for μ. Does this interval contain the true mean of $\mu = 2$?

59. This is a problem for an entire class. Let each member obtain one or more confidence intervals as described in Problem 58, so that 100 such intervals are available for analysis. Find the percentage of these

intervals that actually contain the true mean $\mu = 2$. Compare this percentage to the confidence level of 95% used in constructing the 100 confidence intervals.

60. If the entries in Table 9 in the Appendix are treated as 5-place decimal numbers, then these values may be considered to be a random sample of size 2000 from the population of numbers that make up the interval $0 < x < 1$. Beginning with a column and row of your choice, read off 25 successive numbers from Table 9. This procedure simulates the selection of a random sample of size 25 from this population. Make an approximate 95% confidence interval for the mean of the population using your sample data. (See the discussion on confidence intervals for the means of nonnormal populations in Problem 58.) Does your confidence interval contain the population mean $\mu = .5$?

61. The maximum value that can be attained by the variance of the sample fraction, f, is $\frac{1}{4}n$; hence, the maximum width of a $100(1 - \alpha)\%$ confidence interval for p, using the normal approximation, is $Z_{\alpha/2}/\sqrt{n}$. How large a sample should we take from a binomial population in order to be about 95% confident that the sample fraction will not differ from p by more than .02?

62. The width of the approximate 95% confidence interval for the binomial parameter p is

$$2 \times 1.96 \sqrt{f(1 - f)/n}.$$

If it is possible to make a reasonably good guess of the value of f prior to the sample being taken, then an approximation for the value of n necessary to produce a confidence interval of width w may be determined by setting the expression above equal to w and solving for n. The maximum value of n will occur when $f = \frac{1}{2}$. Find the value of n when

a. $w = .05, f = .1$. b. $w = .05, f = .5$.
c. $w = .01, f = .05$. d. $w = .01, f = .20$.

63. A medical investigator wants to estimate the proportion of patients in the population who will benefit from a certain drug therapy. It is believed that this proportion is around .75. If he is to estimate the true proportion to within .03, how large a sample must he take? (See Problem 62.)

64. In selecting a random sample of size n from a finite population of size N, an approximate $100(1 - \alpha)\%$ confidence interval for the mean of the population is

$$\bar{X} \pm t_{\alpha/2,\, n-1} \frac{s}{\sqrt{n}} \sqrt{\frac{N - n}{N - 1}}.$$

See Section 6.2 for a discussion of the finite-population correction factor $(N - n)/(N - 1)$. Suppose that the data in Problem 34 of Chapter 2 are treated as a population of 106 numbers. Select 30 numbers at random from this population and make an approximate 95% confidence interval for the mean using the formula given above. Does your confidence interval contain μ?

Summary and Keywords

The **problem of estimation** (p. 230) in statistics is to estimate an unknown population parameter by means of a quantity that can be calculated from a sample from the population; such a quantity is an **estimator** (p. 231), and its numerical value for a given sample is an **estimate** (p. 231). An estimator is **unbiased** (p. 231) if its expected value is the parameter (unknown) being estimated. The sample mean and variance are unbiased estimators of the population mean and variance. The precision of the sample mean (as an estimate of the population mean) is measured by its standard deviation, which can be approximated from the sample by the **standard error** (p. 232). A widely used form of estimation is by **confidence intervals** (p. 236) constructed for given **confidence levels** (p. 237). In the case of known variance, confidence intervals for the mean of a normal population involve the standardized mean and the **percentage points** (p. 237) of the normal distribution. In the case of unknown variance, confidence intervals for the mean of a normal population are based on the **t-statistic** (p. 241) and the **t-distribution** (p. 241) with the appropriate **degrees of freedom** (p. 241). Confidence intervals for a binomial proportion can in an analogous way be constructed from the **proportion of successes** (p. 252) in a sample.

REFERENCES

7.1 Fisher, R. A. *Statistical Methods for Research Workers*. 13th ed. New York: Hafner, 1963.

7.2 Hogg, Robert V., and Craig, Allen T. *Introduction to Mathematical Statistics*. 3d ed. New York: Macmillan, 1970. Chapters 6, 7, and 8.

7.3 Mosteller, Frederick; Rourke, Robert E. K.; and Thomas, George B., Jr. *Probability with Statistical Applications*. 2d ed. Reading, Mass.: Addison-Wesley, 1970. Chapter 12.

7.4 Snedecor, George W., and Cochran, William G. *Statistical Methods*. 6th ed. Ames, Iowa: Iowa State University Press, 1967. Chapters 1, 2, and 3.

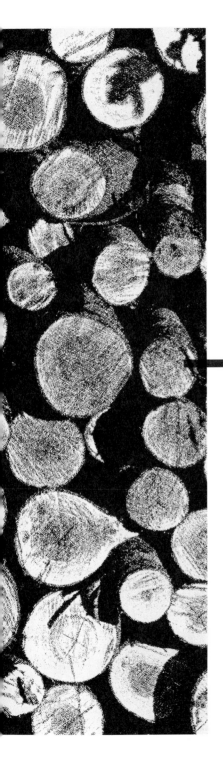

8

Tests of Hypotheses

8.1 Introduction

hypothesis-testing problem

Turning from estimation to the theory of testing hypotheses, let us consider an example that exhibits in concrete form the problems that arise in the general **hypothesis-testing problem**.

A *triangle test* provides one way of checking whether a candidate for a food technician's evaluation panel can distinguish subtle differences in taste. In a single trial of this test the subject is presented with three food samples, two of which are alike, and is asked to select the odd one. Except for the taste difference, the samples are as similar as possible, and the order of presentation is random. In the absence of any ability at all to distinguish tastes, the subject has a one-third chance of correctly distinguishing the odd food sample, and the question is whether he can do better than that.

To check the subject's ability, we present him with a series of triangle tests, and the order of presentation is randomized within each trial to eliminate any consistent bias and to make the trials independent of one another. Let p be the probability that the subject correctly identifies the odd food sample in a single trial. Then $p > \frac{1}{3}$ if the subject has an ability in taste discrimination, and $p = \frac{1}{3}$ if he has none. Since it is hard to see how he could do worse than chance, we rule out the possibility $p < \frac{1}{3}$ on a priori grounds.

To place the problem in a standard framework, we label two hypotheses, a **null hypothesis** and an **alternative hypothesis**:

null hypothesis
alternative hypothesis

$$\text{Null hypothesis } H_0: \qquad p = \tfrac{1}{3}$$

$$\text{Alternative hypothesis } H_a: p > \tfrac{1}{3}$$

We confront the problem of testing the null hypothesis H_0 against the alternative H_a. The hypotheses H_0 and H_a together make up what we call the **model**; we assume a priori that either H_0 is true or else H_a is true.

model

If we make n trials with the subject, and if X is the number of correct identifications he makes, then X has the binomial distribution with parameters n and p:

$$P(X = r) = \binom{n}{r} p^r (1-p)^{n-r}, \qquad r = 0, 1, 2, \ldots, n \qquad 8.1$$

We want to use the statistic X to decide whether or not the alternative hypothesis H_a is more reasonable or plausible than the null hypothesis H_0—that is, whether or not the subject has ability. And we want to set up in advance of experimentation a rule for making the decision; we want to set up a statistical procedure in the form of a **test**. This parallels what we do in constructing confidence intervals. For a given confidence-interval problem,

test

X := a test statistic

SECTION 8.1 INTRODUCTION

we settle on confidence limits $\bar{X} \pm 2.262(s/\sqrt{n})$, say, in advance of sampling, and we let the sample give to these limits actual numerical values. In the present problem we want to set up a **rejection region** or **critical region**, a set R of X-values that will lead us to reject H_0 and prefer H_a to H_0. If the experiment gives to X a numerical value lying in R, we reject H_0; otherwise, we do not.

rejection region
critical region

To illustrate, suppose that n is 10. Consider two rejection regions:

$$R_5 = \{5, 6, 7, 8, 9, 10\}$$

and

$$R_7 = \{7, 8, 9, 10\}$$

If we use R_5, the rule or test is: If $X \geq 5$, reject H_0 and accept H_a (decide that the subject has ability); if $X < 5$, accept H_0 (decide that the subject has no ability). If we use R_7, the rule is: If $X \geq 7$, reject H_0 and accept H_a; if $X < 7$, accept H_0. (Of course, we could use some cutoff point other than 5 or 7.)

Each of these two rules makes sense: We decide in favor of the hypothesis H_a that $p > \frac{1}{3}$ when X is large. The **level** or **size** of the test (or rejection region or critical region) is the *probability of rejecting the null hypothesis H_0 when in fact H_0 is true.* The level is customarily denoted by α. If H_0 is true and an application of the test leads to a rejection of H_0, then an error has been committed; α is the probability of this error.

level
size

For the rejection region R_5, this error occurs when the subject has no skill but scores five or more successful identifications just by luck. The chance of this is

$$\alpha \text{ for } R_5 = P(X \geq 5) = \sum_{r=5}^{10} \binom{10}{r} (\tfrac{1}{3})^r (\tfrac{2}{3})^{10-r} = .213$$

For the rejection region R_7, the chance of this error is

$$\alpha \text{ for } R_7 = P(X \geq 7) = \sum_{r=7}^{10} \binom{10}{r} (\tfrac{1}{3})^r (\tfrac{2}{3})^{10-r} = .020$$

Note that $P(X \geq 7)$ is smaller than $P(X \geq 5)$—the level for R_7 is smaller than the level for R_5.

Thus the test R_7 is more stringent than R_5—with R_7 there is a lesser chance of erroneously attributing skill to a subject who has none. Since errors are to be avoided and since this error is less likely with R_7 than with R_5, it might seem that the test R_7 is preferable to R_5. But that is not the whole story, because with R_7 there is a *greater* chance that a subject *with* skill will erroneously be classified as *devoid* of skill. For this reason, neither

R_5 nor R_7 is preferable to the other in any absolute sense. See Section 8.5 for a more detailed comparison of the behavior of the tests R_5 and R_7.

In what follows we shall encounter many testing problems. They share certain features with the example above, and their analysis can conveniently be set out in a standard sequence of steps.

(1) Formulate the null hypothesis H_0 in statistical terms.
(2) Formulate the alternative hypothesis H_a in statistical terms.
(3) Set the level (or size) α and the sample size n.
(4) Select the appropriate statistic and the rejection region R.
(5) Collect the data and calculate the statistic.
(6) If the calculated statistic falls in the rejection region R, reject H_0 in favor of H_a; if the calculated statistic falls outside R, do not reject H_0.

Each of these steps calls for comment.

Step 1 The null hypothesis must be stated in statistical terms. It would not suffice in our example merely to hypothesize that the subject cannot distinguish subtle differences in taste; we need the specific binomial model, together with the identification of $p = \frac{1}{3}$ as representing absence of skill.

Step 2 The alternative hypothesis is essential to the problem. Suppose that in order to check alleged ESP powers in a subject we present him at each of a series of trials with three cards—one diamond and two spades—presented face down and in random order. He is to pick out the diamond. The form is that of the triangle test, and $H_0: p = \frac{1}{3}$ again represents the null hypothesis of no ability. Should the alternative hypothesis again be $H_a: p > \frac{1}{3}$? Although that would seem to represent positive ESP prowess, it is sometimes claimed that certain subjects have ESP power but disbelieve in its existence and hence, subconsciously and perversely, change from a right answer to a wrong one. Since p is less than $\frac{1}{3}$ for such a subject, the alternative hypothesis should include the case $p < \frac{1}{3}$ as well as the case $p > \frac{1}{3}$; we should take $H_a: p \neq \frac{1}{3}$. In the taste experiment an excessively large X leads us to reject H_0 in favor of H_a; in the present case an excessively small X should also lead us to reject H_0 in favor of H_a. (In 3000 ESP trials, 1900 correct guesses, which is 900 over the 1000 to be expected at chance, would be terrifying—but so would 100 correct guesses, which is 900 under expectation.) Since the form of the alternative hypothesis affects the form of the appropriate test, the alternative hypothesis must correctly represent the problem at hand.

Step 3 <u>In practice the α-level selected is in the range from about .001 to about .1.</u> The smaller α is, the less will be the chance of mistakenly

1% → 10 %

rejecting H_0 (in our example the smaller will be the number of subjects without skill to whom skill is mistakenly attributed). But the smaller α is, the greater will be the chance of mistakenly accepting H_0 (in the example the larger will be the number of subjects with skill who are mistakenly judged to have none). Section 8.5 has further discussion of this trade-off as well as of the effect of the sample size n.

Step 4 In the taste experiment the statistic is the number X of successes, and the rejection region has the commonsense form $\{r, r + 1, \ldots, n\}$. In all our testing problems, common sense will dictate the proper form of the rejection region; choosing the exact region (choosing the actual value of r in our example) requires knowing the distribution of the statistic (the binomial distribution, in our example).

Step 5 The sampling procedure must accord with the model, the binomial model in our example.

Step 6 In our example H_0 is rejected if $X \geq r$ (where $r = 5$ for R_5 and $r = 7$ for R_7). If $X \geq r$, then H_0 does not *well explain* the observed value of X (if H_0 is indeed true, the observation is strangely large). On the other hand, if $X < r$, one cannot with full confidence accept H_0 as true. Suppose the subject has skill, but only to a minute degree which corresponds to a value of p only minutely greater than $\frac{1}{3}$. The test is in this case unlikely to reveal his skill; the test cannot well distinguish between a subject completely devoid of skill and one almost devoid of it. For this reason, if the test fails to reject H_0, we accept H_0 only in a provisional way. This is illustrated by line 3 of Table 1.5 (p. 5): although the data there do not indicate that the drug is effective in preventing coronary heart disease, a later study with a larger sample showed that it *was*. See the final paragraph in Section 9.8.

In what follows we will construct a number of tests of hypotheses. In each case we must actually construct the rejection region R, and these constructions will clarify the general principles set out above. Tests of hypotheses are used extensively in research, though not necessarily in rigid accord with the theory. The theoretical framework, however, makes it possible to understand what test is appropriate to what problem and to see what the strength of the test will be.

8.2 Hypotheses on a Normal Mean

Consider the null hypothesis that a normal population has a specified mean μ_0, the variance σ^2 being unknown. Under the null hypothesis $H_0: \mu = \mu_0$, the statistic

$$t = \frac{\bar{X} - \mu_0}{s/\sqrt{n}} \qquad 8.2$$

has a t-distribution with $n - 1$ degrees of freedom (see Section 7.4), which can be made the basis of a test.

Suppose first that the alternative hypothesis is $H_a: \mu > \mu_0$. It is intuitively clear that we ought to reject H_0 in favor of H_a when $\bar{X}$ is too large—in fact, when $\bar{X}$ exceeds μ_0 by too much. Since the t-statistic (8.2) increases when $\bar{X}$ increases, we can just as well reject H_0 when t is excessively large. How large? Recall that the upper α percentage point $t_{\alpha, n-1}$ is defined so that $P(t > t_{\alpha, n-1})$ is equal to α. Therefore, if we adopt the rule of rejecting H_0 when t exceeds $t_{\alpha, n-1}$, then the chance of rejecting H_0 when it is true is just α. The testing problem and the rule are

$$H_0: \mu = \mu_0$$
$$H_a: \mu > \mu_0 \qquad 8.3$$
$$R: t > t_{\alpha, n-1}$$

The rejection region is specified by the inequality $t > t_{\alpha, n-1}$. If we substitute the ratio in Equation (8.2) for t here, multiply by $s/\sqrt{n}$, and then add μ_0, the inequality becomes $\bar{X} > \mu_0 + t_{\alpha, n-1}(s/\sqrt{n})$. Thus we reject the null hypothesis $\mu = \mu_0$ when $\bar{X}$ exceeds μ_0 by too much, $t_{\alpha, n-1}(s/\sqrt{n})$ providing the proper measure of *too much*.

If μ is equal to μ_0, the distribution of the ratio (8.2) corresponds to the density curve (a) in Figure 8.1. If instead μ is equal to μ_1, say, where $\mu_1 > \mu_0$, then the ratio (8.2) has the density curve (b); it has the same shape as (a) but is displaced to the right, its mean being $\mu_1 - \mu_0$ instead of 0. The cutoff point $t_{\alpha, n-1}$ is where the area of the right tail of curve (a) equals the level α of the test. The area of the left tail of curve (b) is the

Figure 8.1

Distributions corresponding to $H_0: \mu = \mu_0$ and $H_a: \mu = \mu_1$

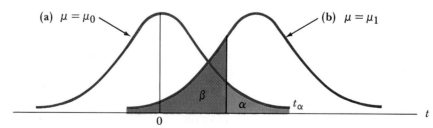

SECTION 8.2 HYPOTHESES ON A NORMAL MEAN

probability of erroneously accepting H_0 if the mean μ is really μ_1 (this probability is labeled β in the figure).

In using the t-test—a test involving the t-statistic as defined by (8.2)—it is essential to use the standard error $s/\sqrt{n}$ in the denominator. It is a common mistake to calculate $(\bar{X} - \mu_0)/s$ instead of $(\bar{X} - \mu_0)/(s/\sqrt{n})$.

Example 1

Problem: A manufacturer of small electric motors asserts that on the average they will not draw more than .8 amperes under normal load conditions. A sample of 16 of the motors was tested and it was found that the mean current was .96 amperes with a standard deviation of .32 amperes. Are we justified in rejecting the manufacturer's assertion? Assume normality.

Solution: Here the null hypothesis that the assertion is correct is $H_0: \mu = .8$. If we are willing to take a risk of 1 in 20 of rejecting his assertion if it is true, we set α at .05. Since we want to reject his assertion only if the evidence indicates that the mean current consumption (the population value) is greater than .8, we have a one-sided alternative $H_a: \mu > .8$. Under the assumption of normality, we use a t-statistic with 15 degrees of freedom and a one-tailed test; the cutoff point is $t_{.05, 15}$, or 1.753:

$$H_0: \mu = .8$$
$$H_a: \mu > .8$$
$$R: t > 1.753$$

For our data the calculated t is

$$t = \frac{\bar{X} - \mu_0}{s/\sqrt{n}} = \frac{.96 - .80}{.32/4} = 2.00$$

Since this exceeds 1.753, we reject the null hypothesis and conclude that the mean current consumption is greater than .8 amperes.

Example 2

Problem: In an attempt to determine whether or not special training will increase IQ, 25 children are given a standard IQ test. The children are then given a course designed to increase their IQ scores and are tested once more at the end of the course. The differences between the second and first scores are recorded; the mean difference for the 25 children is found to be 3 points, and the sample standard deviation is 9 points. Has the training increased IQ? Assume normality.

Solution: The null hypothesis that the training has no effect is that the population mean of differences is $\mu = 0$. We will reject H_0 only if we think the mean difference is positive, so the alternative is $H_a: \mu > 0$. We need a one-tailed test, and if we set α at .05 the cutoff point is $t_{.05, 24}$, or 1.711:

$$H_0: \mu = 0$$
$$H_a: \mu > 0$$
$$R: t > 1.711$$

The calculated t is

$$t = \frac{\bar{X} - 0}{s/\sqrt{25}} = \frac{3 - 0}{9/5} = 1.667$$

The data do not support the hypothesis that the training is effective.

Similarly, if the alternative is $\mu < \mu_0$, the sensible thing is to reject $\mu = \mu_0$ when t is small:

$$H_0: \mu = \mu_0$$
$$H_a: \mu < \mu_0 \qquad\qquad 8.4$$
$$R: t < -t_{\alpha, n-1}$$

Notice again that $t < -t_{\alpha, n-1}$ has probability α if μ does equal μ_0; also notice that $t < -t_{\alpha, n-1}$ is the same thing as $\bar{X} < \mu_0 - t_{\alpha, n-1}(s/\sqrt{n})$.

one-sided The alternative hypotheses in (8.3) and (8.4) are **one-sided,** and so are **two-sided** the corresponding rejection regions. A **two-sided** alternative, including both possibilities $\mu > \mu_0$ and $\mu < \mu_0$, calls for a two-sided rejection region:

$$H_0: \mu = \mu_0$$
$$H_a: \mu \neq \mu_0 \qquad\qquad 8.5$$
$$R: t > t_{\alpha/2, n-1} \text{ or } t < -t_{\alpha/2, n-1}$$

The cutoff points are $\frac{1}{2}\alpha$ percentage points, since the test is two-tailed; as Figure 8.2 indicates, the total area of the two tails taken together is α. The rejection rule can be restated as $|\bar{X} - \mu_0| > t_{\alpha/2, n-1}(s/\sqrt{n})$. We reject H_0 if $\bar{X}$ is excessively far from μ_0 in *either* direction.

Figure 8.2

A two-tailed critical region on the *t*-distribution

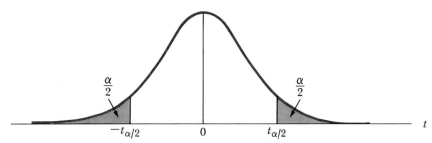

Example 3

Problem: Last year's retail sales records show the average monthly expenditure per person for a certain food product was $5.50. We want to know whether there has been any significant change in this average during the first quarter of this year.

Analysis: To make a comparison, we sample 30 families and find that the mean expenditure is $5.10 with a standard deviation of $.90. Here the null hypothesis of no change is $H_0: \mu = 5.50$. The two-sided alternative, $H_a: \mu \neq 5.50$, represents a change in either direction, up or down. Therefore we want a two-tailed test; if α is set at .01, the cutoff is the $\frac{1}{2}\alpha$, or .005, percentage point for 29 degrees of freedom: $t_{.005,\,29}$ is 2.756. Hence

$$H_0: \mu = 5.50$$
$$H_a: \mu \neq 5.50$$
$$R: t > 2.756 \text{ or } t < -2.756$$

Since

$$t = \frac{\bar{X} - 5.50}{s/\sqrt{30}} = \frac{5.1 - 5.50}{.9/\sqrt{30}} = -2.434$$

the data do not indicate any change in the amount spent.

In each of our three examples we observe the following features.

(1) The null hypothesis always contains the equality statement, and the alternative hypothesis is determined by the question implicit in the statement of the problem.

(2) The calculated statistic is based on the difference between the observed mean $\bar{X}$ and the mean μ_0 under the null hypothesis. Since the difference between them is meaningful only in relation to the amount of variation in the data, the statistic is the ratio of the difference to its estimated standard deviation—that is, to the standard error.

(3) When the difference between $\bar{X}$ and μ_0 is large enough that the calculated *t*-value falls into one of the tails of the *t*-distribution (the appropriate tail or tails determined by the alternative hypotheses), we reject the null hypothesis because the probability of such a *t*-value, given that the null hypothesis is true, is too small to be attributed to normal sampling variation.

Whenever we use the *t*-distribution for calculating confidence intervals or for testing hypotheses, we are assuming that the data are a random sample from a normal population. The first assumption, randomness, is necessary if we want to make probability statements about the results we obtain; therefore it cannot be relaxed.

The second assumption, normality, may be relaxed. Because of the central limit theorem we may say that even though the population has a distribution that is not normal, the *t*-statistic may be used and the probabilities will not be greatly affected, provided we have a sufficiently large sample. The sample size that may be considered sufficiently large depends upon how much the population concerned departs from normality.

Problems

1. Given each of the following sets of values, test the indicated hypothesis.
   a. $n = 16$, $\bar{X} = 40$, $s = 4$, $H_0: \mu \leq 38$, $H_a: \mu > 38$, $\alpha = .01$.
   b. $n = 9$, $\bar{X} = 5.2$, $s = 3$, $H_0: \mu > 4.9$, $H_a: \mu \leq 4.9$, $\alpha = .05$.
   c. $n = 25$, $\bar{X} = -2.1$, $s = 2$, $H_0: \mu = -3.2$, $H_a: \mu \neq -3.2$, $\alpha = .01$.

2. For each of the following sets of values test the indicated hypothesis.
   a. $n = 16$, $\bar{X} = 1550$, $s^2 = 12$, $H_0: \mu = 1500$, $H_a: \mu \neq 1500$, $\alpha = .01$.
   b. $n = 9$, $\bar{X} = 10$, $s^2 = .81$, $H_0: \mu \geq 12$, $H_a: \mu < 12$, $\alpha = .005$.
   c. $n = 49$, $\bar{X} = 19$, $s^2 = 1$, $H_0: \mu \leq 18$, $H_a: \mu > 18$, $\alpha = .05$.

3. Given the following information, test the stated hypothesis at the given level.

SECTION 8.2 HYPOTHESES ON A NORMAL MEAN 273

a. $n = 9$, $\bar{X} = 76$, $\Sigma(X_i - \bar{X})^2 = 32$, $H_0: \mu = 75$, $H_a: \mu \neq 75$, $\alpha = .05$.

b. $n = 25$, $\Sigma X_i = 500$, $\Sigma X_i^2 = 12{,}400$, $H_0: \mu \leq 17$, $H_a: \mu > 17$, $\alpha = .10$.

4. A real-estate broker trying to sell a business claims that the average daily receipts are $3200. An interested buyer conducts an investigation and finds that during the next 16 days the average amount is $2984 with a standard deviation of $320. Does this finding contradict the broker's claim at the
    a. 10% significance level?
    b. 5% significance level?
    c. 1% significance level?

5. Answer Problem 4 for a buyer who is concerned only with the possibility of *lower* daily receipts.

6. The estimated variance based on four measurements of a spring tension was .25 grams. The mean was 26 grams. Test the hypothesis that the true value is 25 grams. Use a 5% level.

7. On the basis of some approximate calculations an engineer concluded that 150 man-hours would be required to produce a certain airplane part. To test this, he followed the production of the first 10 such parts and found the average production time to be 146 man-hours with an estimated <u>standard deviation of the mean</u> of 2.5 man-hours. At the 5% level, will the engineer reject his previous conclusion?

8. The manufacturer of a new drug claims it will lower blood pressure by 14 points on the average. When the drug was administered to 10 patients the following drops in blood pressure were registered: 12, 7, 15, 9, 8, 16, 10, 11, 12, 10. Is this claim sustained at the .005 and .05 levels of significance?

9. In a study of court administration the following times to disposition were found in 20 randomly selected cases of a certain type. Test the null hypothesis that the average time to disposition is 80 days against the alternative that the average time is greater than 80 days. Use the 1% level for the test.

43	90	84	87	116
95	86	99	93	92
121	71	66	98	79
102	60	112	105	98

10. The following are the logarithms of direct microscopic counts on ten

samples of raw skim milk. Is there enough evidence to conclude that the mean log DMC is greater than 6.00? Use $\alpha = .05$.

| 6.97 | 6.83 | 6.56 | 6.15 | 6.99 |
| 7.08 | 6.11 | 5.95 | 6.54 | 6.30 |

11. The drained weights in ounces for a sample of 15 cans of fruit are as follows. At the 2% level of significance, test the hypothesis that on the average a 12-ounce drained-weight standard is being maintained.

12.1	12.1	12.3	12.0	12.1
12.4	12.2	12.4	12.1	11.9
11.9	11.8	11.9	12.3	11.8

12. Use the data given in Problem 6 of Chapter 2 to test the null hypothesis that the mean time lapse to filing the motion to review the sentence is 18 months against the alternative that it is greater than 18 months. Use $\alpha = .05$.

13. A certain type of truck engine has an average of 150 brake horsepower (BHP) at 2000 revolutions per minute. An engineer adds a turbocharger to 21 of these engines and finds an average BHP of 160 with a standard deviation of 10 BHP. Will the addition of the turbocharger increase the mean BHP for the population of engines of this type? Use $\alpha = .05$.

14. Pine trees of a certain variety have an average growth of 10.1 inches in 3 years. A forestry biologist claims that a new variety will have greater average growth in this length of time. A sample of 15 trees of the new variety has an average growth of 10.8 inches with a standard deviation of 2.1 inches. Is the biologist's claim substantiated? Use $\alpha = .05$.

15. The average score on a state-wide twelfth-grade practical-skills examination is 75.5. A group of 35 randomly selected high-school students received special tutoring and scored an average of 82.1 with a standard deviation of 4.7. Do these results provide sufficient evidence to indicate that tutoring increases examination scores? Use $\alpha = .01$.

16. Specifications for the manufacture of a certain brand of cough syrup require an average of 20 mg of the active ingredient for each milliliter of the syrup. Four randomly selected 1-ml samples of the syrup yield the following measurements of the active ingredient: 20.1, 20.2,

20.0, 20.3. Is there sufficient evidence to indicate that the product is not meeting specifications? Use $\alpha = .10$.

17. A newspaper advertisement states that the application of a special reflective coating to windows of a house will reduce air-conditioning costs by 20%. A consumer group measured the actual cost reductions in five randomly selected homes and found an average reduction of 14% with a standard deviation of 8%. Do these data refute the statement in the advertisement? Use $\alpha = .05$.

18. The Seacoast Railway claims that its trains block crossings in Gatorville, Florida, no more than 5 minutes per train on the average. The actual times that ten randomly selected trains blocked the crossings are given below. Do these data refute the company's claim? Use $\alpha = .05$.

| 10.4 | 9.7 | 6.5 | 9.5 | 8.8 |
| 11.2 | 7.2 | 10.5 | 8.2 | 9.3 |

8.3 Hypotheses on a Binomial p

If we take a sample from a binomial population with the intention of using the information it contains to test the hypothesis that p has a specified value p_0, we can use an exact test or an approximate test. To use the exact test we must have tables of the binomial probabilities or be willing to compute them. Ordinarily, then, we use an approximate test based on the central limit theorem.

Under the null hypothesis $H_0: p = p_0$, the ratio

$$Z = \frac{X - np_0}{\sqrt{np_0(1 - p_0)}} \qquad 8.6$$

has approximately a standard normal distribution, where X is the number of successes in the sample (see Section 5.9). The denominator in the ratio (8.6) is the standard deviation of X; we know its value under the null hypothesis, and there is no need to estimate it from the sample.

Testing H_0 is like testing for a normal mean when there are infinitely many degrees of freedom—that is, with the normal distribution in place of the t-distribution. The rejection region will be one-tailed or two-tailed depending on whether the alternative hypothesis is one-sided or two-sided.

Dividing numerator and denominator in the ratio (8.6) by the sample size n gives

$$Z = \frac{X - np_0}{\sqrt{np_0(1 - p_0)}} = \frac{f - p_0}{\sqrt{p_0(1 - p_0)/n}} \qquad 8.7$$

where f is X/n, the fraction of successes in the sample. Clearly, the test may be performed either with X or with f, as convenience dictates. We must, however, be sure to use the right standard deviation in the denominator.

Example 1

Problem: In a sample of 400 bushings there were 12 whose internal diameters were not within the tolerances. Is this sufficient evidence for concluding that the manufacturing process is turning out more than 2% defective bushings? Let α be .05.

Solution: The null hypothesis that the process is in control is $H_0: p \leq .02$. The alternative hypothesis that it is out of control is $H_a: p > .02$. The upper .05 percentage point $Z_{.05}$ for the standard normal distribution is 1.645. We should reject for larger Z-values, since a one-sided test calls for a one-sided rejection region:

$$H_0: p \leq .02$$
$$H_a: p > .02$$
$$R: \frac{X - 400 \times .02}{\sqrt{400 \times .02 \times .98}} > 1.645$$

Our data give

$$\frac{X - 400 \times .02}{\sqrt{400 \times .02 \times .98}} = \frac{12 - 8}{2.8} = 1.429$$

The data do not favor the hypothesis that the process is out of control; 12 defective bushings out of 400 is not excessively large.

Example 2

Problem: To check whether a pair of dice are balanced, we want to test the null hypothesis that the chance of rolling a 7 is $\frac{1}{6}$, as the computations in Chapter 4 predict.

Analysis: The alternative to $p = \frac{1}{6}$ we take to be $p \neq \frac{1}{6}$, since we are interested in discovering deviations in either direction. If α is .05, the

SECTION 8.3 HYPOTHESES ON A BINOMIAL p

cutoff points are $\pm Z_{.025}$, or ± 1.96, on a two-tailed test. Suppose that n is 500.

$$H_0: p = \tfrac{1}{6}$$
$$H_a: p \neq \tfrac{1}{6}$$
$$R: \frac{f - \tfrac{1}{6}}{\sqrt{\tfrac{1}{6} \times \tfrac{5}{6}/500}} > 1.96 \text{ or } < -1.96$$

If in 500 rolls we get 7 a total of 66 times, then

$$\frac{f - \tfrac{1}{6}}{\sqrt{\tfrac{1}{6} \times \tfrac{5}{6}/500}} = \frac{\tfrac{66}{500} - \tfrac{1}{6}}{\sqrt{\tfrac{1}{6} \times \tfrac{5}{6}/500}} = -2.08$$

The statistic falls into the lower tail of the critical region, and so we reject the null hypothesis. We reject the hypothesis that the dice are balanced, that the chance of rolling a 7 is $\tfrac{1}{6}$.

In the chapters that follow we will encounter many statistics that arise from standardizing some random variable by its mean and standard deviation and then replacing the standard deviation by its natural estimate. A standard notation will help in this connection. The standard deviation of a statistic Y we denote σ_Y; an estimate of this standard deviation we denote s_Y. Thus, the standard deviation of $\bar{X}$ is $\sigma_{\bar{X}}$, equal to $\sigma/\sqrt{n}$, of which the estimate $s_{\bar{X}}$ is $s/\sqrt{n}$. The number X of successes in binomial trials has standard deviation σ_X, equal to $\sqrt{np(1-p)}$, and the estimate s_X of this is $\sqrt{nf(1-f)}$. Finally, the fraction f of successes has standard deviation σ_f, equal to $\sqrt{p(1-p)/n}$, and the estimate s_f of this is $\sqrt{f(1-f)/n}$.

Problems

19. A manufacturer claims that at least 95% of the items he produces are failure-free. Examination of a random sample of 600 items showed 39 to be defective. Test the claim at a significance level of .05.

20. On 384 out of 600 randomly selected farms it was discovered that the farm operator was also the owner. Is there reason to believe that *more* than 60% of the operators are also owners?

21. In a sample of 144 voters it was found that 90 were in favor of a bond issue. Test the hypothesis that opinion is equally divided on this issue. Let $\alpha = .01$.

22. In a laboratory experiment a monkey was trained to pull lever A rather

than lever B upon a certain stimulus. To test the effect of the training, the monkey was then subjected to 100 tests in which he pulled lever A 63 times. On the basis of these data can the experimenter reject the hypothesis that the stimulus had no effect? Use the
   a. 5% level of significance.
   b. 1% level of significance.
   c. .5% level of significance.
23. Your friend is a strong believer in horoscopes and is willing to put his beliefs to a test. You suggest that for a one-year period you will present him each week with three horoscopes for that week, one of them being his own. At the end of each week he is to pick out the horoscope best describing his experiences. At the 5% level, how many correct identifications will be required before you will concede that his faith has some justification?

24. A multiple-choice test, consisting of 100 questions, has for each question four choices, only one of which is correct. At the 1% level, how many questions must a student answer correctly in order that his exam score can't be attributed to mere guessing?

25. A bank has three branch offices. A random sample of 200 customers showed 75 who banked at the office on 30th Street. Do these data indicate that more than $\frac{1}{3}$ of the bank's customers prefer to use this office? Use $\alpha = .05$.

26. The incidence of pulmonary embolism in a certain population of high-risk patients is 20%. From this population patients were randomly selected and were treated with a special anticoagulant; of these, 8 developed pulmonary embolism. Do these data indicate that the anticoagulant is effective in reducing the incidence of pulmonary embolism? Use $\alpha = .025$.

27. A bottle-manufacturing process is "under control" if no more than 1% of the bottles are defective. A random sample of 120 bottles showed 5 to be defective. Do these data indicate that the process is "out of control"? Use $\alpha = .10$.

8.4 *p*-Values

The results of tests of hypotheses are frequently given in a form different from that used in Sections 8.2 and 8.3. Consider Example 1 of Section 8.3. There the test statistic is

SECTION 8.4 p-VALUES

$$\frac{X - 400 \times .02}{\sqrt{400 \times .02 \times .98}} \qquad 8.8$$

For the data at hand, $X = 12$ and this statistic has the value 1.429.

The rule for a one-sided, .05-level test is to reject H_0 if the statistic (8.8) has a value exceeding 1.645, which does not happen for our data. But suppose the level α had been .10 instead of .05. The rule would then be to reject H_0 if the value of (8.8) exceeds 1.282, which does happen for $X = 12$.

For which values of α would the corresponding one-sided rule lead us on the basis of our data ($X = 12$) to reject H_0? If $p = .02$ (the largest value in H_0), the statistic (8.8) is approximately normally distributed; from Table 2 we get the following cutoff points:

For $\alpha =$	Reject H_0 if $X >$
.05	1.645
.06	1.555
.07	1.476
.08	1.405
.09	1.341
.10	1.282

p-value
significance
level

Thus the observed value 1.429 of the statistic (8.8) leads to rejection of H_0 if α is .08 or more but not if α is .07 or less. In fact, the area under the normal curve to the right of 1.429 is about .076 (Table 2 again), so our data lead to rejection of H_0 if $\alpha > .076$ but not if $\alpha < .076$. The number .076 is the **p-value**, or **significance level**, of the data (with respect to the test in question).

In reporting the results of a test, rather than stating the level ($\alpha = .05$, or $\alpha = .10$, or whatever) and stating whether or not H_0 is rejected by the data at that level, we can instead simply give the *p*-value. This in effect allows the reader of the report to set his own level. If the *p*-value is .076, the reader knows the data reject H_0 for every α greater than .076 but not for any α less than .076. The *p*-value also has a direct appeal. A large value for the statistic (8.8) is unusual if H_0 is true; the *p*-value measures the unusualness—if $p = .02$, the probability is only .076 that a repeat of the experiment would give a larger (more unusual) value of (8.8).

Example 2 of Section 8.3 calls for a two-sided test. The rule for the .05 level is to reject H_0 if the statistic

$$\frac{f - \frac{1}{6}}{\sqrt{\frac{1}{6} \times \frac{5}{6}/500}} \qquad 8.9$$

is outside the range from -1.96 to $+1.96$. The data ($f = \frac{66}{500}$) give (8.9) the value -2.08, so H_0 is rejected. By Table 2, the area under the normal curve outside the range from -2.08 to $+2.08$ is about .038, and this is the p-value. For $\alpha > .038$, the data reject H_0, but not for $\alpha < .038$.

A hypothesis test is usually called a *test of significance* if the result is given in the form of a p-value. This is a convenient form of testing, but we must not lose sight of the role of the alternative hypothesis. The area to the *left* of -2.08 under the normal curve is about .019. It would be wrong in connection with Example 2 of Section 8.3 to give the p-value as .019 rather than .038. The alternative hypothesis is two-sided, and so a value of (8.9) is unusual (more unusual than the one actually observed) if it falls to the left of -2.08 *or to the right of* $+2.08$.

It is the form of the alternative hypothesis H_a that dictates which sets of data (or values of the statistic) are not *well explained* by the null hypothesis H_0. For this reason we must keep H_a in mind when constructing tests of significance and calculating p-values.

The t-tests of Section 8.2 can also be restated as tests of significance. To find the exact p-value requires a full table of the t-distribution, however; the table of percentage points (Table 4) does not suffice.

Problems

28. Find the p-values in
   a. Problem 19.   **b.** Problem 20.   **c.** Problem 21.
   d. Problem 22.

29. It has been claimed that the probability p of a head in a toss of a coin is not $\frac{1}{2}$ because the coin is not symmetrical. Suppose that a test of the null hypothesis ($H_0: p = \frac{1}{2}$) against the alternative ($H_0: p \neq \frac{1}{2}$) is conducted, and assume that the observed fraction of heads in n tosses of the coin is $f = .51$. Find the p-value if
   a. $n = 100$.   **b.** $n = 10{,}000$.

30. In the test of the null hypothesis ($H_0: p = p_0$) against the one-sided alternative hypothesis ($H_a: p < p_0$), suppose that the observed fraction f is 50% less than the hypothesized probability p_0. That is, $f = .5p_0$. Compute the p-value for $n = 36$ and
   a. $p = .5$.   **b.** $p = .1$.

31. In the test of the null hypothesis ($H_0: p = \frac{1}{4}$) against the one-sided

alternative hypothesis ($H_a: p > \frac{1}{4}$), which of the following circumstances yields the more statistically significant result?
a. $f = .27$, $n = 1000$.   b. $f = .31$, $n = 100$.

8.5 The Theory of Testing

For a further understanding of the theory underlying hypothesis testing, return to the triangle test of Section 8.1. The hypotheses were

$$H_0: p = \tfrac{1}{3}$$
$$H_a: p > \tfrac{1}{3}$$

The subject makes X correct identifications in n trials, and (8.1) gives the distribution of X.

Suppose once again that $n = 10$, and consider the regions $R_5 = \{5, 6, 7, 8, 9, 10\}$ and $R_7 = \{7, 8, 9, 10\}$. As noted in Section 8.1, rejecting H_0 when X is large makes sense. It is contrary to common sense to use a rule that says to decide in favor of H_a in case $X \leq 3$, for example, or a rule that says to decide in favor of H_a in case X is even. But of the sensible regions R_5 and R_7, which is better? Applying a test may lead to the wrong conclusion. There are in fact *two* kinds of error possible—**Type I and Type II errors.**

Type I and Type II errors

Type I error: We reject H_0 but H_0 is true

Type II error: We accept H_0 but H_0 is false

We assess the strength of a test or rejection region by the probabilities of these two kinds of error.

$$\alpha = P(\text{reject } H_0 | H_0 \text{ is true}) = P(\text{Type I error})$$
$$\beta = P(\text{accept } H_0 | H_0 \text{ is false}) = P(\text{Type II error})$$

Naturally we want α and β to be small. The probability α of a Type I error is what in the preceding sections we called the *level*, or *size*, of the test. And $1 - \beta$ is called the **power**.

power

True Situation	Conclusion	
	H_0 is true	H_0 is false
H_0 is true	No error; Probability $= 1 - \alpha$	Type I error; Probability $=$ size $= \alpha$
H_0 is false	Type II error; Probability $= \beta$	No error; Probability $=$ power $= 1 - \beta$

In Section 8.1 the values of α for R_5 and R_7 were calculated by means of the formula (8.1):

$$\alpha \text{ for } R_5 = P(X \geq 5) = .213$$
$$\alpha \text{ for } R_7 = P(X \geq 7) = .020$$

Since $P(X \geq 7)$ is smaller than $P(X \geq 5)$, R_7 is a *better* rejection region than R_5 as far as Type I error (level) is concerned.

But what about β, the chance of a Type II error? If H_0 is false, so that $p > \frac{1}{2}$, then β depends on which alternative value of p we check. We should write β_p to indicate this dependence on p. Let us check for a p of .65. For the region R_5,

$$\beta_{.65} \text{ for } R_5 = P(X < 5) = \sum_{r=0}^{4} \binom{10}{r}(.65)^r(.35)^{10-r} = .095$$

This time we use .65 as the value for p and .35 as the value for $1 - p$ in the binomial formula because we are computing the chance that we will erroneously accept the null hypothesis that the subject is devoid of skill when in fact he can guess right 65% of the time. For R_7,

$$\beta_{.65} \text{ for } R_7 = P(X < 7) = \sum_{r=0}^{6} \binom{10}{r}(.65)^r(.35)^{10-r} = .486$$

Now $P(X < 7)$ is greater than $P(X < 5)$. As far as Type II error is concerned, R_7 is a *worse* rejection region than R_5.

We can summarize our computations in a table:

	α	$\beta_{.65}$
R_5	.213	.095
R_7	.020	.486

The α column makes us prefer R_7, but the β column makes us prefer R_5. This is inevitable—as α goes down, β goes up, and vice versa. If we refuse to allow a Type I error at all, so that $\alpha = 0$, we must always accept H_0, whatever X may be (so the rejection region R is empty); but then if H_0 should be false, we always make an error of Type II, so that $\beta = 1$. Or if we demand that $\beta = 0$, we must always reject H_0, whatever X may be (so the rejection region R is $\{0, 1, 2, \ldots, 10\}$); in this case if H_0 is true, we always make an error of Type I, so that $\alpha = 1$. It is impossible to arrange that both α and β are equal to 0. For whatever consolation it may be, it is equally impossible to arrange that both α and β are equal to 1.

The procedure in a court of law affords an illuminating parallel. Let the null hypothesis be that the defendant is innocent and let the alternative hypothesis be that he is guilty. To condemn an innocent man is to commit a Type I error; to acquit a guilty man is to commit a Type II error. The rules that

Figure 8.3

OC curves: the probability of accepting $H_0: p = \frac{1}{3}$ as a function of p for $n = 10$ trials and for rejection regions $R_5 = \{5, 6, 7, 8, 9, 10\}$ and $R_7 = \{7, 8, 9, 10\}$

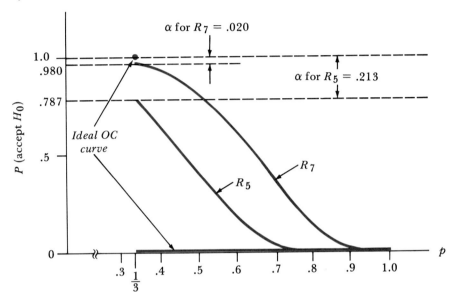

govern a trial are loosely analogous to a statistical procedure—to a statistical test or a rejection region. Any rule that decreases the chance α of a Type I error (a rule, say, that the defendant need not testify against himself) necessarily increases the chance β of a Type II error. And any rule that decreases the chance β of a Type II error (a rule allowing a split jury to convict, say) necessarily increases the chance α of a Type I error. It is impossible to arrange that both α and β are equal to 0. But it is equally impossible to arrange that both α and β are 1 (the contrary appearance of legal reality notwithstanding).

To return to our taste experiment, since the probability of a Type II error depends on which particular p (exceeding $\frac{1}{3}$) holds, a complete understanding of the test requires a graph of β_p. Such a graph is called the *operating characteristic curve*, or OC curve. The OC curve for our experiment is shown in Figure 8.3.

Sometimes, $1 - \beta_p$ is graphed instead of β_p; this graph is called the *power curve* for the test. Of course, the OC curve and the power curve are merely different representations of the same information.

In Figure 8.3 the height of the OC curve labeled R_5 over a point p on the horizontal scale is the probability

$$\beta_p \text{ for } R_5 = P(X < 5) = \sum_{r=0}^{4} \binom{10}{r} p^r (1-p)^{10-r}$$

of accepting H_0 when the rejection region R_5 is used and p is the actual chance that the subject can correctly identify the odd food sample on a single trial. If $p > \frac{1}{3}$, this height is β_p, the probability of a Type II error—the probability of incorrectly accepting H_0. If $p = \frac{1}{3}$ (this corresponds to the left end of the graph), this height is $1 - \alpha$, the probability of correctly accepting H_0. Notice that the curve drops off to the right. Naturally, the more skill the subject has—that is, the greater p is—the less likely the test or experiment is to indicate that he has no skill. Similar remarks apply to the OC curve labeled R_7.

Will the OC curves for our experiment help us to determine the best rejection region? The ideal OC curve would have height 1 at $p = \frac{1}{3}$ and would immediately jump down to 0 for $p > \frac{1}{3}$, as indicated in Figure 8.3. But no corresponding rejection region exists. At the left end the R_7 curve is nearer this ideal than the R_5 curve is; everywhere to the right of this the R_5 curve is nearer this ideal than the R_7 curve is. That the R_7 curve lies above the R_5 curve *at the left end* makes us prefer R_7 to R_5; that the R_7 curve lies above the R_5 curve *everywhere else* makes us prefer R_5 to R_7. Neither region is better than the other. The rejection region R_7 is optimal in this sense: R_7 has an α of .02, and among rejection regions (for sample size $n = 10$) with an α this small, R_7 has the smallest β-values. And R_5 is optimal as well: R_5 has an α of .213, and among rejection regions with an α of this size, R_5 has the smallest β-values. This supports our commonsense feeling that all rational rejection regions must have the form

$$\{r, r + 1, \ldots, 10\}$$

For a proof and further theoretical considerations, see Reference 8.2.

To get an OC curve nearer the ideal shown in Figure 8.3, we must increase n. In addition to increased accuracy, this means, of course, increased effort, time, and money. Figure 8.4 repeats the upper OC curve from Figure 8.3 ($n = 10$ and $R = \{7, 8, 9, 10\}$). The middle curve is the OC curve for $n = 50$ and $R = \{24, 25, \ldots, 50\}$; the bottom one is the OC curve for $n = 200$ and $R = \{82, 83, \ldots, 200\}$. In the case where $n = 50$ and the case where $n = 200$ (curves (b) and (c), respectively, in Figure 8.4), the cutoff point of the rejection region was chosen to make $\alpha = .02$ again. The higher n is, the more rapidly the curve drops from $1 - \alpha$, or .98, toward the horizontal scale.

In connection with Step 6 (see p. 267) in Section 8.1 it was pointed out that the hypotheses H_0 and H_a are not treated in a symmetric fashion. Suppose our test corresponds to the middle OC curve in Figure 8.4: $n = 50$ and $R = \{24, 25, \ldots, 50\}$. If p is $\frac{1}{3}$, then $P(X \geq 24)$, or α, is small—namely, .02— and thus H_0 does not *well explain* an observed X of 24 or more, so we reject H_0^* if $X \geq 24$. But if $X < 24$, can we confidently accept H_0? Since $P(X < 24)$, or $1 - \alpha$, is equal to .98 if $p = \frac{1}{3}$, H_0 *does* well explain an observed X of less than 24. But since the OC curve falls off continuously, $P(X < 24)$ is almost .98 for p-values just a little greater than $\frac{1}{3}$ (for example, $P(X < 24) = .97$ if $p = .34$), and these p-values also well explain an ob-

Figure 8.4

OC curves: the probability of accepting $H_0: p = \frac{1}{3}$ as a function of p for (a) $n = 10$ trials and $R = \{7, 8, 9, 10\}$, (b) $n = 50$ trials and $R = \{24, 25, \ldots, 50\}$, and (c) $n = 200$ trials and $R = \{82, 83, \ldots, 200\}$

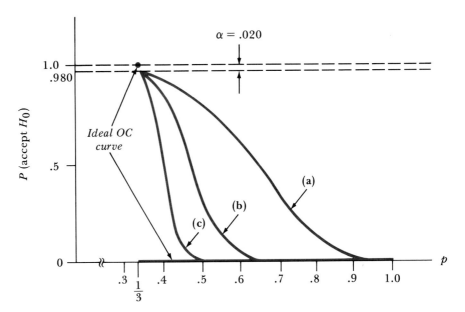

served X that is less than 24. In the event that $X < 24$, then, we do not confidently accept H_0, because we cannot confidently reject *all* of H_a. We can reject the values of p far exceeding $\frac{1}{3}$ but not those barely exceeding $\frac{1}{3}$. Hence the lack of symmetry between H_0 and H_a. We control α by choosing a rejection region that gives a prescribed value for α. The β-values are not completely under our control; they are small for large p but nearly $1 - \alpha$ for p near $\frac{1}{3}$. A larger sample has the desirable effect of making the OC curve drop away more rapidly.

The principles illustrated here in terms of the triangle test of Section 8.1 apply to the other testing problems as well. For example, for the testing problem (8.3) on page 268, the Type I error is the area to the right of t_α under the curve (a) in Figure 8.1. The area to the left of t_α under the curve (b) is the probability β of a Type II error if the mean μ has value μ_1. Suppose μ_0 is fixed, so the curve (a) remains fixed, and suppose μ_1 increases. so the curve (b) moves to the right. It can be seen geometrically that the area β decreases (t_α remains fixed), so the OC curve will fall off in the same way as those in Figures 8.3 and 8.4.

Problems

32. In each of the following situations (i) formulate the null hypothesis H_0 in statistical terms, (ii) formulate the alternative hypothesis H_a in statistical terms, (iii) state what the test statistic could be, and (iv) state what form the rejection region R should have.
    a. A quality-control manager wants to test his company's production scheme against a consumer group's claim that the company's one-pound product weighs less than one pound.
    b. A statistician wants to test whether a given coin is balanced.
    c. A campaign manager for a presidential candidate wants to test his contention that his candidate is ahead in an election.
    d. A prospective buyer of a product wants to test the manufacturer's claim that the proportion of defective items is below 2%.

33. A gambler believes a certain die to be loaded in favor of the side marked six. Calculate the probability of rejecting the null hypothesis $p = \frac{1}{6}$ for each of the seven possible rejection regions of the form $R_0 = \{0, 1, \ldots, 6\}$, $R_1 = \{1, 2, \ldots, 6\}$, $R_2 = \{2, 3, \ldots, 6\}$, etc., if he rolls the die six times.

34. A box contains four marbles, some of which are white. The others are black. To test the hypothesis that there are two of each color, we select two *without* replacement and conclude that there are not two of each if both marbles selected are of the same color.
    a. What is the probability we will come to the wrong conclusion if, in fact, there are two of each color in the box?
    b. Suppose that there is but one white marble in the box. What is the probability that we will falsely conclude that there are two white and two black?

35. Assume the same situation as in Problem 34, but assume that a sample of two is selected *with* replacement. Let θ be the number of white marbles in the box. We reject $H_0: \theta = 2$ if both marbles are the same color. What is the level of the test?

36. A woman asserts that she can tell by taste alone whether tea has been made with tea bags or with loose tea. To test her claim, a statistician presents her with ten pairs of cups; in each pair one cup has been made with a tea bag and the other with loose tea of the same kind. She is asked for each pair to identify the tea-bag cup. If p is the probability of correct identification, the null hypothesis of no ability is $H_0: p = \frac{1}{2}$. Find the exact level of the test that rejects the null hypothesis when the number X of correct identifications satisfies
    a. $X > 6$. b. $X > 7$. c. $X > 8$. d. $X > 9$.

SECTION 8.5 THE THEORY OF TESTING

37. Two boxes, labeled A and B, contain the following numbers of red and white marbles.

	Red	White	Total
Box A	75	25	100
Box B	25	75	100

Jones knows that one box contains a 75-to-25 split between red and white marbles and that the other box contains a 25-to-75 split. What he does not know is which box it is that contains the high proportion of red marbles. He is to select this box (it is in fact box A, but he does not know that) after examining at most two marbles. For each of the following rules find the probability that Jones correctly selects the box with the most reds.

   a. Toss a balanced coin and select A for heads and B for tails.
   b. Toss two balanced coins; select A if two heads appear, and select B otherwise.
   c. Pick a marble (at random) from box A; select A if it is red, and select B if it is white.
   d. Pick a marble from box B; select B if it is red, and select A if it is white.
   e. Pick two marbles with replacement from box A; select A if both are red, and select B otherwise.
   f. Pick two marbles with replacement from box B; select B if both are red, and select A otherwise.
   g. Pick two marbles without replacement from box A; select A if both are red, and select B otherwise.

38. A diving company receives a shipment of 5 air tanks. Let θ be the number of defectives in the shipment. The null hypothesis $H_0: \theta = 0$ is to be tested against the alternative hypothesis $H_a: \theta > 0$. In test plan A, one tank is randomly selected, and H_0 is rejected if it is defective. In test plan B, two tanks are randomly selected, and H_0 is rejected if one or both are defective.

   a. What is the probability of a Type I error for each plan?
   b. Intuitively plan B is "better" than plan A because more data are taken. Demonstrate this fact by comparing the probabilities of a Type II error for the two plans for $\theta = 1, 2, 3, 4, 5$.

39. The probability that a person has an allergic reaction to a certain experimental drug is $p = .30$. The compound is modified and retested on ten people. The null hypothesis $H_0: p = .30$ is rejected if two or fewer have an allergic reaction.

a. What is the probability of a Type I error?
b. Find the probability of a Type II error for $p = .20$ and $p = .10$.

40. A manager of an employment service told a newspaper reporter that at least 60% of its customers are satisfied with the service they received. The reporter decided to test the manager's claim by selecting 25 customers at random and rejecting the manager's claim if 10 or fewer were satisfied with the service.
a. Find the probability of a Type I error.
b. Find the probability of a Type II error if, in fact, only 40% of the customers were satisfied with the service.
c. Which type of error is damaging to the manager?

41. Suppose that the woman in Problem 36 has some ability to distinguish between the two types; suppose in fact that $p = \frac{3}{4}$, where p is the probability of a correct identification. Find β, the chance of a Type II error, in each of the four tests.

42. Use the normal approximation to repeat Problems 36 and 41 and find α and β for $n = 100$ and rejection regions
a. $f > .60$.   b. $f > .70$.   c. $f > .80$.   d. $f > .90$.

CHAPTER PROBLEMS

43. The t-test in Section 8.2 is used to test a hypothesis about the mean of a normal population when the variance is unknown. If the population variance is known, the test statistic Z defined by $Z = (\bar{X} - \mu_0)/(\sigma/\sqrt{n})$ is best. Critical values for this statistic may be found in either the standard normal table (Table 2) or the "∞" row of the t-table (Table 4). Rework Problems 9 and 17 treating the sample variances as if they were the true variances. What is the advantage of using the true variance when it is known?

44. Suppose a Z-test as defined in Problem 43 is used to test the null hypothesis $H_0: \mu = \mu_0$ against the alternative hypothesis $H_a: \mu > \mu_0$ at the 5% level. Show that the probability of a Type II error is just the area under the standard normal curve to the left of $1.645 - (\mu - \mu_0)/(\sigma/\sqrt{n})$. Find the analogous expressions for the lower-tail and two-tail tests.

45. A sample of size 16 is selected from a normal population with variance 100. The null hypothesis $H_0: \mu = 77$ is tested against the alternative hypothesis $H_a: \mu > 77$ at the 5% level. Use the procedure described in Problem 44 to find the probability of a Type II error when $\mu = 83$.

CHAPTER PROBLEMS

46. The standard deviation of a normal population is known to be $\sigma = 40$, and the Z-test in Problem 43 is to be used to test the null hypothesis $H_0: \mu = 32$ against the alternative hypothesis $H_a: \mu > 32$ at the 5% level. If the true mean of the population is $\mu = 34$, how large a sample must be taken so that β is no more than .01? See Problem 44 for the procedure for finding β.

47. With data obtained from the procedure described in Problem 58, Chapter 7, test the null hypothesis $H_0: \mu = .5$ against the alternative hypothesis $H_a: \mu \neq .5$ at the 5% level. Note that the true mean of the population in question is $\mu = .5$.

48. Toss a coin 25, 100, and 400 times, and record the number of heads in each case. Compute the lower-tail p-values for the test of the (false) null hypothesis $H_0: p = .55$, where p is the probability of a head. Comment on the effect of sample size on rejecting a false null hypothesis.

49. Let μ denote the mean of a Poisson population (see Chapter 5, Section 10). A test statistic for testing the null hypothesis $H_0: \mu = \mu_0$ against either one-sided or two-sided alternatives is $Z = (\bar{X} - \mu_0)/\sqrt{\mu_0/n}$. Critical values may be found in either the standard normal table (Table 2) or the "∞" row of the t-table (Table 4). As a rule of thumb, the test may be applied whenever $n\mu_0 > 30$. If $n = 36$ and $\bar{X} = 4.1$, test the null hypothesis $H_0: \mu = 3.0$ against the alternative hypothesis $H_0: \mu > 3.0$ at the 5% level.

50. The number of cars exceeding the speed limit on a certain highway in any given daylight hour was found to be a Poisson random variable with mean $\mu = 8$. After an intensive enforcement effort, the number of speeding cars was recorded for 40 randomly chosen daylight hours, and the sample mean was found to be $\bar{X} = 6.8$ speeding cars per hour. Do these data indicate that the enforcement campaign was effective in reducing the number of speeding cars? Use $\alpha = .05$. (See Problem 49.)

51. The mean and standard deviation of a sample of size 15 from a normal population are $\bar{X} = 11.3$ and $s = 4.1$, respectively. Find all values of μ_0 for which the null hypothesis $H_0: \mu = \mu_0$ is accepted in testing against the alternative hypothesis $H_a: \mu \neq \mu_0$ at the 5% level. Show that this interval of values is the same as the 95% confidence interval for μ.

Summary and Keywords

The **hypothesis-testing problem** (p. 264) is to choose, on the basis of a sample, between a **null hypothesis** and an **alternative hypothesis** (p. 264). The two hypotheses together make up the **model** (p. 264), which is a description of the form of the population in question (for example, normal or binomial). A **test** (p. 264) consists of a rule for deciding, on the basis of the sample, between the null and alternative hypotheses, and it is specified by the **critical** or **rejection region** (p. 265). The **level** or **size** (p 265) of the test is the probability that it will falsely lead to a rejection of the null hypothesis. Hypotheses concerning the mean of a normal population are based on the t-statistic, and the form of the rejection region depends on whether the problem is **one-sided** or **two-sided** (p. 270). For tests on a binomial proportion, based on the standardized number of successes in a sample, the form of the test again depends on the form of the testing problem. A widely used alternative approach to testing is through **p-values** or **significance levels** (p. 279). The general theory of testing concerns balancing the level, or the probability of a **Type I error** (p. 281), against the **power** (p. 281), or (what amounts to the same thing) against the probability of a **Type II error** (p. 281).

REFERENCES

8.1 Hoel, Paul G. *Elementary Statistics*. 3d ed. New York: John Wiley & Sons, 1971. Chapter 8.

8.2 Hogg, Robert V., and Craig, Allen T. *Introduction to Mathematical Statistics*. 3d ed. New York: Macmillan, 1970. Chapter 9.

8.3 Mosteller, Frederick; Rourke, Robert E. K.; and Thomas, George B., Jr. *Probability with Statistical Applications*. 2d ed. Reading, Mass.: Addison-Wesley, 1970. Chapter 9.

8.4 Snedecor, George W., and Cochran, William G. *Statistical Methods*. 6th ed. Ames, Iowa: Iowa State University Press, 1967. Chapters 1–3.

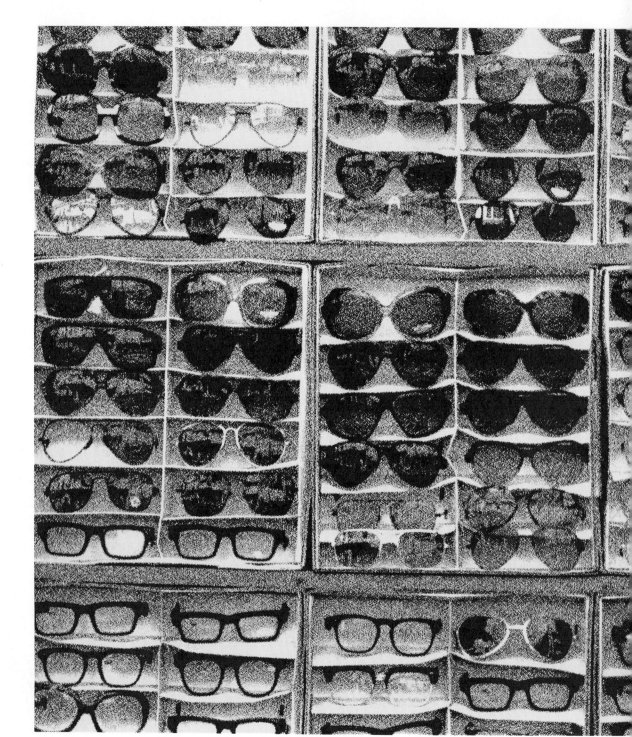

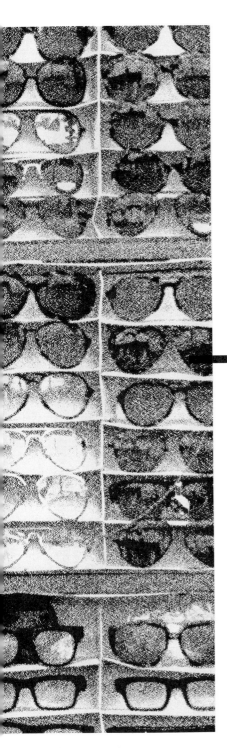

9

Two-sample Techniques and Paired Comparisons

9.1 Introduction

We have looked at some of the inferences that are possible when we have a single random sample from a normal population, and also when our random sample is assumed to have been taken from a binomial population. We now consider some of the inferences that may be made when we have two independent random samples, one from each of two populations.

For example, for the Public Health Service high-blood-pressure study, Table 1.2 (page 3) shows that there were 37 who experienced hypertensive events in the sample of 193 patients on the active drug, and there were 89 who experienced hypertensive events in the sample of 196 patients on placebo. The problem is to compare the two ratios, $\frac{37}{193}$ and $\frac{89}{196}$, to see whether the drug really does reduce the incidence of hypertensive events. This is an inference problem involving samples from two binomial populations. Before studying problems of that kind, we will look into inference problems involving samples from two normal populations.

9.2 Samples from Two Normal Populations

Table 2.1 (page 13) lists the change in blood pressure of 171 patients given a blood-pressure-reducing drug. Denote these numbers by $X_{11}, X_{12}, \ldots, X_{1n_1}$, where n_1 is the sample size, 171. Let $\bar{X}_1$ be the mean: $\bar{X}_1 = n_1^{-1} \sum_{j=1}^{n_1} X_{1j}$. In fact, $\bar{X}_1 = -9.85$. Table 2.2 lists the change in blood pressure of 176 patients given a placebo. Denote these numbers by $X_{21}, X_{22}, \ldots, X_{2n_2}$, where $n_2 = 176$ is the sample size, and let $\bar{X}_2 = n_2^{-1} \sum_{j=1}^{n_2} X_{2j}$ be their mean; in fact, $\bar{X}_2 = .24$. The problem is to draw inferences about the difference between **the means of the two different populations**, and the way to do this is to compare $\bar{X}_1$ and $\bar{X}_2$.

the means of the two different populations

The general setup is this: Let $X_{11}, X_{12}, \ldots, X_{1n_1}$ represent the observed values in a sample of size n_1 taken from a normal population with mean μ_1 and variance σ^2, and let $X_{21}, X_{22}, \ldots, X_{2n_2}$ be the observations in a sample of size n_2 taken from a second normal population with mean μ_2 and variance σ^2 (the same σ^2 as for the first population). When we have two samples, an observation X_{ij} has two subscripts; the first denotes the sample, and the second designates the particular element within the sample. Thus X_{24} is the fourth value in the second sample.

A study of the two-sample situation outlined in the preceding paragraph reveals three basic assumptions:

SECTION 9.2 SAMPLES FROM TWO NORMAL POPULATIONS

Assumption 1   Independent random samples.

Assumption 2   Normal populations.

Assumption 3   A common variance for the two populations.

The methods to be presented here are valid only if these assumptions are satisfied. However, we later consider briefly the consequences of relaxing these assumptions.

Data that satisfy our three assumptions and that fit exactly into the framework of the two-sample situations may be obtained as the result of an experiment for comparing the effects of two treatments. In statistical usage, **treatments** are any procedures, methods, or stimuli whose effects we want to estimate and compare. Treatments in one situation might represent different machines, but in another they could be different operators; they could be different chemicals, or different rates of application of one chemical; and so on. In the Public Health Service hypertension study, one treatment is the blood-pressure-reducing drug and the second treatment is the placebo (really no treatment at all).

treatments

Suppose that we want to compare the effect of two treatments when they are applied to similar experimental material. We can divide the experimental material into a number of *experimental units* and then assign a treatment to each experimental unit at random by tossing a coin or by using a table of random numbers. The purposes of randomization will be discussed in Chapter 12; see also Chapter 1, p. 5.

After the treatments are applied, we obtain an objective measure of the effect of a given treatment from each unit to which the treatment was applied, so that we have a set of values that may be classified into two groups according to the treatments that produced them. If we assume (1) that the responses of all experimental units to which a particular treatment could be applied have a normal distribution, (2) that the variance among the experimental units treated alike is the same for both treatments, and (3) that the responses constitute random samples from the distributions of all such responses, then our three assumptions are satisfied and the methods appropriate to the two-sample situation may be used to estimate the difference between the effects of the treatments and to test the hypothesis that there is no difference in their effects.

Simple experimental designs of this sort are of the class known as *completely randomized* designs. These will be covered in more detail in Chapter 12.

After the data have been obtained, either by selecting the samples or by performing an experiment, the information they contain may be summarized in tabular form by means of the **summary table** shown in Table 9.1.

summary table

Table 9.1 Summary Table for Two Samples from Continuous Populations

Sample	Size	df	Mean	Sum of Squares
1	n_1	$n_1 - 1$	$\bar{X}_1$	$\Sigma (X_{1j} - \bar{X}_1)^2$
2	n_2	$n_2 - 1$	$\bar{X}_2$	$\Sigma (X_{2j} - \bar{X}_2)^2$
Totals	$n_1 + n_2$	$n_1 + n_2 - 2$		Pooled SS

pooled sum of squares

The **pooled sum of squares**, or the pooled SS, is obtained by adding together the sums of squares for the individual samples in the last column of the table. In practical problems these sums of squares would ordinarily be obtained by means of the machine formulas:

$$\Sigma (X_{1j} - \bar{X}_1)^2 = \Sigma X_{1j}^2 - \frac{(\Sigma X_{1j})^2}{n_1}$$

$$\Sigma (X_{2j} - \bar{X}_2)^2 = \Sigma X_{2j}^2 - \frac{(\Sigma X_{2j})^2}{n_2}$$

After the information contained in the data has been summarized in the table, we turn our attention to the estimation of the parameters of interest. As one might suppose, the best unbiased estimators for the population means μ_1 and μ_2 are the respective sample means $\bar{X}_1$ and $\bar{X}_2$, and the best unbiased estimator for the difference between the means, $\delta = \mu_1 - \mu_2$, is the difference between the sample means, $d = \bar{X}_1 - \bar{X}_2$. The latter statistic is of particular interest because two populations are involved, and our primary concern is making inferences about the difference between the means. In experimental situations the difference between the means is also the difference between the effects of the two treatments.

Since the variance is assumed to be the same for both populations, we could obtain an unbiased estimate for σ^2 from each sample separately, but the best unbiased estimate is obtained by pooling the information contained in the two samples to get the pooled estimate of variance:

$$s^2 = \frac{\text{Pooled SS}}{\text{Pooled df}} = \frac{\Sigma (X_{1j} - \bar{X}_1)^2 + \Sigma (X_{2j} - \bar{X}_2)^2}{n_1 + n_2 - 2} \qquad 9.1$$

It is easy to show that this pooled estimate of variance is the weighted mean of the individual sample estimates, where the weights are the degrees of freedom; that is,

SECTION 9.2 SAMPLES FROM TWO NORMAL POPULATIONS

$$s^2 = \frac{(n_1 - 1)s_1^2 + (n_2 - 1)s_2^2}{n_1 + n_2 - 2} \qquad 9.2$$

where

$$s_1^2 = \frac{\Sigma(X_{1j} - \bar{X}_1)^2}{n_1 - 1}$$

$$s_2^2 = \frac{\Sigma(X_{2j} - \bar{X}_2)^2}{n_2 - 1}$$

To find confidence intervals for the means or test hypotheses about them, we must first find estimates for the variances and standard deviations of the estimators. As is always the case, the variances of the means are estimated by dividing the estimate for the population variance by the appropriate sample size; that is,

$$s_{\bar{X}_1}^2 = \frac{s^2}{n_1}$$

$$s_{\bar{X}_2}^2 = \frac{s^2}{n_2} \qquad 9.3$$

pooled variance estimator   where s^2 is our **pooled variance estimator** as defined in Equation (9.1). The estimated standard deviations of the sample means are the corresponding square roots.

According to Equation (5.10), page 176, if X and Y are independent random variables, then their variances add:

$$\text{Var}(X + Y) = \text{Var}(X) + \text{Var}(Y)$$

Since the distribution curve for $-Y$ is the mirror image of that for Y, their measures of spread are the same:

$$\text{Var}(-Y) = \text{Var}(Y)$$

Hence $X - Y$, or $X + (-Y)$, has for its variance the sum

$$\text{Var}(X) + \text{Var}(-Y) = \text{Var}(X) + \text{Var}(Y)$$

Thus we have the formula

$$\text{Var}(X - Y) = \text{Var}(X) + \text{Var}(Y)$$

One might expect a minus on the right, but that is incorrect (for one thing, it could even give a negative variance for $X - Y$).

Since $\bar{X}_1$ and $\bar{X}_2$ are the means of independent random samples, they are independent of each other. Under the assumption of common variance, they have variances σ^2/n_1 and σ^2/n_2, respectively; therefore, the variance of their difference $d = \bar{X}_1 - \bar{X}_2$ is

$$\sigma_d^2 = \frac{\sigma^2}{n_1} + \frac{\sigma^2}{n_2} \qquad 9.4$$

The *estimated* variance of $\bar{X}_1 - \bar{X}_2$ is

$$s_d^2 = s^2\left(\frac{1}{n_1} + \frac{1}{n_2}\right) \qquad 9.5$$

where s^2 is the pooled estimate of the common variance σ^2. If the sample sizes are equal (if $n_1 = n_2 = n$), the estimated variance of d, Equation (9.5), reduces to

$$s_d^2 = \frac{2s^2}{n} \qquad 9.6$$

Example 1

To compare the durabilities of two paints for highway use, 12 four-inch lines of each paint were laid down across a heavily traveled road. The order was decided at random. After a period of time, reflectometer readings were obtained for each line (the higher the readings, the greater the reflectivity). The data are as follows:

Paint A	12.5	11.7	9.9	9.6	10.3	9.6
	9.4	11.3	8.7	11.5	10.6	9.7
Paint B	9.4	11.6	9.7	10.4	6.9	7.3
	8.4	7.2	7.0	8.2	12.7	9.2

For these sets of data we find

Paint A	Paint B
$\Sigma X_{1j} = 124.8$	$\Sigma X_{2j} = 108.0$
$\bar{X}_1 = 10.4$	$\bar{X}_2 = 9.0$
$\Sigma X_{1j}^2 = 1312.00$	$\Sigma X_{2j}^2 = 1010.64$
$(1/n_1)(\Sigma X_{1j})^2 = 1297.92$	$(1/n_2)(\Sigma X_{2j})^2 = 972.00$
$\Sigma(X_{1j} - \bar{X}_1)^2 = 14.08$	$\Sigma(X_{2j} - \bar{X}_2)^2 = 38.64$

SECTION 9.2 SAMPLES FROM TWO NORMAL POPULATIONS

The summary table is

Paint	n	df	Mean	Sum of Squares
A	12	11	10.4	14.08
B	12	11	9.0	38.64
Sums	24	22		52.72

Hence   *d = difference between the means.*

$$d = \bar{X}_1 - \bar{X}_2 = 1.40$$

The pooled estimate of variance is

$$s^2 = \frac{52.72}{22} = 2.40$$

The estimated variance of the difference between the means is

$$s_d^2 = s^2\left(\frac{1}{n} + \frac{1}{n}\right) = 2.40\left(\frac{2}{12}\right) = 0.40$$

and the estimated standard deviation of d is

$$s_d = \sqrt{0.40} = 0.63$$

$s_d = s\sqrt{\frac{1}{n_1} + \frac{1}{n_2}}$

In the next two sections we'll see how to use these statistics for inference—for estimation and for hypothesis testing.

Problems

1. Given the following summary table, and assuming common variance, find the best estimates for μ_1, $\mu_2 - \mu_1$, σ^2, $\sigma_{\bar{X}_1}^2$, σ_d^2.

Sample	n	$\bar{X}_i$	$\Sigma(X_{ij} - \bar{X}_i)^2$
1	10	14	240
2	16	42	480

1 Jun 83 notes

2. Given the following summary table, and assuming common variance, find the best estimates for

a. $\mu_1, \mu_2, \mu_1 - \mu_2$. **b.** $\sigma^2, \sigma_{\bar{X}_2}^2, \sigma_d^2$.

Sample	n	$\bar{X}_i$	$\Sigma(X_{ij} - \bar{X}_i)^2$
1	21	12	170
2	6	9	80

3. Find the pooled variance for the following two sets of data.

Sample	Data
1	2 3 4 5
2	3 5 7

4. The following data are the ages (to the nearest $\frac{1}{2}$ year) of four randomly selected college freshmen and six randomly selected college seniors. Find the estimated standard deviation of $\bar{X}_1 - \bar{X}_2$.

	Ages
Freshmen	17.5 18.0 18.5 18.0
Seniors	21.0 20.5 21.5 22.0 21.5 21.0

9.3 Confidence Intervals for $\mu_1 - \mu_2$ and μ_i

It is assumed that the populations from which our data were obtained have normal distributions; therefore, the sample means and the difference between them will be normally distributed. Let d be the difference between the sample means: $d = \bar{X}_1 - \bar{X}_2$; let δ be the difference between the population means: $\delta = \mu_1 - \mu_2$. Then d is an unbiased estimator of δ, and s_d^2, defined by (9.5), is an unbiased estimator for the variance of d. Furthermore,

$$t = \frac{d - \delta}{s_d} = \frac{(\bar{X}_1 - \bar{X}_2) - (\mu_1 - \mu_2)}{s\sqrt{\frac{1}{n_1} + \frac{1}{n_2}}} \qquad 9.7$$

has a t-distribution with $n_1 + n_2 - 2$ degrees of freedom. Recall (see Sec-

SECTION 9.3 CONFIDENCE INTERVALS FOR $\mu_1 - \mu_2$ AND μ_i

tion 7.4) that the percentage point t_{h, n_1+n_2-1} for the t-distribution for $n_1 + n_2 - 1$ degrees of freedom is defined by the requirement

$$P(t < -t_{h, n_1+n_2-1}) = P(t > t_{h, n_1+n_2-1}) = h.$$

If $h = \alpha/2$, then

$$P(-t_{(\alpha/2), n_1+n_2-1} \leq t \leq t_{(\alpha/2), n_1+n_2-1}) = 1 - \alpha$$

Therefore a $100 \times (1 - \alpha)\%$ confidence interval for δ is that interval between the following quantities:

$$L = d - t_{(\alpha/2), n_1+n_2-2} s_d$$
$$R = d + t_{(\alpha/2), n_1+n_2-2} s_d$$

9.8

In Example 1 in the preceding section, the degrees of freedom are 22. For a 95% confidence interval, $\alpha = .05$ and $\alpha/2 = .025$; the corresponding percentage point on the t-distribution (Table 4 in the Appendix) is $t_{.025, 22} = 2.074$. Therefore the 95% confidence limits for the difference between the mean reflectivities of the two paints are

$$L = 1.40 - (2.074)(.63) = .09$$

and

$$R = 1.40 + (2.074)(.63) = 2.71$$

We can be 95% confident that the difference between the actual means is between .09 and 2.71.

Note that this confidence interval is for $\delta = \mu_1 - \mu_2$. If we wanted the interval for $\mu_2 - \mu_1$, the estimator would be $\bar{X}_2 - \bar{X}_1$, or -1.40, and the confidence limits would be -2.71 and $-.09$ (the variance of $\bar{X}_2 - \bar{X}_1$ is the same as the variance of $\bar{X}_1 - \bar{X}_2$).

We find a confidence interval for the means of either or both of the populations exactly as we found a confidence interval for the mean of a normal population in Section 7.5, except that the pooled estimate of variance is used to calculate the standard error of $\bar{X}_i$ (where i is 1 or 2), rather than the separate estimate of variance that could be obtained from the data in the ith sample. The t-statistic,

$$t = \frac{\bar{X}_i - \mu_i}{s_{\bar{X}_i}}$$

9.9

will therefore have the pooled degrees of freedom $n_1 + n_2 - 2$, and the confidence limits for μ_i are

$$L = \bar{X}_i - t_{(\alpha/2), n_1+n_2-2} s_{\bar{X}_i}$$
$$R = \bar{X}_i + t_{(\alpha/2), n_1+n_2-2} s_{\bar{X}_i}$$

9.10

For Example 1 in the preceding section, 99% confidence limits for the mean reflectivity of paint B are

$$L = 9 - (2.819)\sqrt{\frac{2.40}{12}} = 7.74$$

and

$$R = 9 + (2.819)\sqrt{\frac{2.40}{12}} = 10.26$$

where $2.819 = t_{.005,\ 22}$.

Problems

5. Find 95% confidence intervals for
   a. $\mu_2 - \mu_1$ in Problem 1.   **b.** $\mu_2 - \mu_1$ in Problem 2.

6. In an experiment designed to compare the mean service lives of two types of tires it was found that the difference between the means was 2,250 miles, and that the pooled estimate of variance was equal to 625,000. There were 20 tires of each type. Find a 95% confidence interval for the difference between the mean lives of these types of tires.

7. Fifteen of each of two types of fabricated wood beams were tested for breaking load with the following results. Find a 98% confidence interval for the difference between the mean breaking loads.

Type	n	$\bar{X}_i$	$\Sigma (X_{ij} - \bar{X}_i)^2$
I	15	1,560	48,000
II	15	1,600	36,000

8. Measurements of tensile strength of two types of rubber were taken on ten pieces of each kind. The mean strengths of the two samples were 3,475 and 3,000 pounds per square inch, respectively, with corresponding estimated standard deviations of 695 and 385 pounds per square inch. Find a 99% confidence interval for the difference in strength. Would this interval lead you to conclude that there is a real difference?

9. Given the following summary table,
   a. Find 90% confidence intervals for μ_1, μ_2, and $\mu_2 - \mu_1$.

SECTION 9.3 CONFIDENCE INTERVALS FOR $\mu_1 - \mu_2$ AND μ_i **303**

b. Would the confidence interval for $\mu_2 - \mu_1$ lead you to conclude that the means are different? Why?

Sample	n	$\bar{X}_i$	$\Sigma (X_{ij} - \bar{X}_i)^2$
1	9	112	48
2	16	118	136

10. The table below gives for a certain crime the time (in minutes) spent in court per appearance there for each of two jurisdictions. Thus the $\bar{X}_i$ of 8.9 means that for the 50 appearances in the first jurisdiction, the average time in court was 8.9 minutes.
   a. Find 90% confidence intervals for the two population means.
   b. On the basis of a 95% confidence interval, would you conclude that the two jurisdictions are different in respect of time in court per appearance?

Jurisdiction	n	$\bar{X}_i$	$\Sigma (X_{ij} - \bar{X}_i)^2$
1	50	8.9	105
2	52	10.6	95

11. Find 98% confidence intervals for μ_1, μ_2, and $\mu_1 - \mu_2$ of Problem 2.

12. Samples of ore were taken from two locations, and the percentage of copper in each sample was measured. The data are given below. Find a 90% confidence interval for the mean difference in the percentage of copper in the ore from the two locations.

Location	Copper (%)				
1	5.5	5.9	6.2	5.8	
2	4.2	3.9	4.1	3.8	4.1

13. Nineteen plants of a certain species were divided into two groups. One group received normal lighting, while the other received special lighting that had certain wavelengths filtered out. At the end of a specified growing period, the plant biomass in grams was measured. The results are summarized below. Find a 99% confidence interval for the mean difference in plant biomass for the two groups.

Group	n	$\bar{X}$	s
Normal lighting	9	5.3	1.10
Filtered light	10	2.1	.69

14. A company considering one of two makes of automobiles for its fleet road-tested four automobiles of each make. The gas mileage of the eight automobiles is given below. Find a 95% confidence interval for the mean difference in gas mileage.

Auto Make	Miles per Gallon			
1	18.5	19.2	17.6	17.9
2	21.4	23.6	20.5	23.2

15. Seven subjects were randomly selected for a study of pain-relievers. Four were given pain-reliever A, and three were given pain-reliever B. The amount of pain reliever in the bloodstream was measured 10 minutes after ingestion of the drug. The data are given below. Find a 95% confidence interval for the difference between the mean amounts of drug A and drug B in the bloodstream.

Drug	Amount of Pain-Reliever (mg/100 ml)			
A	32	35	36	33
B	25	27	28	

9.4 Testing the Hypothesis $\mu_1 = \mu_2$

To test the hypothesis that the difference between the means is equal to a given value, δ_0, we again rely on the quantity defined in Equation (9.7), for under the hypothesis $H_0: \mu_1 - \mu_2 = \delta_0$, the statistic

$$t = \frac{\bar{X}_1 - \bar{X}_2 - \delta_0}{s_d} \qquad 9.11$$

SECTION 9.4 TESTING THE HYPOTHESIS $\mu_1 = \mu_2$

has a t-distribution with $n_1 + n_2 - 2$ degrees of freedom. The test is a simple t-test.

A hypothesis of particular interest in the two-sample case, or in the case of comparing two treatments in a completely randomized experiment, is the one that states that there is no difference between the means—$H_0 : \mu_1 = \mu_2$. Under this hypothesis, $\delta_0 = 0$ and the test statistic as given in Equation (9.11) reduces to

$$t = \frac{\bar{X}_1 - \bar{X}_2}{s_d} \qquad 9.12$$

Here, as with other t-tests, the critical region depends upon the alternative hypothesis. If it states simply that the means are not equal ($H_a : \mu_1 \neq \mu_2$), we use the two-tailed critical region $|t| \geq t_{(\alpha/2), n_1+n_2-2}$. If the alternative is one-sided, i.e., $H_a : \mu_1 > \mu_2$ or $H_a : \mu_1 < \mu_2$, we use the one-tailed critical region $t > t_{\alpha, n_1+n_2-2}$ or $t < -t_{\alpha, n_1+n_2-2}$.

For a one-tailed test, there is often some doubt as to which tail should be used. This difficulty is resolved if we always write our t, Equation (9.12), and the alternative hypothesis in such a way that the subscripts are in the same order. If the alternative is $H_a : \mu_2 > \mu_1$, and if we write the numerator of our t-statistic as $\bar{X}_2 - \bar{X}_1$, then the inequality in H_a can be looked on as an arrowhead pointing to the right and we use the upper tail. If the alternative hypothesis is written as $H_a : \mu_1 < \mu_2$, we use $\bar{X}_1 - \bar{X}_2$, and the inequality sign then points to the left, showing that we use the lower tail for the critical region.

The determination of the proper tail to use can also be seen this way. For the testing problem

$$H_0 : \mu_1 = \mu_2$$

$$H_a : \mu_1 > \mu_2$$

we ought to reject H_0 in favor of H_a if $\bar{X}_1$ (which estimates μ_1) is greater than $\bar{X}_2$ (which estimates μ_2) by an excessive amount. Since this happens exactly when $\bar{X}_1 - \bar{X}_2$ is greater than 0 by an excessive amount, we reject H_0 if the value of the statistic (9.12) is greater than t_{α, n_1+n_2-2}. Similarly, for the problem

$$H_0 : \mu_1 = \mu_2$$

$$H_a : \mu_1 < \mu_2$$

we should prefer H_a to H_0 if $\bar{X}_1$ is greatly less than $\bar{X}_2$, or if $\bar{X}_1 - \bar{X}_2$ is greatly less than 0, so we reject H_0 if the value of the statistic (9.12) is less than $-t_{\alpha, n_1+n_2-2}$.

Example 1

Problem: The weight gains (in ounces) of infants over a six-week period were recorded for each of two diets. The following table summarizes the data. Is there reason to believe that the second diet leads to greater weight gains than the first—that μ_2 is larger than μ_1? Let $\alpha = .05$.

Diet	n	df	$\bar{X}_i$	$\Sigma (X_{ij} - \bar{X}_i)^2$
1	15	14	21	1224
2	9	8	29	756
		22		1980

$$s^2 = \frac{1980}{22} = 90, \quad s_d^2 = 90\left(\frac{1}{15} + \frac{1}{9}\right) = 16$$

Solution: The hypothesis is $H_0: \mu_1 = \mu_2$. The alternative hypothesis is $H_a: \mu_2 > \mu_1$; therefore, we calculate

$$t = \frac{\bar{X}_2 - \bar{X}_1}{s_d} = \frac{29 - 21}{4} = 2$$

The critical region is $t > t_{.05, 22} = 1.717$. The calculated t is greater than the tabular value; therefore, we reject the hypothesis and conclude that $\mu_2 > \mu_1$.

Example 2

To test the hypothesis that in the example of Section 9.2 the two paints are equal in reflectivity, we test the hypothesis $H_0: \mu_1 = \mu_2$ against the alternative $H_a: \mu_1 \neq \mu_2$. The t-statistic is

$$t = \frac{\bar{X}_1 - \bar{X}_2}{s_d} = \frac{1.40}{.63} = 2.222$$

The critical region for $\alpha = .05$ is $|t| \geq t_{.025, 22} = 2.074$. Our calculated t falls into the upper tail of the critical region; therefore, we reject the null hypothesis and conclude that the means are not equal.

Example 3

Consider once again the data in Tables 2.1 and 2.2 (pages 13 and 14). The sample means and variances are:

SECTION 9.4 TESTING THE HYPOTHESIS $\mu_1 = \mu_2$

Treatment	n	df	Mean	Variance
1. Active drug	171	170	$\bar{X}_1 = -9.85$	$s_1^2 = 133.40$
2. Placebo	176	175	$\bar{X}_2 = .24$	$s_2^2 = 123.88$
Sums	347	345		

Hence

$$d = \bar{X}_1 - \bar{X}_2 = -9.85 - .24 = -10.09$$

$$s_1 = 11.55, \quad s_2 = 11.13$$

The pooled estimate of variance is

$$s^2 = \frac{170(133.40) + 175(123.88)}{170 + 175} = 128.57$$

The estimated variance of $\bar{X}_1 - \bar{X}_2$ is

$$s_d^2 = s^2(\tfrac{1}{171} + \tfrac{1}{176}) = 1.48$$

and $s_d = 1.22$. The degrees of freedom are 345, which is so large that we can use the bottom line of Table 4 in the Appendix. In other words, the t-statistic (9.7) has nearly the standard normal distribution.

Let us test the null hypothesis $\mu_1 = \mu_2$ (the hypothesis that the drug is no more effective than the placebo in reducing blood pressure) against the alternative hypothesis that $\mu_1 < \mu_2$ (the hypothesis that the drug does help). This is a one-sided hypothesis, and the cutoff point for the 5%-level test is $-Z_{.05} = -1.645$. The value of the t-statistic is

$$t = \frac{\bar{X}_1 - \bar{X}_2}{s_d} = \frac{(-9.85) - (.24)}{1.22} = -8.27$$

This is below -1.645, and so we reject the null hypothesis and conclude that the drug does help. For a 0.5%-level test, the cutoff point is $-Z_{.005} = -2.576$, and the data lead to the rejection of the null hypothesis even at this smaller level.

In fact the t-value -8.27 is so highly significant that no tables are really necessary: it is "obvious" that the data in the summary table above lead to the conclusion that the drug is effective in reducing blood pressure. But it is "obvious" only because the difference $\bar{X}_1 - \bar{X}_2$ is obviously large in comparison with the standard error s_d. What is not obvious is a fact which statistical theory tells us, namely, the fact that s_d is the right measure of the variability in $\bar{X}_1 - \bar{X}_2$. It is statistical theory that tells us that the magnitude of $\bar{X}_1 - \bar{X}_2$ is to be judged by comparing it with s_d.

Problems

16. Five students were taught touch typing by method 1 and four were taught by method 2. The numbers of words per minute for these nine randomly selected students are given below. Is there a statistically significant difference in the mean words per minute for the two teaching methods? Use $\alpha = .05$.

Method	Words per Minute
1	41 33 38 42 44
2	56 45 50 51

17. Independent random samples from two normal populations with common variance gave the following results. At the 5% level of significance, can we conclude that the mean of the first population is greater than that of the second?

Sample	n	$\bar{X}_i$	$\Sigma (X_{ij} - \bar{X}_i)^2$
1	15	35.2	35
2	12	34.0	25

18. Two techniques for a certain laboratory test are compared to determine which one is more efficient. The average number of tests per day that a person can perform is recorded for each of the two methods. Given the following summary data, would you be justified in concluding that the second method is preferable? Let $\alpha = .01$.

Method	Number of Days	$\bar{X}_i$	$\Sigma (X_{ij} - \bar{X}_i)^2$
1	10	43	168
2	15	48	200

19. Given the following data for two samples from normal populations with the same variance, test the following hypotheses.
   a. $H_0: \mu_1 = 25$, $H_a: \mu_1 < 25$, $\alpha = .05$.
   b. $H_0: \mu_2 = 22$, $H_a: \mu_2 > 22$, $\alpha = .01$.
   c. $H_0: \mu_1 = \mu_2$, $H_a: \mu_1 \neq \mu_2$, $\alpha = .05$.

SECTION 9.4 TESTING THE HYPOTHESIS $\mu_1 = \mu_2$ **309**

Sample	n	$\bar{X}_i$	$\Sigma (X_{ij} - \bar{X}_i)^2$
1	15	22	1224
2	9	25	756

20. A business consultant suggested to a company that its sales would increase if the company's salesmen would wear more conservative clothes. As a test, 15 randomly selected salesmen were asked to wear conservative business suits on the job for one month while 12 others were to wear their usual casual attire. The numbers of sales for the two groups were recorded, and the data are summarized below. Does it appear that conservative attire for the salesmen will increase the average monthly sales volume? Use $\alpha = .05$.

Group	n	$\bar{X}$	s
Conservative	15	37	5.5
Casual	12	28	6.1

21. The following tables summarize some of the background characteristics of two groups of judges. The first group consists of those selected by their colleagues as commanding their respect, and the second group consists of those not so selected. For group 1, $n = 17$; for group 2, $n = 54$. For each of the four characteristics, test at the 5% level the hypothesis that the population means are the same.

Age

Group	$\bar{X}_i$	s_i
1	59.737	9.244
2	50.300	7.031

Years at School

Group	$\bar{X}_i$	s_i
1	18.700	3.302
2	14.321	3.592

Years on Bench

Group	$\bar{X}_i$	s_i
1	12.358	4.319
2	14.123	4.930

Number of Visits to Prisons

Group	$\bar{X}_i$	s_i
1	1.354	.812
2	.503	.756

310   CHAPTER 9   TWO-SAMPLE TECHNIQUES AND PAIRED COMPARISONS

22. The following are yields in bushels per acre for two oat varieties. Each was tried on eight plots. Can we conclude that the yields are the same for the two varieties, A and B? Use $\alpha = .1$.

A	71.2	72.6	47.8	76.9	42.5	49.6	62.8	48.2
B	56.6	60.7	45.4	73.0	42.8	65.2	41.7	57.3

23. The following are percentages of fat found in samples of two types of meat. Do the meats have different fat contents?

Meat A	30	26	30	19	25	37	27	38	26	31
Meat B	40	34	28	29	26	36	28	37	35	42

24. To test the effect of controlled grazing versus continuous grazing, 32 steers were tested: 16 steers were treated each way for a period of time. The gain in weight in pounds for each animal is given below. Can we conclude there is a difference in the meat gain for the two procedures? Use $\alpha = .1$.

Controlled Grazing		Continuous Grazing	
45	62	94	12
96	128	26	89
120	99	88	96
28	50	85	130
109	115	75	54
39	96	112	69
87	100	104	95
76	80	53	21

25. Two different acids were tested for their effects on the yield of a dye.

Acid 1	Acid 2
134.5	142.0
135.0	150.0
141.5	151.5
143.5	152.0
146.0	152.5
135.0	134.5
134.5	138.5
144.0	146.0

Eight runs were made with each type. The yields above were obtained. Test the hypothesis that there is no difference in their effects. Use $\alpha = .05$.

26. Of 22 college-age women athletes selected for a study on the effect of weight training in distance running, 12 were given a weight-training program while the 10 others served as a control group. At the end of the program both groups were timed in a two-mile run; the data are summarized below. Has the weight-training program been effective in reducing the mean running time for two miles? Use $\alpha = .05$.

Group	n	$\bar{X}$	s
Weight-trained	12	14.5	1.3
Control	10	16.6	1.7

27. A sociologist asked randomly selected workers in two different industries to fill out a questionnaire on job satisfaction. The answers were scored 1 to 20, with higher scores indicating greater job satisfaction. The data are given below. Does it appear that there is a significant difference in the job satisfaction of the workers in the two industries? Use $\alpha = .05$.

Industry	n	$\bar{X}$	s
A	30	15.3	2.2
B	28	12.1	3.1

9.5 Which Assumptions Can Be Relaxed?

In Section 9.2 we stated three assumptions under which the methods of Sections 9.3 and 9.4 are valid. Let us now reconsider those assumptions and see which, if any, can be relaxed. The first, randomness, must be satisfied if we are to make probability statements in connection with our inferences.

As is frequently the case, the second assumption, normality, is not a strong assumption, for, because of the central limit theorem, if the samples are sufficiently large, even large departures from normality will not affect the probabilities to any great extent.

The third assumption, common or homogeneous variance, cannot be relaxed. If the population variances are not the same, a significant result for the *t*-test may be due to the different variances and not to different means. If this assumption is not satisfied, there is no exact test of the hypothesis $H_0: \mu_1 - \mu_2 = \delta_0$; however, there are several approximate tests available (see Reference 9.4). It is possible to assess this third assumption by comparing the standard deviations s_1 and s_2 for the two samples. In Example 3 of the preceding section these are 11.55 and 11.13, an indication that the variance is the same in the two populations.

9.6 Paired Comparisons

In the completely randomized experiment for comparing two treatments, the treatments are randomized over, or randomly assigned to, the whole set of experimental units; that is, a random device is used to determine which of the two treatments is applied to any given experimental unit. It is common to restrict the randomization so that each treatment is applied to the same number of units. As we have seen, the data from this type of experiment can be analyzed by using the two-sample techniques. These are also called the method of **group comparisons**.

group comparisons

For this type of experiment the efficiency and the precision of estimation are inversely proportional to the *pooled* estimate of variance, s^2; the smaller the variance, the greater the precision. To increase the precision of an experiment, with a fixed amount of experimental material, we have to alter the design of the experiment in order to reduce the unexplained variation in the data as measured by s^2. Since s^2 is based on the variation among experimental units treated alike, either we must reduce this variation by using more homogeneous experimental material, or we must eliminate some of this variation by using *a priori* information about the responses of the experimental units. It is the latter approach we now consider.

If the experimental units occur in pairs or can be grouped into pairs in such a way that the variation in the responses between the members of any pair is less than the variation between members of different pairs, we can improve the efficiency of our experiment by randomizing the two treatments over the two members of each pair. We restrict our randomization so that each treatment is applied to one member of each pair; hence, we obtain a separate estimate of the difference between the treatment effects from each pair, and the variation between, or among, the pairs is not included in our estimate of the variance. If the variation among the pairs of units is large relative to the variation within the pairs, the variance will be smaller than if a completely randomized design were used.

SECTION 9.6 PAIRED COMPARISONS

For example, suppose we want to compare the efficiencies of two different barnacle-resistant paints and there are ten ships available. We could randomly assign the two paints to five ships each and then, after a suitable length of time, obtain a measure of the weight of barnacles clinging to each ship. Since the variance would be based on the variation among ships painted alike, we could expect it to be large, because the ships would very likely be sailing in different waters for different periods of time.

On the other hand, where the port side of a hull goes, the starboard goes also; therefore, we would probably have a much more precise experiment if we were to paint one side of each hull with one paint, and the other side with the second paint, tossing a coin to decide which side gets which paint. Each ship would then provide a measure of the difference in effectiveness of the two paints, and the variation due to other factors, such as time at sea and the parts of the world where the ships sailed, would be eliminated from our estimate of variance.

These same considerations lead to the use of identical twins or of littermates in animal experiments, and to grouping experimental units into pairs according to some factor such as age, weight, chemical composition, environment, etc. If the pairing is successful, much of the variation due to the factor on which it is based is eliminated from the estimate of variance, and the efficiency of the experiment is increased accordingly. The paired experiment is the simplest example of a class of experimental designs known as the *randomized complete block* designs.

paired comparisons The analysis of **paired comparisons** reduces to the single-sample techniques of Chapters 7 and 8. The data for a paired comparison may be represented as follows:

Pair	Treatment 1	Treatment 2	Difference
1	Y_{11}	Y_{12}	X_1
2	Y_{21}	Y_{22}	X_2
3	Y_{31}	Y_{32}	X_3
⋮	⋮	⋮	⋮
n	Y_{n1}	Y_{n2}	X_n

In this table a single observation is denoted by Y_{ij}, where the first subscript corresponds to the pair, the second to the treatment. For example, Y_{32} is the response of that member of the *third* pair that had the *second* treatment applied to it.

The assumptions are the same as for the group comparisons—independent random observations, normal populations of responses, and common within-treatment variance.

To estimate the mean difference and to test hypotheses about the difference between the treatment effects, we use the differences between the members of each pair,

$$X_i = Y_{i1} - Y_{i2} \qquad 9.13$$

and treat these as a random sample from the population of differences. We then have

$$\bar{X} = \frac{1}{n} \sum X_i \qquad 9.14$$

and

$$s_X^2 = \frac{1}{n-1} \sum (X_i - \bar{X})^2 \qquad 9.15$$

as unbiased estimates for μ_X, the mean difference, and for σ_X^2, the variance of the differences, respectively.

The variance of the mean difference $\bar{X}$ is estimated unbiasedly by

$$s_{\bar{X}}^2 = \frac{s_X^2}{n} \qquad 9.16$$

Since the differences X_i constitute a single random sample from a normal population, the appropriate methods for finding confidence intervals and for testing hypotheses have already been considered in Chapters 7 and 8. Thus $100(1 - \alpha)\%$ confidence limits for μ_X are given by

$$\begin{array}{c} L = \bar{X} - t_{(\alpha/2),\, n-1} s_{\bar{X}} \\ R = \bar{X} + t_{(\alpha/2),\, n-1} s_{\bar{X}} \end{array} \qquad 9.17$$

and the test of the hypothesis $H_0: \mu_X = \mu_0$ is the single-sample t-test with $n - 1$ degrees of freedom based on the statistic

$$t = \frac{\bar{X} - \mu_0}{s_{\bar{X}}} \qquad 9.18$$

The test of the hypothesis of no difference in the effects of the two treatments is the special case $H_0: \mu_X = 0$.

Example 1

In order to determine whether or not a particular heat treatment is effective in reducing the number of bacteria in skim milk, counts were made before and after treatment on 12 samples of skim milk, with the following results.

SECTION 9.6 PAIRED COMPARISONS

The data are in the form of log DMC, the logarithms of direct microscopic counts.

Sample	log DMC Before Treatment	log DMC After Treatment	Difference $X_i = Y_{i1} - Y_{i2}$
1	6.98	6.95	.03
2	7.08	6.94	.14
3	8.34	7.17	1.17
4	5.30	5.15	.15
5	6.26	6.28	−.02
6	6.77	6.81	−.04
7	7.03	6.59	.44
8	5.56	5.34	.22
9	5.97	5.98	−.01
10	6.64	6.51	.13
11	7.03	6.84	.19
12	7.69	6.99	.70
		Total	3.10

For these data,

$$\Sigma X_i = 3.10, \quad \bar{X} = .258$$
$$\Sigma X_i^2 = 2.199, \quad (\Sigma X_i)^2/n = .8008$$
$$\Sigma (X_i - \bar{X})^2 = 2.199 - .8008 = 1.3982$$
$$s_X^2 = \frac{1.3982}{11} = .12711$$
$$s_{\bar{X}}^2 = \frac{.12711}{12} = .0106$$
$$s_{\bar{X}} = .103$$

For the hypothesis of no effect, $H_0: \mu_X = 0$, against $H_a: \mu_X > 0$,

$$t = \frac{.258}{.103} = 2.50$$

From the tables, $t_{.05, 11}$ equals 1.796; therefore, we reject the null hypothesis at the 5% level and conclude that the heat treatment has reduced the number of bacteria.

Problems

28. In order to compare wear in two types of tires, ten pairs were mounted on the rear wheels of ten cars. Each pair consisted of one tire of each type, and the assignment to the left or right wheel was randomized. After the cars were driven under controlled conditions, the amount of wear was recorded for each tire. For each pair of rear tires, the amount of wear on the tire of the second type was subtracted from the amount of wear on the tire of the first type. The mean difference of wear was .83, and the standard deviation of the differences was .1032.
    a. Find a 95% confidence interval for the mean difference in wear for the two types of tires.
    b. Test the hypothesis $H_0: \delta = .80$ at the 1% level.

29. A new reducing diet was tested on nine women selected at random, with the results in the table below.
    a. Give an 80% confidence interval for the mean weight difference in the population represented by the nine women.
    b. At a 1% significance level, would you conclude that the diet is effective?

Woman	Weight Before	Weight After
1	133	132
2	146	143
3	135	137
4	168	160
5	148	151
6	152	148
7	180	171
8	126	120
9	122	124

30. Each of four technicians replaced the picture tubes in each of two brands of television sets. The times to make the replacements are given below. Does it appear that the picture tube for brand A takes less

	Time (minutes)			
Technician	1	2	3	4
Brand A	67	83	93	101
Brand B	75	91	97	115

SECTION 9.6 PAIRED COMPARISONS

time on the average to replace than the picture tube for brand B? Use $\alpha = .05$.

31. To compare the average weight gains of pigs fed two different rations, nine pairs of pigs were used. The pigs within each pair were littermates, the rations were assigned at random to the two animals within each pair, and they were individually housed and fed. The gains, in pounds, after 30 days are given in the following table.
    a. Test the hypothesis that the mean gains are the same against the alternative that feed A produces a larger gain.
    b. Find a 98% confidence interval for the difference between the average gains.

Ration	Litter									Sum
	1	2	3	4	5	6	7	8	9	
A	60	38	39	49	49	62	53	42	58	450
B	53	39	29	41	47	50	56	47	52	414

32. To test the effect on workers of continuous music in a factory, each of ten workers was observed for one month without music and for one month with music. The table below gives the average number of items produced per day.
    a. Test the hypothesis that there was no change in production.
    b. Find a 99% confidence interval for the production difference.

Worker	Without Music	With Music	Worker	Without Music	With Music
1	8.37	7.38	6	4.73	4.90
2	5.01	6.96	7	7.82	8.53
3	7.24	8.03	8	5.67	6.28
4	6.35	6.40	9	8.01	8.59
5	6.24	6.02	10	6.98	7.32

33. The weather bureau measured the ozone levels at five locations in Orange County before and after a cool front moved through the area.
    a. Was there a significant drop in the ozone level? Use $\alpha = .05$.
    b. Make a 90% confidence interval for the mean change in the ozone level.

	Ozone (ppm)				
Location	1	2	3	4	5
Before	.13	.15	.09	.14	.10
After	.10	.12	.07	.10	.08

34. Five randomly selected patients who underwent knee operations were placed in a new rehabilitation program. Data were obtained before and after the program by an instrument designed to measure knee strength. A standard rehabilitation program gives a mean gain in knee strength of $\mu = 10$. Is there evidence to indicate that the new program is an improvement over the standard program? Use $\alpha = .05$.

Patient	1	2	3	4	5
Before	10	17	20	11	28
After	25	35	33	27	49

9.7 Groups versus Pairs

Paired comparisons can give greater precision of estimation than group comparisons, but only if the pairing is effective. To be effective the pairs must be such that the variation among the pairs is greater than the variation between the units within the pairs. The degrees of freedom for the t-test based on paired comparisons are equal to $n - 1$ compared with $2(n - 1)$ for the group comparison based on the same number of experimental units, and since degrees of freedom are like money in the bank, they should not be invested unless a suitable return can be expected.

With pairing, then, as compared with grouping, we lose $n - 1$ degrees of freedom. But if the variance is reduced enough to more than compensate for their loss, then a gain in efficiency is achieved. If, however, the experimental material is nearly homogeneous, we have no justification for pairing, because the variation among pairs will be but little greater (if at all) than that within the pairs. We would squander degrees of freedom without getting a sufficient reduction in the variance.

For data from experiments involving two treatments, it is important that the proper method of analysis be employed. If the data are from a paired

SECTION 9.7 GROUPS VERSUS PAIRS

experiment and if we ignore the pairing and use the group analysis of Sections 9.3 and 9.4, the estimate of variance is inflated by the variation among pairs and will be too large. A difference between the treatment effects that is actually significant might not be detected because of the inflated variance estimate.

To illustrate what can happen when the two-sample techniques are used to analyze paired data, we use the example of Section 9.6 concerning the reduction due to a heat treatment of the number of bacteria in skim milk. The summary table is as follows:

	n	df	$\bar{Y}$	Sum of Squares
Before	12	11	6.721	8.0997
After	12	11	6.463	4.7750
Sums	24	22		12.8747

Here

$$s^2 = \frac{12.8747}{22} = .5852$$

$$s_d^2 = \frac{2(.5852)}{12} = .0975$$

$$s_d = .312$$

For the hypothesis that $\mu_X = 0$,

$$t = \frac{\bar{Y}_1 - \bar{Y}_2}{s_d} = \frac{.258}{.312} = .827$$

which is not significant.

Using this analysis we would conclude that the treatment had no effect. Notice that the standard deviation of the difference between the means is three times as large as when these paired data were analyzed by the proper techniques.

Just as we have run into trouble using group techniques to analyze paired data, we are also in trouble if the data from a group experiment are paired. Random pairing would result in a loss of degrees of freedom, and at the same time the variance estimate could be either larger or smaller than it should be and could lead to the wrong conclusion.

Suppose we were to systematically pair the results of a completely randomized experiment by ranking both sets of response from high to low and then pairing the largest of each group, the next largest, and so on.

Such a procedure would generally result in a variance estimate that is much smaller than it should be, and therefore in *t*-values that are too large. We might well detect a difference that does not exist.

The design of the study dictates the method of analysis that should be used.

Problems

35. Analyze the data of Problem 29 by the method of group comparisons, answering the same questions. Discuss the different results.
36. Repeat Problem 32 ignoring the pairing. Discuss the different results.
37. Ten egg timers were timed in two positions—vertical, and at 20 degrees from vertical. The values in seconds are in the following table.
    a. Treat the data as two groups and calculate the *t* for testing the hypothesis that the mean time is the same in both positions.
    b. Use paired-comparison methods to find the value of *t* for testing the same hypothesis. Note that when differences among timers are eliminated from the variance, as is the case with paired comparisons, the *t*-value is larger.
    c. What conclusions would be reached in these two analyses? Use $\alpha = .01$.

	Timer									
Position	1	2	3	4	5	6	7	8	9	10
Vertical	170	191	205	181	210	192	183	205	185	216
Tipped	160	197	175	181	163	172	177	185	183	177

38. In the experiment described in Problem 22, suppose that plots of ground had been matched by soil types and that each variety of oats had been grown on each type of soil. With the data paired by soil types as given below, would it now appear that variety A has the higher mean yield? Use $\alpha = .05$.

Soil Type	1	2	3	4	5	6	7	8
Variety A	71.2	72.6	47.8	76.9	42.5	49.6	62.8	48.2
Variety B	65.2	60.7	42.8	73.0	41.7	56.6	57.3	45.4

9.8 Two Samples from Binomial Populations

We now suppose that we have two independent random samples, one of size n_1 from a binomial population with a proportion of successes equal to p_1, the other of size n_2 from a binomial population with proportion of successes equal to p_2. We let X_1 and X_2 be the observed number of successes in the first and second samples, respectively.

Our primary objectives will be the estimation of the difference $p_1 - p_2$ and the testing of the hypothesis that $p_1 = p_2$. To achieve these objectives, we make use of the normal approximation to the binomial.

The proportion p_i is estimated unbiasedly by the corresponding sample fraction f_i, that is, X_i/n_i; and the difference $p_1 - p_2$ is estimated unbiasedly by $f_1 - f_2$. The variance of f_i is estimated by

$$s_{f_i}^2 = \frac{f_i(1 - f_i)}{n_i} \qquad 9.19$$

Since f_1 and f_2 are independent, the variance of their difference, $f_1 - f_2$, is the sum of their variances and is estimated by

$$s_{f_1-f_2}^2 = \frac{f_1(1 - f_1)}{n_1} + \frac{f_2(1 - f_2)}{n_2} \qquad 9.20$$

For large n_1 and n_2, $f_1 - f_2$ is approximately normally distributed, so that

$$\frac{(f_1 - f_2) - (p_1 - p_2)}{\sqrt{\frac{p_1(1 - p_1)}{n_1} + \frac{p_2(1 - p_2)}{n_2}}} \qquad 9.21$$

has approximately a standard normal distribution. If we replace the p_i in the denominator by their estimates f_i, we obtain

$$\frac{(f_1 - f_2) - (p_1 - p_2)}{\sqrt{\frac{f_1(1 - f_1)}{n_1} + \frac{f_2(1 - f_2)}{n_2}}}$$

and this too has approximately a standard normal distribution. This leads to the approximate confidence limits

$$L = f_1 - f_2 - t_{(\alpha/2), \infty} \sqrt{\frac{f_1(1 - f_1)}{n_1} + \frac{f_2(1 - f_2)}{n_2}}$$
$$R = f_1 - f_2 + t_{(\alpha/2), \infty} \sqrt{\frac{f_1(1 - f_1)}{n_1} + \frac{f_2(1 - f_2)}{n_2}} \qquad 9.22$$

To test the hypothesis $H_0: p_1 = p_2$ that the proportions of successes in the two populations are the same, we use a test based on the standard normal distribution. The hypothesis states that the proportions are equal to each other but does not specify the common value; therefore, we must estimate the variance of the difference under the assumption that the proportions are the same.

If $p_1 = p_2 = p$, both f_1 and f_2 are unbiased estimates for p, but the best estimate will be obtained by pooling the two samples into one sample of size $n_1 + n_2$ with the pooled number of successes $X_1 + X_2$. The observed fraction for the combined samples,

$$f = \frac{X_1 + X_2}{n_1 + n_2} \qquad 9.23$$

is our best estimate for the common value of p.

If $p_1 = p_2$, the estimated variance of the difference $d = f_1 - f_2$ is

$$s_d^2 = \frac{f(1-f)}{n_1} + \frac{f(1-f)}{n_2} = f(1-f)\left(\frac{1}{n_1} + \frac{1}{n_2}\right)$$

The estimated standard deviation of $d = f_1 - f_2$ is

$$s_d = \sqrt{f(1-f)\left(\frac{1}{n_1} + \frac{1}{n_2}\right)} \qquad 9.24$$

Under the null hypothesis, the quantity

$$\frac{f_1 - f_2}{s_d} = \frac{f_1 - f_2}{\sqrt{f(1-f)\left(\frac{1}{n_1} + \frac{1}{n_2}\right)}} \qquad 9.25$$

has approximately a standard normal distribution if n_1 and n_2 are large. This, then, provides a basis for an approximate test procedure. Once again, we have a t-test with infinite degrees of freedom.

Example 1

Problem: Two different methods of manufacture, casting and die forging, were used to make parts for an appliance. In service tests of 100 of each type it was found that ten castings failed during the test, but only three forged parts failed. Find 95% confidence limits for the difference between the proportions of the cast and forged parts that would fail under similar conditions.

SECTION 9.8 TWO SAMPLES FROM BINOMIAL POPULATIONS

Solution: Here

$$f_1 = \frac{10}{100} = .10$$

$$f_2 = \frac{3}{100} = .03$$

$$s_{f_1}^2 = \frac{.10(.90)}{100}$$

$$s_{f_2}^2 = \frac{.03(.97)}{100}$$

$$s_{f_1-f_2}^2 = \frac{1}{100}[.10(.90) + .03(.97)] = .001191$$

$$s_{f_1-f_2} = \sqrt{.001191} = .0345$$

From the t-table, $t_{.025, \infty}$ is found to be 1.96, and approximate confidence limits for $p_1 - p_2$ are given by

$$L = f_1 - f_2 - t_{.025, \infty} \, s_{f_1-f_2} = .07 - 1.96(.0345) = .002$$

and

$$R = f_1 - f_2 + t_{.025, \infty} \, s_{f_1-f_2} = .07 + 1.96(.0345) = .138$$

We are approximately 95% confident that the difference between the proportions of failures is between .00 and .14.

Example 2

Problem: In order to test the effectiveness of a vaccine, 120 experimental animals were given the vaccine and 180 were not. All 300 animals were then infected with the disease. Among those vaccinated 6 died as a result of the disease. Among the control group there were 18 deaths. Can we conclude that the vaccine decreases the mortality rate?

Solution: Here we want to test the hypothesis $H_0: p_1 = p_2$ against $H_a: p_1 > p_2$, where p_1 is the proportion of deaths among the control animals and p_2 is the proportion of deaths among the animals given the vaccine:

$$f_1 = \frac{18}{180} = .10$$

$$f_2 = \frac{6}{120} = .05$$

$$f = \frac{18 + 6}{180 + 120} = \frac{24}{300} = .08$$

$$s^2_{f_1-f_2} = .08 \times .92 \times \left(\frac{1}{180} + \frac{1}{120}\right) = .001022$$

$$s_{f_1-f_2} = .03197$$

Our calculated t is

$$t = \frac{f_1 - f_2}{s_{f_1-f_2}} = \frac{.10 - .05}{.03197} = 1.564$$

The critical region is $t > t_{.05, \infty} = 1.645$. Therefore, we have no reason at the .05 level of significance to reject the hypothesis that $p_1 = p_2$—no reason to believe that the mortality rate for the second group is less than that for the first.

As an interesting sidelight, the preceding example illustrates the dangers of the uncritical acceptance of numerical facts. If on the basis of the data of Example 2 it were reported that, compared with the control group, only one-third as many of the vaccinated animals died, this would be a factual statement. However, before accepting the conclusion implicit in this statement we should look for additional information. Not only are the actual numbers of deaths in each group important, it is also necessary that the sample sizes be taken into account since the observed numbers are meaningful only in relation to the size of the experiment.

Suppose that the number of animals in each group had been equal to 20 but that the number of deaths, 6 and 18, remained the same. In this case the calculated statistic would equal 3.87, and the hypothesis $H_0: p_1 = p_2$ would be rejected. There is, however, no significant difference in the mortality rates when 6 out of 120 is compared with 18 out of 180.

Example 3

For a final example, let us return to the data in Chapter 1 on heart disease among patients in the Public Health Service hypertension study, the data given here in Table 9.2. In the discussion following the corresponding table in Chapter 1 (Table 1.5), several observations were made:

(1) Line 1 makes it clear that the active drug helps in the prevention of extreme elevation of blood pressure, and the evidence is so strong as to require no statistical analysis.

SECTION 9.8 TWO SAMPLES FROM BINOMIAL POPULATIONS

Table 9.2 Heart Disease

	Active	Placebo
Number of patients	193	196
(1) Number of cases where blood pressure increased to 130 or more	0	24
(2) Number of hypertensive events	37	89
(3) Number of cases of heart attack and other coronary heart disease	35	38
(4) Number of cases of enlarged heart	12	20

(2) Line 2 is strong evidence that the drug helps prevent hypertensive events. But exactly *how* strong is the evidence?

(3) Line 3 seems to indicate that the drug does not help in the prevention of heart attack and other coronary heart disease. How is the evidence to be weighed?

(4) Line 4 is unclear. Does the drug help prevent enlargement of the heart?

Let us analyze the lines 2, 3, and 4. In each case, the subscript 1 will refer to the active drug and the subscript 2 will refer to the placebo. And in each case the null hypothesis will be $p_1 = p_2$ and the alternative will be $p_1 < p_2$.

Line 2: In this case

$$f_1 = \frac{37}{193} = .192, \quad f_2 = \frac{89}{196} = .454, \quad f = \frac{126}{389} = .324, \quad s_d = .047$$

Here s_d is the estimated standard deviation, defined by (9.24). The statistic (9.25) has the value

$$\frac{f_1 - f_2}{s_d} = \frac{.192 - .454}{.047} = -5.6$$

Since $-Z_{.005} = -2.576$, this is significant at the 0.5% level. Indeed, -5.6 is off the normal table (Table 2 in the Appendix): $P(Z < -5.52) \leq P(Z < -3.59) = .0002$, so that even at the 0.02% level the evidence in line 2 of Table 9.2 favors the alternative hypothesis that the drug helps to prevent hypertensive events.

Line 3: In this case

$$f_1 = \frac{35}{193} = .181, \quad f_2 = \frac{38}{196} = .194, \quad f = \frac{73}{389} = .188, \quad s_d = .04$$

The statistic (9.25) has the value

$$\frac{f_1 - f_2}{s_d} = \frac{.181 - .194}{.04} = -.3$$

Since $-Z_{0.1} = -1.282$, this is not significant even at the 10% level. In fact, since $P(Z \leq -.3) = .38$, the level of significance, or p-value, in the sense of Section 8.4, is merely 38%. This is what common sense told us, too: Line 3 of Table 9.2 is no evidence that the drug helps in the prevention of heart attacks and other coronary heart disease.

Line 4: Here

$$f_1 = \frac{12}{193} = .062, \quad f_2 = \frac{20}{196} = .102, \quad f = \frac{32}{389} = .082, \quad s_d = .028$$

and

$$\frac{f_1 - f_2}{s_d} = \frac{.062 - .102}{.028} = -1.4$$

Since $-Z_{.1} = -1.282$ and $-Z_{.05} = -1.645$, this is significant evidence against the null hypothesis at the 10% level but not at the 5% level. The p-value is 8%. As evidence that the drug helps prevent enlargement of the heart, line 4 of Table 9.2 carries a little weight, but not much.

In connection with line 3 of Table 9.2 as analyzed in the preceding example, recall (see p. 3) that a subsequent study involving a much larger sample of patients showed that reducing blood pressure in mild hypertensives by means of drugs *does* help to prevent coronary heart disease. It would be wrong to say that the data in line 3 of the table show that we should accept the null hypothesis that the drug is without effect in the prevention of coronary heart disease. The data only tell us that we should reserve judgment—that we should *not reject* the null hypothesis of no effect in favor of the alternative hypothesis that the drug does help prevent coronary heart disease. This is a general principle of testing, one that doesn't apply just to this particular example alone: The data may tell us to reject the null hypothesis, or they may tell us to reserve judgment (not to reject the null hypothesis), but they never tell us to accept the null hypothesis unequivocally.

SECTION 9.8 TWO SAMPLES FROM BINOMIAL POPULATIONS

Problems

39. In a sample of 100 from one binomial population there were 32 successes. In a sample of 200 from a second binomial population there were 88 successes.
  a. Find estimates for p_1, p_2, and $p_2 - p_1$.
  b. Estimate the standard deviation of $f_2 - f_1$.
  c. Find a 95% confidence interval for $p_2 - p_1$.
  d. Test the hypothesis that $p_1 = p_2$. Use the two-sided alternative and $\alpha = .01$.

40. Samples of 200 were taken from each of two binomial populations. They contained 104 and 96 successes, respectively.
  a. Find the estimated standard deviation of $f_1 - f_2$.
  b. Find a 98% confidence interval for $p_1 - p_2$.
  c. Test the hypothesis that $p_1 = p_2$. Use the two-sided alternative and $\alpha = .05$.

41. A sample of 500 males and a sample of 500 females were taken. Of the males, eight were found to be color-blind, while only one of the females was color-blind. From these data, can we conclude that the proportion of color-blind females is smaller than the proportion of color-blind males? Use the following levels of significance.
  a. $\alpha = .05$.   b. $\alpha = .025$.   c. $\alpha = .01$.

42. A sample of 500 people were classified as being *athletic* or *nonathletic*. Among 300 classified as *athletic* it was found that 48 regularly eat a certain breakfast food. Among the 200 *nonathletic* persons 52 did so.
  a. Find a 99% confidence interval for the difference in the proportions that use the product.
  b. Test the hypothesis that there is no difference. Let $\alpha = .01$.

43. Two different alloys were used in the manufacture of parts for an appliance. One hundred samples of each were subjected to shock testing. Defects developed in 18 of those made from alloy I and in 26 of those made from alloy II. Can we conclude that those made of alloy I stand up to shock better than those made of alloy II? Let $\alpha = .05$.

44. In a sample of 50 turnings from an automatic lathe five were found to be outside specifications. The cutting bits were then changed and the machine restarted. A new sample of 50 contained three defective turnings.
  a. Find a 95% confidence interval for the decrease in the proportion defective after changing bits.
  b. Can we conclude that the proportion defective has been reduced? Let $\alpha = .05$.

328   CHAPTER 9   TWO-SAMPLE TECHNIQUES AND PAIRED COMPARISONS

45. During the five-year period 1961–65, if 300 out of 1000 juveniles committed to training schools were girls, and if over the last five years 540 out of 2000 were girls, can we conclude that the proportion of girls among those committed has decreased from the 1961–65 period? Let $\alpha = .05$.

46. A random sample of the records of 50 students attending a university in 1960 showed that 5 made the dean's list for academic achievement. In 1979, 15 students in a random sample of 75 made the dean's list. Make a 95% confidence interval for the difference in the proportions of students making the dean's list in 1960 and 1979.

47. A candidate for mayor of Springfield had a polling company test his candidacy first in March and then in September. The results of the two polls are given below. Has the proportion of voters in favor of his candidacy increased significantly in the period from March to September? Use $\alpha = .05$.

	n	x
March	255	110
September	278	136

48. An experiment was conducted to test visual memory under ordinary and under distractive conditions. Each subject was shown a photograph and one hour later was asked to pick it out of four similar photographs. No pressure was placed on group 1, but group 2 was subjected to planned distractions while the first photograph was examined. The data given below are the numbers of correct identifications. Does it appear that the distractions had an effect on visual memory? Use $\alpha = .05$.

Group	n	f
1	101	51
2	125	47

CHAPTER PROBLEMS

49. A physiologist claimed that the difference between the average resting pulse rates of baseball and basketball players exceeds 4 beats per

minute. Random samples of 10 basketball players and 40 baseball players gave the following results. Do the data indicate that the physiologist's claim is incorrect? Use $\alpha = .05$.

Group	n	$\bar{X}_i$	$\Sigma (X_{ij} - \bar{X}_i)^2$
Basketball	10	68	82
Baseball	40	70	302

50. The methods of this chapter may be used to analyze the data of Problem 34, Chapter 2. Test for the equality of the means of the cranial widths of the two populations of skulls at the 5% level of significance. (You may use the fact that the mean and standard deviation of the 106 cranial widths are 141.6 and 5.3, respectively, but you must compute the mean and standard deviation of the other 4 cranial widths.)

51. For the data of Problem 33, Chapter 2, make 95% confidence intervals for
    a. The mean weight at age 20.
    b. The mean weight at age 35.
    c. The difference in mean weight between age 35 and age 20.
    d. Can the confidence interval for part c be obtained from the confidence intervals for parts a and b?

52. Record the heights of four randomly selected male students and four randomly selected female students in the class. Consider these data to be random samples from the populations of college-age males and females.
    a. Test the (false) null hypothesis that the means of the two populations are equal. Use an appropriate one-tailed test and $\alpha = .05$.
    b. Make a 95% confidence interval for the mean difference between the heights of the two populations.
    c. What effect would increasing the sample size have on the probability of rejecting the false null hypothesis and on the length of the confidence interval?

53. Two methods of teaching were tested on two groups of 200 students each. At the conclusion of the period, tests were given, with the results

Group	n	$\bar{X}_i$	$\Sigma (X_{ij} - \bar{X}_i)^2$
1	200	82	16,200
2	200	79	20,000

in the table. Is the difference between the two means significantly different from zero? Find the *p*-value. (See Section 8.4.) Use the normal distribution tables (i.e., there are infinitely many degrees of freedom).

54. A statistic for testing for the equality of two Poisson means is

$$Z = \frac{\bar{X}_1 - \bar{X}_2}{\sqrt{\bar{X}(1/n_1 + 1/n_2)}}$$

where $\bar{X} = (n_1\bar{X}_1 + n_2\bar{X}_2)/(n_1 + n_2)$. (See Section 5.10 and Problem 82 of Chapter 5.) Critical values may be found in either the standard normal table (Table 2) or the "∞" row of the *t*-table (Table 4). As a rule of thumb, the statistic may be applied when $n_i\bar{X}_i > 30$. If $n_1 = 20$, $n_2 = 40$, $\bar{X}_1 = 4$, and $\bar{X}_2 = 2$, test the null hypothesis $H_0: \mu_1 - \mu_2 = 0$ against the alternative hypothesis $H_a: \mu_1 - \mu_2 \neq 0$ at the 5% level of significance.

55. The number of defectives per day in the manufacturing of aluminum cans is a Poisson random variable. Data were taken for 7 days at one plant and 14 days at another plant, and the average numbers of defectives were found to be $\bar{X}_1 = 35.5$ and $\bar{X}_2 = 30.1$, respectively. Is there a statistically significant difference between the mean number of defective cans produced by the two plants? Use $\alpha = .05$ (see Problem 54).

Summary and Keywords

An important problem is to draw inferences about the relation between the **means of two different normal populations** (p. 294) in order to understand the differences (if there are any) between two **treatments** (p. 295). The procedures are based on the **summary table** (p. 295), the **pooled sum of squares** (p. 296), and the **pooled variance estimator** (p. 297). These procedures make it possible to estimate the difference of population means and to test for equality of population means. Where there is great variability among the experimental units, **group comparisons** (p. 312) can be replaced by **paired comparisons** (p. 313) to increase efficiency.

Another important problem is to compare the proportions in two binomial populations by comparing the proportions of successes in samples from these populations.

REFERENCES

9.1 Croxton, Frederick E.; Cowden, Dudley J.; and Bolch, Ben W. *Practical Business Statistics*. 4th ed. New York: Prentice-Hall, 1969. Chapters 11 and 12.

REFERENCES

9.2 Li, Jerome C. R. *Statistical Inference,* I. Ann Arbor: Edwards Brothers, 1964. Chapters 10, 11, and 21.

9.3 Ostle, Bernard. *Statistics in Research*. 2d ed. Ames, Iowa: Iowa State University Press, 1963. Chapter 7.

9.4 Snedecor, George W., and Cochran, William G. *Statistical Methods*. 6th ed. Ames, Iowa: Iowa State University Press, 1967. Chapter 4.

10

Approximate Tests: Multinomial Data

10.1 The Multinomial Distribution

In earlier chapters we discussed the binomial distribution in some detail (Section 5.8); recall that it is the appropriate model when we are sampling from a population in which the elements belong to either one of two classes, and the sample is taken in such a manner that the probability of obtaining an element from a given class remains constant—that is, is not affected by the sampling process. We now consider the more general case where the elements of the population are classified as belonging to one of k classes, where $k \geq 2$. We refer to such a population as a *multinomial* population, and if the proportions of the elements belonging to each class are not changed by the selection of the sample, the appropriate model is the **multinomial distribution**.

multinomial distribution

The multinomial distribution is the *joint* distribution for the random variables $Y_1, Y_2, \ldots, Y_k$, the numbers of elements in a sample of size n that belong to each of the k classes of the population, Y_i being the number that belong to the ith class. The corresponding sample fractions Y_i/n we denote by f_i. These quantities satisfy the conditions

$$\sum_{i=1}^{k} Y_i = n \qquad \qquad 10.1$$

$$\sum_{i=1}^{k} f_i = 1 \qquad \qquad 10.2$$

The parameters of the multinomial distribution are the sample size n and the proportions $p_1, p_2, \ldots, p_k$ of the elements in the population that belong to each of the k classes, p_i being the proportion that belong to the ith class. Since the k classes contain all of the elements,

$$\sum_{i=1}^{k} p_i = 1 \qquad \qquad 10.3$$

As one might expect, each p_i is estimated in an unbiased way by the corresponding sample fraction

$$f_i = \frac{Y_i}{n}$$

10.2 A Hypothesis about the Multinomial Parameters

The basic hypothesis to be considered here is that the proportions belonging to the k classes are equal to a set of specified values; thus, $H_0: p_i = p_{i0}$, where $i = 1, 2, \ldots, k$. An exact test for this hypothesis would be difficult to apply, particularly for large sample sizes; therefore, it is usually tested by a procedure involving approximations.

If the hypothesis is true, the mean, or *expected,* number of elements of the ith class in a sample of size n is

$$E_i = np_{i0}$$

The *observed* number in the ith class is Y_i. If we calculate for each class the quantity

$$\frac{(Y_i - E_i)^2}{E_i}$$

we have the squared difference between the observed and expected numbers in that class relative to its expected number. The sum of these quantities over all classes, denoted by χ^2 (chi-square, pronounced *kai-square*),

$$\chi^2 = \frac{(Y_1 - E_1)^2}{E_1} + \frac{(Y_2 - E_2)^2}{E_2} + \cdots + \frac{(Y_k - E_k)^2}{E_k}$$

or simply

$$\chi^2 = \sum_{i=1}^{k} \frac{(Y_i - E_i)^2}{E_i} \qquad 10.4$$

is a measure of the lack of agreement between the data and the hypothesis. The idea is that if the null hypothesis is true, then the **observed frequencies**

observed frequencies

$$Y_1, Y_2, \ldots, Y_k$$

expected

ought not deviate too much from their respective **expected** values

$$E_1, E_2, \ldots, E_k$$

chi-square statistic

The **chi-square statistic** (10.4) gathers together the discrepancy between Y_i and E_i for all the values of i. To test the null hypothesis, we ask whether the statistic (10.4) has a larger value than can reasonably be accounted for by the workings of chance.

It can be shown that for sufficiently large sample sizes the distribution of the statistic (10.4) can be approximated by a chi-square distribution with $k - 1$ degrees of freedom. Note that the degrees of freedom are equal to one less than the number of *classes* and are not related to the size of the sample.

chi-square distribution

The χ^2-distribution, or **chi-square distribution,** is a continuous distribution ordinarily derived as the sampling distribution of a sum of squares of independent standard normal variables. It is a skewed distribution such that only nonnegative values of the variable χ^2 are possible, and it depends upon a single parameter, the degrees of freedom. The χ^2-distributions for 1, 4, and 10 degrees of freedom are shown in Figure 10.1. It can be seen that the skewness decreases as the degrees of freedom increase. In fact, it can be shown that as the degrees of freedom increase without limit, the χ^2-distribution approaches a normal distribution.

Tables of the χ^2-distributions have been computed. Let χ^2 denote a random variable with the χ^2-distribution with d degrees of freedom. Since $\chi^2 \geq 0$, we have $P(\chi^2 < 0) = 0$, and the distribution of χ^2 can be completely described by giving the values of $P(0 \leq \chi^2 \leq a)$ for the various positive values of a. The distribution of χ^2 can just as well be described by giving the values of $P(\chi^2 > a)$ for the various positive values of a. Table 6 in the Appendix instead gives the percentage points. For certain values of d,

Figure 10.1

Chi-square distributions with 1, 4, and 10 degrees of freedom

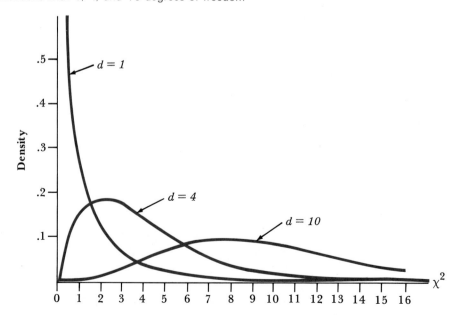

SECTION 10.2 A HYPOTHESIS ABOUT THE MULTINOMIAL PARAMETERS

the degrees of freedom, and for certain values of α, Table 6 gives the number $\chi^2_{\alpha, d}$ such that for the χ^2-distribution with d degrees of freedom, the area to the *right* of $\chi^2_{\alpha, d}$ is α. In other words, if χ^2 has the chi-square distribution with d degrees of freedom, then

$$P(\chi^2 > \chi^2_{\alpha, d}) = \alpha.$$

The probability is α that the random variable χ^2 will have a value *larger* than $\chi^2_{\alpha, d}$; the tabulated values are defined by this requirement. For example, for $\alpha = .05$ and $d = 20$, the tabulated value of $\chi^2_{.05, 20}$ is 31.41—the upper 5% point for the chi-square distribution with 20 degrees of freedom is 31.41. This means that if χ^2 has the chi-square distribution with 20 degrees of freedom, then $P(\chi^2 > 31.41) = .05$.

To test the hypothesis H_0 that $p_i = p_{i0}$ for all i, we calculate the multinomial chi-square sum (10.4) and compare the calculated value with the percentage points of the χ^2-distribution with $k - 1$ degrees of freedom. Since good agreement between the observed and expected numbers would result in a small χ^2-value and perfect agreement would give for χ^2 a value of 0, we are justified in rejecting the hypothesis only when χ^2 is large; hence, this test is always a one-tailed test on the upper tail of the χ^2-distribution. The critical region is $\chi^2 \geq \chi^2_{\alpha, k-1}$. For the theory to apply, the sample size n must be large enough that each expected frequency E_i is at least 5.

Example 1

Problem: A geneticist postulates that in the progeny of a certain dihybrid cross the four phenotypes should be present in the ratio 16 = 9:3:3:1. Examination of 800 members of the progeny generation results in the observed numbers 439, 168, 133, and 60 for the four phenotypes. Are these numbers in agreement with the hypothesized ratio? Let $\alpha = .05$.

Solution: The hypothesis is $H_0: p_1 = \frac{9}{16}, p_2 = \frac{3}{16}, p_3 = \frac{3}{16}, p_4 = \frac{1}{16}$. The calculations may be conveniently arranged in tabular form as follows:

	Phenotype			
	1	2	3	4
Y_i	439	168	133	60
$E_i = np_{i0}$	450	150	150	50
$Y_i - E_i$	−11	18	−17	10
$(Y_i - E_i)^2$	121	324	289	100

$$\chi^2 = \frac{121}{450} + \frac{324}{150} + \frac{289}{150} + \frac{100}{50} = 6.36$$

From Table 6 in the Appendix we find that $\chi^2_{.05,\,3}$ is 7.81; therefore, since the calculated value is smaller than this, we have no reason to reject the hypothesis. The results do not disagree with the hypothetical ratio 9:3:3:1.

In the above example we may be interested in testing the hypothesis that some, but not all, of the proportions are equal to specified values. For instance, we might be interested in testing whether the proportions in the first two phenotypes are $\frac{9}{16}$ and $\frac{3}{16}$, respectively, regardless of the proportions in the last two. In such cases the number of classes is reduced to conform with the hypothesis by combining the unspecified classes into one. For the genetic example they would be: first phenotype, second phenotype, other phenotypes. If we are interested in only one class, the problem falls into the framework of the binomial test, for we would have two classes—the class of interest and the class consisting of *all other classes*.

Problems

1. According to genetic theory, when red and white varieties of snapdragons are crossed, the first generation of hybrids have flowers of intermediate pink, or rose, color. When this generation is self-pollinated, the second hybrid generation shows a segregation of colors in the ratio 1 red : 2 pink : 1 white. When such an experiment was performed with 100 plants, it resulted in 22 red, 55 pink, and 23 white flowers. Are these results consistent with theory? Test at the 5% level of significance.

2. A person is told that a large jar contains red, white, and black marbles either in the proportion 1:1:1 or in the proportion 3:4:3. He draws a sample of 300 marbles that contains 95 red, 111 white, and 94 black marbles. On the basis of this information, decide which of the two proportions is more reasonable by testing each of the two possibilities with χ^2.

3. The following table gives for each day of a one-week period the number of crimes reported to a certain police station. Test the hypothesis that crimes are uniformly distributed over the days of the week. Let $\alpha = .05$.

SECTION 10.2 A HYPOTHESIS ABOUT THE MULTINOMIAL PARAMETERS

Day	No. of Crimes
Monday	38
Tuesday	31
Wednesday	40
Thursday	39
Friday	40
Saturday	44
Sunday	48

4. A supermarket manager wants to find out whether shoppers prefer some store locations over others for a certain product. He sets up this product in four locations with these results.

Location	A	B	C	D
Number of items sold	312	380	420	388

   a. Test at the 5% level the null hypothesis that there is no shopper preference.
   b. Suppose the manager believes location C is preferred. Test at the 5% level the null hypothesis that customers select location C with probability $\frac{1}{4}$ or less. use $\chi^2_{2\alpha, k-1}$

5. The probabilities for the various outcomes when a pair of dice is tossed are given below with the frequencies observed when a pair of dice was tossed 1800 times.
   a. What is the estimated probability of throwing a 7 with these dice? Of throwing a 12?
   b. At the 1% level of significance, are the observed frequencies in agreement with the theoretical values?

						r					
	2	3	4	5	6	7	8	9	10	11	12
$P(X = r)$	$\frac{1}{36}$	$\frac{2}{36}$	$\frac{3}{36}$	$\frac{4}{36}$	$\frac{5}{36}$	$\frac{6}{36}$	$\frac{5}{36}$	$\frac{4}{36}$	$\frac{3}{36}$	$\frac{2}{36}$	$\frac{1}{36}$
Frequency	40	108	175	184	225	330	223	228	128	87	72

6. Someone gave me an interesting pair of dice. I rolled them 36 times

and obtained the following outcomes. Lacking further information, would you conclude it is a fair pair of dice? Let $\alpha = .01$.

Outcome	2	3	4	5	6	7	8	9	10	11	12
Frequency	0	0	4	1	4	0	5	10	9	3	0

7. One can obtain a pair of dice such that one die will produce only a 1, 5, or 6 with a probability of $\frac{1}{3}$ for each, and the other die will produce a 3, 4, or 5 with a probability of $\frac{1}{3}$ for each. Given the results of the 36 rolls tabulated in the preceding problem, would you conclude that the dice that produced those results were of the type just described? Let $\alpha = .05$.

8. Five students worked respectively 3, 4, 5, 6, and 7 hours on a card-punching project. The following table gives the number of cards each produced during the last hour of his working period. Test at the 5% level the null hypothesis that fatigue has no effect on efficiency.

Student	1	2	3	4	5
Number of hours worked	3	4	5	6	7
Number of cards produced during last hour	43	44	38	32	30

9. Test at the 5% level of significance the hypothesis $p_1 = p_2 = p_3 = p_4 = \frac{1}{4}$ on the basis of each of the two following observed frequency distributions. Explain your results.

First sample	12	8	9	11
Second sample	120	80	90	110

10. Suppose you know that four probabilities are either in the ratio 1:2:2:1 or else in the ratio 1:3:3:1. Given the table below, which of the two sets of ratios would you tend to believe?

SECTION 10.3 BINOMIAL DATA

Observed frequency	70	170	170	70
Expected frequency according to 1:2:2:1	80	160	160	80
Expected frequency according to 1:3:3:1	60	180	180	60

11. People who used Sunbelt Van Lines in moving to Florida were classified according to their destination within the state. Do the data indicate that this carrier is as likely to move people to one area of the state as to another? Test at the 5% level of significance.

Area	North	Central	South
Number	62	80	58

12. Smith put forth the theory that the genetic types A, B, and C would occur in the population in the proportions 1:2:1, and he reported the results of two sampling experiments that seemed to support his theory very well. However, Jones pointed out that Smith's data had very little of the random variation one usually finds in real data. It seemed to Jones that the data agreed with the theory too well, and this led him to suspect that Smith simply contrived the data in order to gain support for his theory. Given the χ^2 numbers for Smith's two data sets given below, do you believe that Jones is justified in his concern?

Experiment	A	B	C
1	101	201	98
2	149	299	152

10.3 Binomial Data

In Section 8.3 we studied tests concerning the proportion p in a binomial population. The two-sided test there is a special instance of the chi-square test for the case where k is 2. It is instructive to trace the connection.

We relabel the success category in the population as class 1 and the failure category as class 2 ($k = 2$). If p_0 is the proportion of successes in the population, then class 1 has proportion $p_{10} = p_0$ and class 2 has proportion $p_{20} = 1 - p_0$. And now the null hypothesis

$$H_0: p_1 = p_0 \text{ and } p_2 = 1 - p_0$$

is exactly the same thing as the null hypothesis

$$H_0: p = p_0$$

considered in Section 8.3.

If X is the number of successes in a sample of size n, then the number Y_1 of observations falling in class 1 is X and the number Y_2 of observations falling in class 2 is $n - X$. Now the chi-square sum (10.4) for testing H_0 is

$$\chi^2 = \frac{(Y_1 - np_{10})^2}{np_{10}} + \frac{(Y_2 - np_{20})^2}{np_{20}}$$

$$= \frac{(X - np_0)^2}{np_0} + \frac{[(n - X) - n(1 - p_0)]^2}{n(1 - p_0)} \qquad 10.5$$

But algebra reduces this expression to

$$\chi^2 = \left[\frac{X - np_0}{\sqrt{np_0(1 - p_0)}} \right]^2 \qquad 10.6$$

which is the square of the statistic (8.7) used to test the hypothesis $p = p_0$.

Example 1

Problem: At one time the sex ratio in a small eastern city was known to be 8 to 1: there were eight women for every man. Suppose that in a recent survey we found that a random sample of 450 contained 68 men. Would we be justified in concluding that the ratio had changed?

Solution: If the ratio is 8 to 1, one out of every nine persons is a male; therefore, if the ratio has not changed, the proportion p of males is $\frac{1}{9}$. We test the hypothesis $H_0: p = \frac{1}{9}$ against the two-sided alternative $H_a: p \neq \frac{1}{9}$ at the 1% level of significance. Let X be the number of men in the sample; then

$$Y_1 = X = 68$$
$$Y_2 = n - X = 450 - 68 = 382$$
$$E_1 = \tfrac{1}{9}(450) = 50$$

SECTION 10.3 BINOMIAL DATA

$$E_2 = \tfrac{8}{9}(450) = 400$$

$$\chi^2 = \frac{(68 - 50)^2}{50} + \frac{(382 - 400)^2}{400} = 7.29$$

From Table 6 of the Appendix, $\chi^2_{.01,\,1}$ is found to be 6.63. The calculated χ^2 is greater than the tabular value; therefore, we reject the null hypothesis and conclude that the ratio has changed.

In this calculation we have used Equation (10.5). Using Equation (10.6) instead gives

$$\chi^2 = \left[\frac{68 - 450 \times \tfrac{1}{9}}{\sqrt{450 \times \tfrac{1}{9} \times \tfrac{8}{9}}}\right]^2 = 2.7^2 = 7.29$$

which checks. The value 2.7 that is to be squared here is the value of the statistic (8.7) used in Section 8.3 for a two-sided test of $H_0: p = p_0$. At the 1% level we are to reject if the statistic has absolute value exceeding $Z_{.005}$, or 2.576; this is true of our value 2.7, so again we reject. The two procedures necessarily lead to the same conclusion because 6.63, the cutoff point for the chi-square test, is 2.576^2.

Since it is always true that $\chi^2_{\alpha,\,1}$ is equal to $Z^2_{\alpha/2}$, a chi-square test based on Equation (10.5) is always the same as the two-sided test of Section 8.3.

Problems

13. An insurance company found that 35 of its 680 customers in a particular area of the country filed claims last year. Do these data indicate that the claim rate in this area is different from the nationwide claim rate of 3%? Let $\alpha = .01$.

14. Rework Problem 21 of Chapter 8 using the chi-square test for binomial data. Verify that the value of the chi-square statistic is equal to the square of the value of the Z statistic.

15. A random sample of 500 voters showed that 153 favored candidate A, 240 favored candidate B, and the rest favored candidate C. Test the hypothesis that 50% of the voters favor candidate B. Use $\alpha = .05$.

16. Out of 300 offenders selected at random, 259 pleaded guilty, 3 chose trial by judge, and 38 chose trial by jury. On the basis of this sample could we conclude that the proportion of offenders asking for a jury trial is .10? Let $\alpha = .05$.

17. Five samples were taken from the output of a machine during one day. From the following results would we conclude at the 5% level that the machine was turning out 8% defective items on the average?

Sample	Defectives	Nondefectives
1	1	49
2	9	41
3	12	38
4	5	45
5	2	48
Totals	29	221

10.4 A Test for Goodness of Fit

goodness-of-fit test

In the tests of hypotheses of the preceding chapters, we assumed that we knew the form of the distribution and we tested for the values of parameters. For example, we *assumed* the population was normal and tested the hypothesis $\mu = \mu_0$. But what if we want to check on the assumption of normality itself? The multinomial chi-square **goodness-of-fit test** can be applied.

For an example, suppose we want to test at the 5% level the hypothesis that the 171 blood-pressure differences in Table 2.1 (page 13) come from some normal population. These 171 figures are grouped into bins in Table 2.4 (page 16). If we combine the first three bins and combine the last seven bins, the data reduce to Table 10.1. Now we must check whether these observed frequencies agree well with a normal distribution having *some* mean and standard deviation—a mean and standard deviation not specified in advance. Our first step is to estimate the mean and standard deviation by $\bar{X}$ and s as calculated from the formulas (3.4) and (3.10). These are

$$\bar{X} = -9.85 \quad \text{and} \quad s = 11.55$$

The next step is to compute what probability a normally distributed random variable X with mean -9.85 and standard deviation 11.55 has of falling in each of the six classes represented in Table 10.1. To do this we standardize the boundaries of the bins:

$$[-20.5 - (-9.85)]/11.55 = -.92$$
$$[-15.5 - (-9.85)]/11.55 = -.49$$

SECTION 10.4 A TEST FOR GOODNESS OF FIT

Table 10.1 Change in Blood Pressure (Active Group)

Class	Measurement	Y_i = Number of Observations
1	Less than -20.5	26
2	-20.5 to -15.5	31
3	-15.5 to -10.5	27
4	-10.5 to -5.5	31
5	-5.5 to -0.5	22
6	More than -0.5	34
		$n = 171$

$$[-10.5 - (-9.85)]/11.55 = -.06$$
$$[-5.5 - (-9.85)]/11.55 = .38$$
$$[-0.5 - (-9.85)]/11.55 = .81$$

The probability that such an X falls in class 2, for example, is the probability that the standardized variable $Z = [X - (-9.85)]/11.55$ falls between the first two of these figures, namely $-.92$ and $-.49$; by Table 2 in the Appendix this probability is $.3212 - .1879$, or $.1333$. Thus we obtain the probabilities (rounded off) shown in Figure 10.2.

Now we chi-square the observed frequencies in Table 10.1 against these probabilities. Since the probability of class 2 is $.133$ and the total number of observations is 171, we expect on the average $171 \times .133$, or 22.74, observations to fall in class 2. Table 10.2 shows the remaining calculations.

Figure 10.2

Areas for the standard normal curve

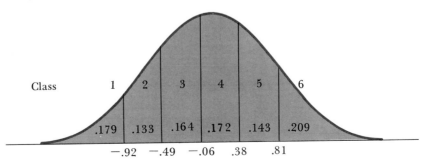

Table 10.2 Calculation for the χ^2 Goodness-of-Fit Statistic for the Data of Table 10.1

Class	Probability	E_i	Y_i	$Y_i - E_i$	$(Y_i - E_i)^2/E_i$
1	.179	30.61	26	−4.61	.69
2	.133	22.74	31	8.26	3.00
3	.164	28.05	27	−1.05	.04
4	.172	29.41	31	1.59	.09
5	.143	24.45	22	−2.45	.25
6	.209	35.74	34	−1.74	.08
Totals	1.000	171.00		0.00	$\chi^2 = 4.15$

The number of degrees of freedom in this example is 3. If we had not had to estimate the mean and standard deviation from the sample (by $\bar{X}$ and s), the degrees of freedom would have been (as in Section 10.2) one less than the number of categories: $k - 1$, or 5. But the rule is that we must subtract one additional degree of freedom for each parameter estimated. We have estimated two parameters, the mean and the standard deviation, and so our degrees of freedom are 5 − 2, or 3. We are to test at the 5% level the hypothesis that the data in Table 10.1 fit a normal curve. The 5% point on a chi-square distribution with 3 degrees of freedom is $\chi^2_{.05, 3} = 7.81$. Our value, 4.15, is less than this; the data do fit with a normal distribution.

The multinomial chi-square may be used to test the goodness of fit of data with other distributions as well. We need only remember that each expected number should be at least 5, so that the chi-square statistic closely follows the chi-square distribution, and that we lose one degree of freedom for every parameter estimated from the sample.

Problems

18. In order to test the representativeness of a nationwide sample of households taken in 1971, the frequency of households in different income brackets in the sample was compared with the U.S. Census figures for that year. Are the sample frequencies in agreement with the census figures? Let $\alpha = .05$.

SECTION 10.4 A TEST FOR GOODNESS OF FIT

Household Income	Percent Households in the Census	Number of Households in the Sample
Under $2000	9.7	181
$2000 to $3999	12.0	228
$4000 to $5999	11.6	235
$6000 to $7999	12.0	265
$8000 to $9999	12.4	270
$10,000 to $14,999	23.2	479
$15,000 and over	19.1	342
Totals	100.0	2000

19. Lacking a table of random numbers, I decided to use the last digits in a set of five-place tables of logarithms. The number of times each integer occurred in a sample of size $n = 50$ is given below. I would expect each integer to occur, on the average, equally often. Are the observed frequencies in agreement with this expectation? Use $\alpha = .05$.

Integer	0	1	2	3	4	5	6	7	8	9
Frequency of occurrence	1	8	3	3	6	6	7	5	7	4

20. The following grades were given to a class of 80 students. The expected numbers are those corresponding to the *curve*. Do you think the instructor used the curve? Let $\alpha = .025$.

	Grade				
	A	B	C	D	F
Given	9	25	35	11	0
Expected	6	19	30	19	6

21. Given the following distribution, test at the 5% level of significance the hypothesis that the data came from a normal distribution.

Class	Observed Frequency
$25 \le X < 50$	15
$50 \le X < 75$	25
$75 \le X < 100$	30
$100 \le X < 125$	20
$125 \le X < 150$	10

22. Apply a goodness-of-fit test to the following data in order to determine if it came from a normal distribution. Let $\alpha = .005$.

Class	Observed Frequency
15 to 19	2212
20 to 24	3503
25 to 29	1313
30 to 34	939
35 to 39	3490
40 to 44	3005
45 to 54	5096
55 to 64	2904
65 to 80	885

23. Test at the 5% level of significance the hypothesis that the lengths of mayfly larvae given in Problem 33, Chapter 7, are normally distributed.

24. Problem 6, Chapter 2, concerns 66 observed time lapses between prison admission and the filing of a motion. Test whether they are normally distributed. Let $\alpha = .05$.

25. Does a normal distribution provide a good fit to the differences in cutoff voltages given in Problem 21, Chapter 2? Let $\alpha = .05$.

26. Apply a goodness-of-fit test to the distribution of the number of citations given in Problem 75, Chapter 3. Can it be concluded that the sample is from a normal population? Let $\alpha = .05$.

27. The following data are the yields obtained from 100 hills of corn. Expected frequencies for the normal distribution, which were computed using $\bar{X} = 914$ and $s = 331.8$, are given along with the data.
   a. Test the data for normality at the 5% level of significance with all eight categories being used in the computation of the chi-square statistic.

SECTION 10.4 A TEST FOR GOODNESS OF FIT

b. Combine the first two categories and combine the last two categories, so that all expected frequencies are 5 or greater. Test again for normality at the 5% level of significance and compare with part a.

Yield (grams)	Number of Hills	Expected Frequency
$100 \leq X < 300$	3	3.2
$300 \leq X < 500$	7	7.4
$500 \leq X < 700$	15	15.3
$700 \leq X < 900$	26	22.4
$900 \leq X < 1100$	22	22.9
$1100 \leq X < 1300$	13	16.5
$1300 \leq X < 1500$	9	8.3
$1500 \leq X < 1700$	5	3.9

28. The following is the distribution for the number of defective items found in 150 lots of a manufactured item. We want to test the hypothesis that X, the number of defectives, has a Poisson distribution, and we proceed as follows:

X = Number Defective	Observed Frequency	Expected Frequency (Rounded)
0	23	20
1	39	41
2	43	41
3	23	27
4	10	14
5	7 ⎫	⎫
6	4 ⎬ 12	⎬ 7
7	1 ⎭	⎭

(1) The mean $\bar{X}$ is 2. This estimates the parameter μ for the Poisson distribution.

(2) Obtain the probabilities for each value of X by using Table 3 in the Appendix.

$$f(r) = \frac{\mu}{r} f(r-1)$$

For example,
$$f(0) = e^{-\mu} = e^{-2} = .135$$
so
$$f(1) = 2(.135) = .270$$
$$f(2) = \frac{2}{2}(.270) = .270$$
$$f(3) = \frac{2}{3}(.270) = .180$$
$$f(4) = \frac{2}{4}(.180) = .090$$

(3) Now calculate the expected number for each class.
(4) Combine the last four observed and expected numbers in order that no expected number will be less than 5. Calculate χ^2. The degrees of freedom equals 4, since we had to estimate μ.

29. Using the procedure outlined in the preceding problem, test the hypothesis that the following have a Poisson distribution. Let $\alpha = .05$.

X	0	1	2	3	4	5
Frequency	40	59	26	17	7	2

30. The following is the distribution for the number of westbound cars arriving at a particular intersection in 60-second intervals. Can it be concluded that a Poisson distribution provides a suitable model? See Problem 28.

Number of cars	0	1	2	3	4	5	6	7	8	9	10
Observed frequency	8	23	39	53	36	30	15	7	5	3	1

10.5 Contingency Tables

Many sets of enumeration data are such that they may be grouped according to two or more criteria of classification, and often we want to know whether or not these various criteria are independent of one another. For instance, we might like to know whether color preference is independent of

SECTION 10.5 CONTINGENCY TABLES

Table 10.3 The $r \times c$ Contingency Table

Rows	Columns					Row Totals
	1	2	3	...	c	
1	Y_{11}	Y_{12}	Y_{13}		Y_{1c}	R_1
2	Y_{21}	Y_{22}	Y_{23}		Y_{2c}	R_2
3	Y_{31}	Y_{32}	Y_{33}		Y_{3c}	R_3
⋮						
r	Y_{r1}	Y_{r2}	Y_{r3}		Y_{rc}	R_r
Column Totals	C_1	C_2	C_3		C_c	n

sex. In this case we could take a sample of people, record their sex and their favorite color, and classify their responses by sex and by color.

If the data are to be classified according to two criteria (the only case we treat here), then the classes based on one of them may be represented by the rows in a two-way table, and the classes based on the other by the columns. A cell of the table is formed by the intersection of a row and column. This is called a **contingency table**. We count the numbers of elements in the sample that belong to each cell of the table. The general two-way table with r rows and c columns is presented in Table 10.3. The notation used in Table 10.3 is as follows:

contingency table

Y_{ij} = number observed to belong to the ith row and jth column. Note that the first subscript denotes the row, the second the column.

R_i = total observed number in the ith row; found by adding across the row.

C_j = total observed number in the jth column; found by adding down the column.

n = sample size = $\Sigma R_i = \Sigma C_j$.

To test the null hypothesis that rows and columns represent independent classifications, we compute an expected number E_{ij} for each cell and use the multinomial chi-square. By **independence** we mean that the proportion of each row total that belongs in the jth column is the same for all rows (or, what is the same thing, that the proportion of each column total that belongs in the ith row is the same for all columns). This is expressed mathematically by saying that the probability that a random element will belong in the (i, j)th cell is equal to the product of the probability that it belongs to the ith row and the probability that it belongs to the jth column. Symbolically, our hypothesis is $H_0: p_{ij} = p_i p_j$; $i = 1, 2, \ldots, r$; $j = 1, 2, \ldots, c$, where

independence

p_{ij} = probability of *i*th row *and* *j*th column,
$p_{i.}$ = probability of *i*th row with columns ignored,
$p_{.j}$ = probability of *j*th column with rows ignored.

The marginal probability $p_{i.}$ is estimated by the observed fraction in the *i*th row (R_i/n), and $p_{.j}$ by the observed fraction in the *j*th column (C_j/n). Under the hypothesis of independence, the value of p_{ij}, or of $p_{i.}p_{.j}$, is estimated by the product of these estimates, R_iC_j/n^2; therefore, the expected number is the sample size *n* multiplied by the estimated probability:

$$E_{ij} = \frac{R_iC_j}{n} \qquad 10.7$$

We can now find the expected number for each cell in the table.

To perform the test, we find the contribution of each cell to the multinomial chi-square. The contribution of the (i, j)th cell is

$$\frac{(Y_{ij} - E_{ij})^2}{E_{ij}}$$

There are a total of $r \times c$ such contributions, and the calculated χ^2 is their sum:

$$\chi^2 = \sum_{i,j} \frac{(Y_{ij} - E_{ij})^2}{E_{ij}} \qquad 10.8$$

This chi-square has $(r-1)(c-1)$ degrees of freedom, the number of rows less one multiplied by the number of columns less one.

If the calculated χ^2 exceeds the tabular value for probability α, we reject the null hypothesis and conclude that rows and columns do not represent independent classifications; if the calculated χ^2 does not exceed the tabular value, we do not reject the null hypothesis.

Example 1

Two polls were taken in a city to investigate opinion about a proposed change in public-aid policy. One was a sample of 240 from the Fifth Ward, the second a sample of 200 from the Ninth Ward. Are the two wards homogeneous with respect to opinion on the issue? Let $\alpha = .05$. The two-way table is shown as Table 10.4.

The figures in the body of the table are the observed frequencies. The expected frequencies are computed by the formula (10.7); for example, $E_{12} = R_1 C_2/n = 240 \times 158/440 = 86.2$. It is convenient to set the

Table 10.4

Rows	Columns			Row Totals
	1 Favor	2 Oppose	3 Undecided	(Sample Sizes)
1 Fifth Ward	126	84	30	240
2 Ninth Ward	95	74	31	200
Column Totals	221	158	61	440

Table 10.5

Rows	Columns			Row Totals
	1	2	3	
1 Fifth Ward	120.5	86.2	33.3	240.0
2 Ninth Ward	100.5	71.8	27.7	200.0
Column Totals	221.0	158.0	61.0	440.0

expected frequencies out in a table (Table 10.5) like that for the observed frequencies. Note that the row and column totals are the same as before (there may be slight differences owing to roundoff error). And now the χ^2 comes out to 1.40:

$$\chi^2 = \frac{(126 - 120.5)^2}{120.5} + \frac{(84 - 86.2)^2}{86.2} + \frac{(30 - 33.3)^2}{33.3}$$
$$+ \frac{(95 - 100.5)^2}{100.5} + \frac{(74 - 71.8)^2}{71.8} + \frac{(31 - 27.7)^2}{27.7} = 1.40$$

Since the number of degrees of freedom is $(3 - 1)(2 - 1)$, or 2, the critical point is $\chi^2_{.05,\,2} = 5.99$. We do not, on the basis of the data, reject the null hypothesis of homogeneity.

Since the sample sizes (240 and 200) are different, the two rows of the table of observed frequencies cannot be compared easily by eye. They are more easily compared if put in percentage form as follows:

	Favor	Oppose	Undecided	Total
Fifth Ward	52.5%	35.0%	12.5%	100.0%
Ninth Ward	47.5%	37.0%	15.5%	100.0%

These percentages look more or less consistent, but an objective judgment requires the statistical test.

Example 2

A sample of 309 veterans upon discharge from the army were asked whether they intended to return to their former civilian jobs; the possible responses were *yes, maybe,* and *no.* A follow-up interview after six months determined whether or not each veteran had in fact returned to his former job. Table 10.6 is the two-way table of data (the column corresponds to

Table 10.6

	Columns			
Rows	1 Yes	2 Maybe	3 No	Row Totals
1 Returned	121	58	21	200
2 Not returned	45	49	15	109
Column Totals	166	107	36	309

Table 10.7

	Columns			
Rows	1	2	3	Row Totals
1 Returned	107.4	69.3	23.3	200.0
2 Not returned	58.6	37.7	12.7	109.0
Column Totals	166.0	107.0	36.0	309.0

intention, the row to status after six months). The expected frequencies (for example, $E_{12} = 107 \times 200/309 = 69.3$) are given in Table 10.7. The χ^2 statistic works out to 10.75:

$$\chi^2 = \frac{(121 - 107.4)^2}{107.4} + \frac{(58 - 69.3)^2}{69.3} + \frac{(21 - 23.3)^2}{23.3}$$
$$+ \frac{(45 - 58.6)^2}{58.6} + \frac{(49 - 37.7)^2}{37.7} + \frac{(15 - 12.7)^2}{12.7} = 10.75.$$

There are 2 degrees of freedom; if $\alpha = .05$ again, then the observed χ^2 exceeds the critical value $\chi^2_{.05, 2} = 5.99$. This time the data lead us to reject the null hypothesis of independence of the rows and columns.

Although the computational procedures are the same in Examples 1 and 2, the points of view are somewhat different. In Example 1 there are two populations (the two wards) corresponding to the two rows of Table 10.4, and a sample is taken from each population. There is nothing random about the row totals 240 and 200; the randomness is in the way these totals split into the categories corresponding to the three columns. We ask whether the two populations have the same composition with respect to the question at hand (public-aid policy). This is testing for **homogeneity** of the two populations.

homogeneity

In Example 2, on the other hand, there is *one* population (the veterans being discharged) from which a sample of 309 is taken. The row totals 200 and 109 in Table 10.6 *are* random; they were of course not set in advance of sampling. This time we ask whether the veteran's intention (the column) is independent of his status after six months (the row). This is testing for **independence** of two attributes in the population.

independence

Despite this difference in point of view, the statistical test appropriate to the problem is the same in each case.

Problems

 31. In a survey of 68 statistics students, only 38 of whom had a mathematical background, the following answers were recorded regarding their preference for course A over course B. Are these opinions homogeneous for the two types of students? Use $\alpha = .01$.

	Prefer A	Prefer B	Undecided
Math background	20	10	8
No math background	10	19	1

32. Since a survey questionnaire remained unanswered by 32% of the surveyed population, it was decided to take a sample of the nonrespondents in order to check whether their opinions are similar to those of the respondents with regard to the most important question in the survey. On the basis of the following data could we assume that the nonrespondents hold the same opinions as the respondents? Let $\alpha = .01$.

	Strongly Agree	Agree	Neutral	Disagree	Strongly Disagree
Respondents	203	150	108	117	102
Nonrespondents	4	9	11	14	12

33. A hospital kept record of the ages and sexes of the patients with a certain disease. From the following data can it be concluded that there is a difference in the way in which this disease attacks the two sexes? Let $\alpha = .05$.

	Age Group				
	10–25	26–35	36–45	46–55	56+
Male	23	34	64	81	98
Female	20	31	55	68	96

34. Students in college were asked to rate the general instruction in the school as poor, fair, or good. The students themselves were classified

	School Instruction		
Students	Poor	Fair	Good
Poor	30	40	30
Fair	75	65	60
Good	50	35	15

SECTION 10.5 CONTINGENCY TABLES

as poor, fair, or good according to their grades. Test at the 2.5% level the hypothesis that the rating of instruction is independent of student performance. Can you detect from the table any specific trend?

35. To assess the impact of the prior record of a prisoner on the decision of a parole board, a sociologist derived the following table from the files of all prisoners (with similar sentences) appearing for consideration by the board. Is parole release dependent upon prior record? Let $\alpha = .005$.

	Released First Time	Released Second Time	Released Third Time	Not Released
No prior record	85	10	5	0
Some prior record	70	60	50	20
Serious prior record	45	80	75	100

36. The blood types for individuals randomly selected from two populations are given below. Do the data indicate that the distributions of blood types for the two populations are different? Use $\alpha = .05$.

	Blood Type			
Population	O	A	B	AB
1	460	359	377	83
2	321	234	186	57

37. The germination rates of 100 seeds of each of four varieties are given below. Are there significant differences among the expected germination rates of the four varieties? Use $\alpha = .05$.

	Variety			
Germinate	A	B	C	D
Yes	91	87	92	93
No	9	13	8	7

38. Seventy-five items selected randomly from each of three assembly lines were classified as good (G), repairable (R), or beyond repair (BR).

Are the expected proportions of items in each of these categories the same across assembly lines? Let $\alpha = .05$.

Line	G	R	BR
1	55	14	6
2	47	13	15
3	52	15	8

39. An accountant randomly sampled accounts receivable at three offices and classified each as either "on time" or "overdue." Are the expected proportions of overdue accounts the same for the three offices? Test at the 5% level of significance.

	Office		
Status	1	2	3
On time	110	108	55
Overdue	20	31	15

CHAPTER PROBLEMS

40. The data of Example 2 of Section 9.8 are displayed below in a 2 × 2 contingency table. Use the chi-square statistic to test for homogeneity of mortality rates across the two populations. Let $\alpha = .05$. Show that the value of the chi-square statistic is equal to the square of the value of the *t*-statistic for these data.

	Fatal	Nonfatal
Vaccinated	6	114
Unvaccinated	18	162

41. Classify members of the class according to sex and eye color. For simplicity, you may want to consider only two categories for eye color (e.g., brown and other). Arrange the data in a contingency table. Do the data show that sex is independent of eye color?

CHAPTER PROBLEMS 359

42. Beginning with a row and a column of your choice, read off 50 successive digits from Appendix Table 9. Test the hypothesis that these digits occur on the average in the proportion $\frac{1}{10}$ each. Let $\alpha = .05$.

43. Generate 100 values of the geometric distribution with $\mu = 2$ ($p = \frac{1}{2}$) as described in connection with Problem 22 of Chapter 6. Subtract 1 from each value so that the smallest value is 0 instead of 1. Test at the 5% level of significance the false null hypothesis that the data are Poisson. Did you reach the correct conclusion with your test?

44. The chi-square test for $r \times c$ contingency tables applies to data having two classification criteria. Data having more than two classification criteria may also be analyzed by this test by considering the criteria two at a time. This problem and the two that follow it are illustrations of this method of analysis for data having 3 classification criteria. See Reference 10.1 for other, generally preferable, approaches to the analysis of multiclassified data.

Random samples of male students and female students listened to recordings of readings by a male and a female reader and were asked to rate the readings as very effective, moderately effective, or ineffective. The numbers of each sex who rated the readings in the three categories follow. The same selection was read by both readers. Test the hypothesis that there is no difference in the responses due to sex of the listener. Sum over sex of reader to get totals.

	Male Reader		Female Reader	
Rating	Male Listener	Female Listener	Male Listener	Female Listener
Ineffective	10	10	5	10
Moderately effective	60	15	25	40
Very effective	30	10	15	50

45. Using the data of Problem 44, test the hypothesis that the male and the female reader are equally effective. Sum over sex of listener.

46. Using the data of Problem 44, test the hypothesis that there is no interaction between sex of listener and sex of reader, that is, that there is no tendency for listeners of one sex to rate higher or lower depending on the sex of the reader. Sum ratings where listener and reader are of the same sex, and where they are of opposite sexes.

47. Two temperatures, T_1 and T_2, and two curing times, C_1 and C_2, were applied in combination to equal numbers of glued joints between pieces of wood. The joints were then tested for tensile strength. The

following data are the numbers of sample joints whose tensile strengths exceeded and did not exceed a prescribed standard.
a. Does time of curing affect strength?
b. Does temperature affect strength?
c. Is there an interaction between time and temperature; i.e., is the effect of temperature at the second time different from that at the first time? HINT: Compare $T_1C_1 + T_2C_2$ with $T_1C_2 + T_2C_1$.

	T_1		T_2	
Test Result	C_1	C_2	C_1	C_2
Failed	30	20	20	10
Passed	20	30	10	40

Summary and Keywords

The **multinomial distribution** (p. 334) describes independent sampling from a population classified into two or more categories or classes. The **chi-square statistic** (p. 335) tests whether a sample comes from a given population by comparing the **observed frequencies** with the **expected values** (p. 335) they should have if they do come from the population in question; the statistic follows the **chi-square distribution** (p. 336). A **goodness-of-fit test** (p. 344), having a similar structure, can be used to check whether a sample came from, for example, a normal population. Tests for **independence** (pp. 351, 355) and for **homogeneity** (p. 355), based on **contingency tables** (p. 351), also have a similar form and can be used to check whether two samples come from the same population or whether two classification schemes are independent.

REFERENCES

10.1 Fienberg, Stephen E. *The Analysis of Cross-Classified Categorical Data.* Cambridge, Massachusetts: The MIT Press, 1970.

10.2 Guttman, Irwin, and Wilks, Samuel S. *Introductory Engineering Statistics,* New York: John Wiley & Sons, 1965. Chapter 12.

10.3 Hoel, Paul G. *Elementary Statistics.* 3d ed. New York: John Wiley & Sons, 1971. Chapter 11.

10.4 Snedecor, George W., and Cochran, William G. *Statistical Methods.* 6th ed. Ames, Iowa: Iowa State University Press, 1967. Chapter 9.

11

Regression and Correlation

11.1 Linear Regression

In most of our statistical procedures so far, we have been concerned with a single observation made on each element of the sample, that is, with a sample of values for a single variable X. We now consider the case in which two measurements are made on each element of the sample—where the sample consists of pairs of values, one for each of the two variables X and Y. For example, consider the heights and weights of individuals. If we take a sample of individuals, from each obtain his height and weight, and then let the height be represented by X and the weight by Y, we obtain from the ith person the pair of numbers (X_i, Y_i). If there are n persons in the sample, we have a sample of size n which consists of the n number pairs

$$(X_1, Y_1), (X_2, Y_2), \ldots, (X_n, Y_n)$$

Our object is to study the relationship between the variables X and Y. One way to study this relationship is by means of regression.

To use a regression analysis we must know or assume the form of the relationship between the variables. This is expressed by a mathematical function in which Y, the **dependent** variable, is set equal to some expression that depends only on X, the **independent** variable, and on certain constants or parameters. Assuming that there is such a relationship does not imply that cause-and-effect is at work (see Section 11.9). We may arrive at the form of the relationship by either of two methods: (1) from analytical or theoretical considerations, or (2) by studying scatter diagrams like those in Figure 11.1.

A **scatter diagram** is obtained by plotting the pairs of values of X and Y as points in a plane, where Y is measured along the vertical axis and X along

Figure 11.1

Scatter diagrams that suggest (a) a linear relationship, (b) a curvilinear relationship, and (c) no relationship

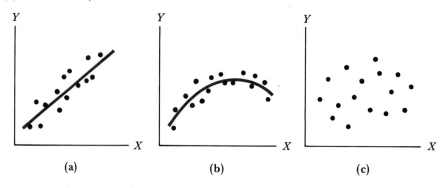

SECTION 11.1 LINEAR REGRESSION

the horizontal axis. After the points have been plotted, an examination of the diagram may reveal that the points form a pattern, a pattern that will indicate what mathematics should be used in the analysis. We shall be concerned here with the case in which the points appear to lie along a straight line in the diagram, or in which theoretical considerations lead us to conclude that the relationship should be linear.

In other words, we will study cases in which the points (X_1, Y_1), (X_2, Y_2), ..., (X_n, Y_n) have a scatter diagram resembling the one in Figure 11.1a. The problem is to describe and study the line that goes through the "middle" of the cloud of points. Before we can do that, we will have to define the line in a precise way.

It is because of sampling variation that the observed points will not all lie exactly on the line in the diagram; they will instead be scattered to some degree about the line. We assume that for each value of X there is a distribution for the values of Y, and we assume that for each i, Y_i is an observation from the distribution corresponding to X_i. The regression line is the line joining the *means* of the distributions corresponding to the different possible values of X. Under these assumptions, the relationship we want to estimate is

$$\mu_{Y|X} = A + BX \qquad 11.1$$

[handwritten annotations: "y intercept" pointing to A, "slope" pointing to B]

This means that the mean value of Y for a fixed value of X is equal to $A + BX$. The line defined by (11.1) is the one that goes through the middle of the cluster of points in Figure 11.1a. The constants A and B are, respectively, the **intercept** and the **slope** of this line. The intercept is the value the mean $\mu_{Y|X}$ has when $X = 0$; if X has the value 0, then the right side of (11.1) has the value A. The slope is the rate at which the mean $\mu_{Y|X}$ changes when X changes; it is the amount by which $\mu_{Y|X}$ increases if X is increased by one unit. If X is increased to $X + 1$, the right side of (11.1) becomes $A + B(X + 1)$, or $(A + BX) + B$, which does exceed $AX + B$ by the amount B.

intercept
slope

These ideas will be made clearer by an example. Our problem is to use the information in the sample of size n to estimate the parameters A and B.*

Example 1

The rate of gasoline consumption by automobiles has become critical because of the prospect of long-range fuel shortages. To check the effect of speed on fuel consumption, the Federal Highway Administration tested au-

---

*In linear regression, α and β are usually used to denote the intercept and slope, respectively. Here these symbols have been used for the probabilities of Type I and Type II errors; to avoid confusion, we use A and B for the regression parameters.

tomobiles of various weights at various speeds.* One car weighing 3975 pounds showed the following results:

Speed (miles per hour)	30	40	50	60	70
Gas rate (miles per gallon)	18.25	20.00	16.32	15.77	13.61

If we take X as speed in miles per hour and Y as gasoline consumption rate in miles per gallon, then $X_1 = 30$ and $Y_1 = 18.25$, $X_2 = 40$ and $Y_2 = 20.00$, and so on. The five points corresponding to these five pairs of measurements are plotted as a scatter diagram in Figure 11.2.

Figure 11.2

Scatter diagram for gasoline consumption rates

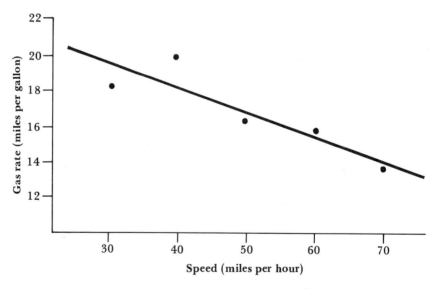

The mileage rate falls off with increased speed, as one expects, and the diagram suggests that the decrease is approximately linear. The points lie along a straight line as in part (a) of Figure 11.1, rather than along a curve as in part (b) or in a formless cloud as in part (c). The line shown in Figure 11.2 is constructed not by eye but by an exact procedure described in the next section.

*U.S. Department of Transportation, "The Effect of Speed on Automobile Gasoline Fuel Consumption Rates" (October 1973).

A randomly selected Y is represented by

$$Y_i = A + BX_i + e_i \qquad 11.2$$

where e_i is the random deviation of the observed Y from the mean $A + BX_i$. The population regression equation, Equation (11.1), is estimated by the *prediction equation* or **regression line**

regression line

$$\widehat{Y} = a + bX \qquad 11.3$$

where $\widehat{Y}$, a, and b are estimators for $\mu_{Y|X}$, A, and B, respectively. To obtain these estimators, we use the *method of least squares*. This method will give the best unbiased estimators for A and B if the following assumptions are satisfied.

Assumption 1 The X-values are known, that is, nonrandom.

Assumption 2 For each value of X, Y is normally and independently distributed with mean $\mu_{Y|X}$ equal to $A + BX$ and variance $\sigma^2_{Y|X}$, where A, B, and $\sigma^2_{Y|X}$ are unknown parameters.

Assumption 3 For each X the variance of Y given X is the same; that is, $\sigma^2_{Y|X} = \sigma^2$ for all X, where σ^2 is an unknown parameter.

Although the second assumption includes normality, this is not required by the least-squares theory. It is included here because we want to make inferences based on our estimates and for these we use normal theory.

In the model specified by Equation (11.2), X_i is not a random variable. The speeds in Example 1, far from being random, were controlled with great care. What is random is e_i, and since e_i is a random variable, Y_i is a random variable (with mean $A + BX_i$ and variance σ^2). If in Example 1 the measurement for 30 miles per hour was made repeatedly with a variety of automobiles (say, with weights near 3975 pounds) or under a variety of driving conditions, the gasoline consumption rate would vary from trial to trial; this is the source of the variability.

11.2 Least Squares

least squares

The principle of **least squares** is illustrated in Figure 11.3. For every observed Y_i there is a corresponding predicted value $\widehat{Y}_i$, equal to $a + bX_i$, given by Equation (11.3). The deviation of the observed Y from the predicted Y is $Y_i - \widehat{Y}_i$, equal to $Y_i - a - bX_i$. The sum of squares of these deviations from the fitted line is

$$\Sigma (Y_i - \widehat{Y}_i)^2 = \Sigma (Y_i - a - bX_i)^2 \qquad 11.4$$

Figure 11.3

Sample points and estimated regression line

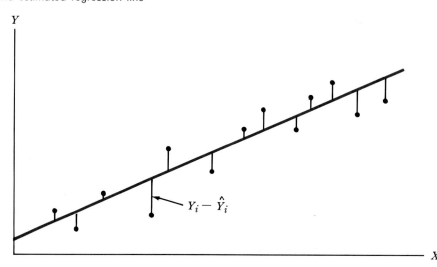

The appropriate estimators a and b are the functions of the sample values that make this sum of squares a minimum. These least-squares estimators are

$$b = \frac{\Sigma (X_i - \bar{X})(Y_i - \bar{Y})}{\Sigma (X_i - \bar{X})^2} \qquad 11.5$$

and

$$a = \bar{Y} - b\bar{X} \qquad 11.6$$

These facts are derived in Section 11.10, but it is not necessary to go through the derivation in order to understand and use the results. Notice, however, that Equation (11.6) implies that the regression line goes through the center of gravity of the points (X_i, Y_i). It can be shown that a and b are unbiased estimates of A and B.

After we have found numerical values for the estimators a and b given by Equations (11.5) and (11.6), we substitute them into the prediction equation $\hat{Y} = a + bX$. There is no other line that could be fitted to the observed points for which the sum of squares of vertical deviations from the line would be smaller.

The machine formulas for the sums of squares and crossproducts are

SECTION 11.2 LEAST SQUARES

$$\sum (X_i - \bar{X})^2 = \sum X_i^2 - \frac{(\sum X_i)^2}{n}$$

$$\sum (Y_i - \bar{Y})^2 = \sum Y_i^2 - \frac{(\sum Y_i)^2}{n}$$

$$\sum (X_i - \bar{X})(Y_i - \bar{Y}) = \sum X_i Y_i - \frac{(\sum X_i)(\sum Y_i)}{n}$$

Example 1

In the following table the calculations outlined previously for obtaining a linear regression equation are illustrated for the data in the fuel consumption example of the preceding section:

	X_i	Y_i	$X_i - \bar{X}$	$Y_i - \bar{Y}$	$(X_i - \bar{X})^2$	$(X_i - \bar{X})(Y_i - \bar{Y})$	$(Y_i - \bar{Y})^2$
	30	18.25	−20	1.46	400	−29.20	2.13
	40	20.00	−10	3.21	100	−32.10	10.30
	50	16.32	0	−.47	0	0.00	.22
	60	15.77	10	−1.02	100	−10.20	1.04
	70	13.61	20	−3.18	400	−63.60	10.11
Sums	250	83.95			1000	−135.10	23.80

$$n = 5, \quad \bar{X} = \frac{250}{5} = 50, \quad \bar{Y} = \frac{83.95}{5} = 16.79$$

$$b = \frac{-135.10}{1000} = -.135, \quad a = 16.79 - (-.135)50 = 23.54$$

The regression equation is

$$\hat{Y} = 23.54 - .135X$$

For $X = 30$,

$$\hat{Y} = 23.54 - .135(30) = 19.49$$

For $X = 70$,

$$\hat{Y} = 23.54 - .135(70) = 14.09$$

The regression line in Figure 11.2 can be constructed by running a line from the point with coordinates (30, 19.49) to the point with coordinates (70, 14.09).

The estimated mean consumption for $X = 55$ is

$$Y = 23.54 - .135(55) = 16.12$$

An approximation to this value can be read off graphically from Figure 11.2. At 55 miles per hour, to travel 100 miles takes 1 hour and 49 minutes, and the gasoline used on the average is 100/16.12, or 6.2 gallons. At 70 miles per hour, to travel 100 miles takes only 1 hour and 26 minutes, but the gasoline used on the average is 100/14.09, or 7.1 gallons. Thus the driver saves 23 minutes at a cost of 0.9 gallons. This result of course is of direct interest only to the owner of this and similar automobiles. To understand the national effect of various speed limits, for example, would require the performance characteristics of other types of automobiles and trucks, together with information on how many miles they are driven per year.

A regression model ordinarily applies only over a limited range of values of X. No one expects the model in the example above to apply to a speed of 180 miles per hour—the car would not go that fast anyway. The predicted gasoline consumption for $X = 180$, however, is $23.54 - .135(180)$, which comes out to $-.76$ miles per gallon. For $X = 0$, $\hat{Y}$ is 23.54 miles per gallon, which is equally absurd. The model cannot be expected to be accurate much outside the 30-to-70 range.

The regression relation (11.3) is often given in a slightly different form. Since $\hat{Y} = a + bX = (a + b\bar{X}) + b(X - \bar{X})$, and since $\bar{Y} = a + b\bar{X}$ by Equation (11.6), the relation between X and Y can be expressed as

$$\hat{Y} = \bar{Y} + b(X - \bar{X})$$

When put in this form, the regression line in Example 1 becomes $\hat{Y} = 16.79 - .135(X - 50)$.

Problems

1. Given the following pairs of values,
   a. Plot the points on graph paper.
   b. Find the regression equation, $\hat{Y} = a + bX$.
   c. Draw the regression line on the graph.

X	1	2	3	4	5
Y	5	1	13	9	17

SECTION 11.2 LEAST SQUARES

2. Given the following pairs of values,
   a. Plot the points on graph paper.
   b. Find the equation for the regression of Y on X.
   c. Draw the regression line on the graph.
   d. Find the equation for the regression of X on Y.
   e. Draw the regression line of X on Y on the graph.
   f. Why are the regression lines different?

X	1	3	5	7	9
Y	2	1	4	3	5

3. Find the equation for the regression of Y on X:

X	12	13	14	15	16	17	18
Y	−8	−6	−4	0	6	12	14

4. Find the equation for the regression of Y on X:

X	−10	−6	−2	2	6	10	14	16	20	23
Y	−8	−5	−4	−3	0	4	7	8	11	10

5. Given the following pairs of values,
   a. Find the regression equation $\hat{Y} = a + bX$.
   b. Find the predicted value of Y given $X = -3$.

X	−3	−2	−1	0	1	2	3
Y	36	33	24	15	9	6	3

6. Below are given the heights (in inches) of eight mother-daughter pairs. Find the regression line for obtaining a predicted value of the daughter's height from a known value of the mother's height. Plot the data and the line on graph paper.

Mother's height (X)	67	64	62	65	69	63	65	66
Daughter's height (Y)	70	69	65	68	66	60	64	66

CHAPTER 11 REGRESSION AND CORRELATION

7. The following table gives the cost per unit of a manufactured piece of equipment as a function of the number of units produced.
   a. Fit a regression line to the data.
   b. Plot the data and the fitted line on graph paper.
   c. Use the regression line to estimate the cost per unit for a production batch of 70 units.

Production units	10	20	50	100	150	200
Cost per unit in $10,000	7.5	7.0	6.8	6.2	6.0	5.6

8. The table below gives, for various years, the percentage of women graduates in education.
   a. Fit a linear regression equation to the data, and plot the points and the line on graph paper.
   b. Give the predicted percentage of females in 1981 according to this regression line.

Year	1935	1940	1945	1950	1955	1960	1965	1970	1975
% women	56.4	58.8	60.7	64.4	64.1	64.2	65.7	67.8	69.4

9. An experiment was conducted to assess the effect allocated shelf space has on sales of a product. The shelf space was fixed at 2, 3, 4, 5, and 6 square feet, each for a period of two weeks, and the number of items sold was recorded for each week of the two-week periods.
   a. Fit a regression line to the data.
   b. Plot the line and the given points on graph paper.

Space allocated	2	3	4	5	6
Units sold	320	381	411	460	482
	344	402	440	471	495

11.3 Variance Estimates

In the preceding section we looked at the estimators for the parameters A and B obtained by the method of least squares, but before we can use these

SECTION 11.3 VARIANCE ESTIMATES

estimators to make further inferences about the linear relationship between Y and X, we must obtain estimators for the variance σ^2 and for the variances of the sampling distributions of a and b.

The variance of Y given X, assumed to be the same for all X, is estimated unbiasedly by the mean squared deviation about the regression line, that is, by

$$s_{Y.X}^2 = \frac{1}{n-2} \sum (Y_i - \hat{Y}_i)^2 \qquad 11.7$$

or

$$s_{Y.X}^2 = \frac{1}{n-2} \left[\sum (Y_i - \bar{Y})^2 - b \sum (X_i - \bar{X})(Y_i - \bar{Y}) \right] \qquad 11.8$$

where the subscripts $Y.X$ indicate that this is the estimator for the variance of Y when we have the regression of Y on X.

The estimated variance of the sample regression coefficient b is the estimate of variance, $s_{Y.X}^2$, divided by the sum of squares for X:

$$s_b^2 = \frac{s_{Y.X}^2}{\sum (X_i - \bar{X})^2} \qquad 11.9$$

The estimated variance of the estimator a of the intercept is more complicated:

$$s_a^2 = s_{Y.X}^2 \left(\frac{1}{n} + \frac{\bar{X}^2}{\sum (X_i - \bar{X})^2} \right) \qquad 11.10$$

In the numerical example in the preceding section we had $n = 5$, $\bar{X} = 50$, $\sum (X_i - \bar{X})^2 = 1000$, $\sum (Y_i - \bar{Y})^2 = 23.80$, $\sum (X_i - \bar{X})(Y_i - \bar{Y}) = -135.10$, $b = -.135$, and $a = 23.54$. Therefore

$$s_{Y.X}^2 = \frac{1}{n-2} \left[\sum (Y_i - \bar{Y})^2 - b \sum (X_i - \bar{X})(Y_i - \bar{Y}) \right]$$

$$= \tfrac{1}{3}[23.80 - (-.135)(-135.10)] = 1.85$$

$$s_{Y.X} = 1.36$$

$$s_b^2 = \frac{s_{Y.X}^2}{\sum (X_i - \bar{X})^2} = \frac{1.85}{1000} = .00185, \qquad s_b = .043$$

$$s_a^2 = s_{Y.X}^2 \left[\frac{1}{n} + \frac{\bar{X}^2}{\sum (X_i - \bar{X})^2} \right] = 1.85 \left[\frac{1}{5} + \frac{2500}{1000} \right] = 5.00,$$

$$s_a = 2.24$$

Problems

10. Given the following pairs of values,
   a. Estimate the population regression line $\mu_{Y|X} = A + BX$.
   b. Estimate $\sigma^2_{Y|X}$, the variance of Y for each fixed X.
   c. Find the unbiased estimates for the variances of a and b.

X	10	20	30	40	50	60	70
Y	−4	−3	−2	0	3	6	7

11. For the data of Problem 7, compute $s^2_{Y \cdot X}$, s^2_a, and s^2_b.

12. Given that $n = 30$, $s^2_{Y \cdot X} = 20$, $s^2_X = 12$, and $\bar{X} = 5$, find s^2_a and s^2_b. Note: $\Sigma(X_i - \bar{X})^2 = (n - 1)s^2_X$.

13. Find $s_{Y \cdot X}$ for the heights of mother-daughter pairs in Problem 6. Interpret the meaning of $s_{Y \cdot X}$ in this case.

14. Find estimated values for $\sigma_{Y|X}$ and σ_B for the data of Problem 8.

15. Compute $s^2_{Y \cdot X}$, s^2_a, and s^2_b for the data of Problem 9.

11.4 Inferences about A and B

inferences

Thus far we have not had to make any assumptions about the form of the distribution of Y given X, since the method of least squares will yield the best linear unbiased estimators for A and B if the Xs are known values and the Ys are independently distributed with common variance. Now, however, we want to make **inferences** about A and B on the basis of the estimators a and b; therefore, we must make some distributional assumptions.

If we assume that the Ys are normally distributed, it follows that our estimators are normally distributed (they are linear functions of the Ys, and a linear function of normal variables is normally distributed; recall that the X_i are nonrandom) and that we may base our confidence intervals and tests of hypotheses on the t-distribution. As usual, it can be shown that even if the distribution of the Ys is not normal, the probabilities based on the t-distribution will be very good approximations if the sample is sufficiently large. Under the normality assumption, the ratios

$$\frac{b - B}{s_b} \quad \text{and} \quad \frac{a - A}{s_a}$$

have t-distributions, each with $n - 2$ degrees of freedom because the esti-

SECTION 11.4 INFERENCES ABOUT A AND B

mates s_b and s_a are based on the sum of squares of deviations from regression, and this has $n - 2$ degrees of freedom.

From the relationship

$$t = \frac{b - B}{s_b} \qquad 11.11$$

it is apparent that confidence intervals for B will have the form common to all those we have calculated, that is,

$$L = b - t_{\alpha/2,\,(n-2)} s_b$$
$$R = b + t_{\alpha/2,\,(n-2)} s_b \qquad 11.12$$

For the gasoline-consumption example, the 90% confidence interval for B is

$$L = -.135 - (2.353)(.043) = -.236$$
$$R = -.135 + (2.353)(.043) = -.034$$

We are 90% confident that the true slope B is between $-.236$ and $-.034$. Similarly, since

$$t = \frac{a - A}{s_a} \qquad 11.13$$

the confidence limits for the true intercept A are given by

$$L = a - t_{\alpha/2,\,(n-2)} s_a$$
$$R = a + t_{\alpha/2,\,(n-2)} s_a \qquad 11.14$$

For our example, the 95% confidence limits for A are

$$L = 23.54 - (3.182)(2.24) = 16.41$$
$$R = 23.54 + (3.182)(2.24) = 30.67$$

We are 95% confident that the intercept of the true regression line (as given by Equation (11.1)) at 0 is between 16.41 and 30.67.

Equations (11.11) and (11.13) also serve as bases for tests of hypotheses concerning the parameters A and B. Under the null hypothesis that the slope is equal to a given value B_0 (that is, $H_0: B = B_0$), the quantity

$$t = \frac{b - B_0}{s_b} \qquad 11.15$$

has a t-distribution with $n - 2$ degrees of freedom. A common hypothesis is that B equals 0. We want to know whether or not there is a linear association between the variables, for if there is not, there is nothing to be gained by using the Xs, as they would contribute nothing to the analysis of the Ys. Under this hypothesis, $H_0: B = 0$, we have

$$t = \frac{b}{s_b} \qquad 11.16$$

We must remember, however, that this quantity is distributed as t only when B is 0. In this situation the alternative hypothesis is usually that B is not equal to zero. There are, of course, situations in which one is interested only in knowing whether or not there is a positive slope (or a negative slope). In these circumstances we would use one-sided alternatives.

To test the hypothesis $H_0: B = 0$ against a two-sided alternative for the gasoline-consumption example, we calculate the value of t as follows:

$$t = \frac{b}{s_b} = \frac{-.135}{.043} = -3.14$$

In the t-table we find that $t_{.05, 3}$ is 2.353; since the calculated t is less than -2.353, we reject the hypothesis that B is 0 at the 10% level. We conclude that there is an underlying linear relationship.

The test of the hypothesis that the intercept A is equal to a specified value A_0 is also an ordinary t-test. Under $H_0: A = A_0$, the function

$$t = \frac{a - A_0}{s_a} \qquad 11.17$$

also has a t-distribution with $n - 2$ degrees of freedom. With a two-sided alternative, H_0 is rejected if the observed value of t falls outside the interval from $-t_{\alpha/2, (n-2)}$ to $+t_{\alpha/2, (n-2)}$.

Problems

16. Given that $b = -1.5$, $n = 10$, $s_{Y.X}^2 = 35$, and $\Sigma (X_i - \bar{X})^2 = 140$, find a 95% confidence interval for B.
17. Give a 98% confidence interval for A in Problem 10.
18. In a regression problem, $n = 30$, $\Sigma X_i = 15$, $\Sigma Y_j = 30$, $\Sigma (X_i - \bar{X})(Y_i - \bar{Y}) = 30$, $\Sigma (X_i - \bar{X})^2 = 10$, and $\Sigma (Y_i - \bar{Y})^2 = 160$.
    a. Find the regression line, $\hat{Y} = a + bX$.

SECTION 11.4 INFERENCES ABOUT A AND B

    b. Estimate the variance, $\sigma^2_{Y|X}$.
    c. Test $H_0: B = 0$ against $H_a: B \neq 0$, $\alpha = .05$.

19. In a regression analysis, $n = 38$, $\bar{Y} = 20$, $\bar{X} = 7$, $\Sigma(Y_i - \bar{Y})^2 = 900$, $\Sigma(X_i - \bar{X})^2 = 60$, and $\Sigma(X_i - \bar{X})(Y_i - \bar{Y}) = 180$.
    a. Find the regression equation.
    b. Test the hypothesis that $A = 0$ against the alternative $A \neq 0$. Let $\alpha = .02$.
    c. Find a 90% confidence interval for B.

20. In a regression analysis, $n = 41$, $\bar{X} = 12$, $\bar{Y} = 10$, $\Sigma(X_i - \bar{X})^2 = 400$, $\Sigma(Y_i - \bar{Y})^2 = 64$, and $\Sigma(X_i - \bar{X})(Y_i - \bar{Y}) = 100$.
    a. Find the equation for the regression of Y on X.
    b. Find the estimated variance.
    c. Find a 90% confidence interval for B.
    d. Test the hypothesis $A = 5$ against the alternative $A > 5$ with $\alpha = .01$.

21. In a regression problem, $n = 18$, $\Sigma(X_i - \bar{X})^2 = 144$, $b = 3.1$, and the estimated variance is 36. Can we conclude that $B > 2$? Let $\alpha = .025$. Find a 95% confidence interval for B.

22. Given that $a = 6$, $n = 20$, $\bar{X} = 3$, $s^2_{Y \cdot X} = 25$, and $\Sigma(X_i - \bar{X})^2 = 90$, test the hypothesis that $A = 10$. Let $\alpha = .05$.

23. Given that $a = -2$, $n = 16$, $\bar{X} = 6$, $s^2_{Y \cdot X} = 12.8$, and $\Sigma(X_i - \bar{X})^2 = 144$, find a 95% confidence interval for A.

24. The data given below are the shoe sizes (X) and the weights (Y) of ten college-age males. Test at the 5% level the hypothesis $H_0: B = 0$ against the alternative hypothesis $H_a: B > 0$. What do the results of the test say about the possibility of predicting weight from a known value of shoe size?

Shoe size X	9.5	9.5	10.5	10.5	11	8.5	8.5	9.5	10	9
Weight Y	140	155	153	150	180	160	155	145	163	150

25. Find 95% confidence intervals for A and B for the data of Problem 7.

26. Do the data in Problem 8 indicate that there has been a statistically significant increase in the percentage of women education graduates over time? Answer by testing the null hypothesis $B = 0$ against the alternative $B > 0$ at the 5% level of significance.

27. Refer to Problem 9. Test the null hypothesis $A = 0$ against the alternative $A \neq 0$ at the 5% level of significance. Explain the meaning of the hypothesis in this case, and explain the result of the test.

11.5 Some Uses of Regression

In addition to the use of a regression analysis to estimate the parameters of the functional relationship between two variables and to test hypotheses concerning these parameters, there are several other applications that are of great importance in practical problems. We briefly consider three of them.

prediction   Having fitted a regression and having obtained the **prediction** equation, we are in a position where, given a value for X, we can predict with some degree of confidence the corresponding mean of Y. We can also predict what a single observed value of Y would be for a given X, but with less confidence. The point estimate for either the mean or a single value is

$$\hat{Y}_0 = a + bX_0 \qquad 11.18$$

where $\hat{Y}_0$ is the predicted value corresponding to the given value X_0.

We can find a confidence interval for the mean $\mu_{Y|X_0} = A + BX_0$ of Y given X_0. The formulas are

$$L = \hat{Y}_0 - t_{\alpha/2,\,(n-2)} s_{\hat{Y}_0}$$
$$R = \hat{Y}_0 + t_{\alpha/2,\,(n-2)} s_{\hat{Y}_0} \qquad 11.19$$

The estimated standard deviation of $\hat{Y}_0$ is

$$s_{\hat{Y}_0} = s_{Y.X} \sqrt{\frac{1}{n} + \frac{(X_0 - \bar{X})^2}{\Sigma(X_i - \bar{X})^2}} \qquad 11.20$$

From the form of $s_{\hat{Y}_0}$ we see that the estimated standard deviation for a predicted mean value depends upon X_0. The farther X_0 gets from the mean of X, the greater will be the standard deviation. As a result, the confidence interval will become wider as the given X is farther displaced from the mean. If we were to calculate a confidence interval for $\mu_{Y|X}$ for every X, the endpoints of these intervals would lie on the two branches of a hyperbola, as shown in Figure 11.4. This indicates that we have more confidence near the center than at the extremes of the range of our X-values. It also reflects the fact that we can interpolate—make predictions for values within the range of the X used to estimate the relationship—with more confidence and greater safety than if we try to extrapolate and make predictions for values outside the range of X in our sample. In addition, it may well be that within the range used for estimating the regression equation, the true relationship can be approximated reasonably well by a straight line, but that outside that region the true relationship breaks away sharply from the regression equation.

A prediction interval for a *single* value of Y, given that X equals X_0, is

Figure 11.4

Confidence intervals for the mean of Y given X

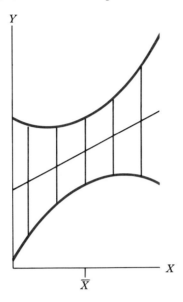

found in the same way as a confidence interval for the corresponding mean, but since the variance must include the variance of a single observed Y, the standard deviation is larger:

$$s_{Y_0} = s_{Y.X}\sqrt{1 + \frac{1}{n} + \frac{(X_0 - \bar{X})^2}{\Sigma(X_i - \bar{X})^2}} \qquad 11.21$$

The prediction interval is

$$\hat{Y}_0 \mp t_{\alpha/2,\,(n-2)} s_{Y_0} \qquad 11.22$$

Here again, because of the dependence on $X_0 - \bar{X}$, the interval will be wider as X_0 departs farther from $\bar{X}$.

Another common use of regression is for statistical control of a variable that cannot be controlled otherwise, or that we do not care to control for practical reasons. Instead, we merely record values of X at the same time as we observe Y and then adjust all the Ys to a common X. Suppose we want to record the time required by electric heating elements to raise the temperature of a specified amount of water a specified number of degrees, and suppose that the line voltage is variable. Certainly, if the voltage is low it will take longer to heat the water than if the voltage is high. We want to eliminate the effect of the varying voltage. We find the cost of voltage-

Figure 11.5

Two points before and after adjustment

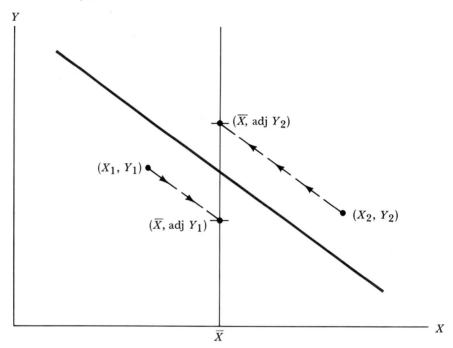

regulating equipment excessive, but we can afford a voltmeter. We now take our readings on the time but also record the voltage during the period of observation. If we let the time be Y and the voltage be X, we fit a regression of Y on X. After this has been done, we adjust each Y to the mean of X according to the formula

$$\text{adj } Y_i = Y_i - b(X_i - \bar{X}) \qquad 11.23$$

Figure 11.5 shows two Ys and their adjusted values. The effect of the adjustment given by Equation (11.23) is to translate the point (X_i, Y_i) parallel to the regression line to a new origin at $\bar{X}$. The adjusted Ys have a common X and may be compared, since the effects of differences among the Xs have been removed. Adjustments of this type are useful in a wide variety of applications.

Sometimes, after fitting a regression of Y on X and then observing values of Y, we want to estimate the value of X that produced the observed Y. This is a discrimination problem, or a problem of classification. In biological assays we fit a curve for response against dose of a drug. After the curve has been obtained we observe a response and want to estimate the dose

SECTION 11.5 SOME USES OF REGRESSION

that produced it. This is an important application of regression techniques, but the formulas will not be given here; see Reference 11.5.

Problems

28. In a regression analysis $n = 18$, $\bar{X} = 8$, $\bar{Y} = 10$, $\Sigma(X_i - \bar{X})^2 = 100$, $\Sigma(Y_i - \bar{Y})^2 = 400$, and $\Sigma(X_i - \bar{X})(Y_i - \bar{Y}) = 160$.
   a. Find the regression equation.
   b. Find the predicted value of Y given that $X = 3$.
   c. Find a 99% confidence interval for the mean of Y for $X = 3$.
   d. Find a 99% prediction interval for a single observation on Y for $X = 3$.

29. Given that $n = 11$, $\bar{Y} = 20$, $\bar{X} = 4$, $\Sigma(X_i - \bar{X})^2 = 64$, $\Sigma(X_i - \bar{X})(Y_i - \bar{Y}) = 256$, and $\Sigma(Y_i - \bar{Y})^2 = 1600$,
   a. Find the regression equation.
   b. Find the predicted value for Y given that $X = 6$.
   c. Find a 95% confidence interval for the mean of Y given that $X = 6$.
   d. Find a 95% prediction interval for a single value of Y given that $X = 6$.

30. Given that $n = 25$, $\Sigma X_i = 75$, $\Sigma Y_i = 50$, $\Sigma X_i^2 = 625$, $\Sigma X_i Y_i = 30$, and $\Sigma Y_i^2 = 228$,
   a. Find the regression equation.
   b. Find $s_{\hat{Y}_0}$, the estimated standard deviation of $\hat{Y}_0$ for a given X_0, as a function of X_0.
   c. Find a 95% prediction interval for a single value of Y given X_0 as a function of X_0.
   d. Find a 95% prediction interval for a single value of Y given $X = 5$.

31. Refer to Problem 7. Find a 95% confidence interval for the mean cost per unit when there are 70 production units.

32. On the basis of the data of Problem 8, between what two limits would you predict the percentage of women education graduates to fall in 1981? Answer by computing a 95% prediction interval.

33. Refer to Problem 9. Make 95% confidence intervals and 95% prediction intervals for the number of units sold with shelf-space 2, 3, 4, 5, and 6 square feet. Sketch the regression line and the intervals on graph paper.

34. The following data are the median sales prices of houses for the first 10 months of 1976. Compute the regression line. Make 95% prediction intervals for the median sales price in November ($X = 11$) and the

median sales price in December ($X = 12$). Do your intervals contain the actual values of 45,800 for November and 45,900 for December?

Month (X)	1	2	3	4	5	6	7	8	9	10
Median sales price in thousands of dollars (Y)	41.6	42.7	43.6	43.3	43.6	46.1	44.6	44.2	44.7	45.3

35. The following data are the logarithms of chemical concentration (X) and fluorescence intensity (Y) which were obtained from a laboratory experiment in chemistry. Obtain a 95% confidence interval and a 95% prediction interval for the logarithm of fluorescence intensity when the logarithm of chemical concentration is $X = -10.0$. Plot the results on graph paper along with the data and the regression line.

log chemical concentration (X)	-11.0	-10.7	-9.9	-9.5	-9.3	-8.9
log fluorescence intensity (Y)	-2.8	-2.6	-1.9	-1.5	-1.4	-1.0

36. Consider the first example in this chapter (p. 365). If you were given, for the same car, the measurement of 10 miles per gallon at a speed of 60 miles per hour, would you consider it consistent with the earlier figures? Or would you decide to drop it as a probable error? Substantiate your answer statistically.

37. The regression line $\hat{Y} = -3 + 0.4X$ was derived from a random sample of 100 income-tax returns from a very specific subpopulation. In this case X represents income and Y the amount claimed as deductible income, both in thousands of dollars. Here X ranged between 15 and 30. These values were obtained: $\bar{X} = 20$, $s_a = 0.64$, and $s_{Y.X} = 0.32$. If a man with an income of $25,000 claims a deduction of $8,000, is that unusual?

11.6 Partitioning the Sum of Squares

For a single variable Y, the variation in Y is measured by the sum of squares, all of which can be considered as being due to random or unexplained

Figure 11.6
Subdivision of $Y_i - \bar{Y}$

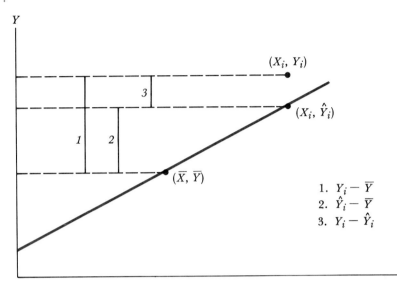

1. $Y_i - \bar{Y}$
2. $\hat{Y}_i - \bar{Y}$
3. $Y_i - \hat{Y}_i$

variation; hence, the estimated variance of Y is based on the total sum of squares. In the regression situation, however, some of the observed variation among the sample Ys is associated with the relationship between Y and X. Figure 11.6 shows one observed point (X_i, Y_i) and the point on the fitted line whose coordinates are the means $(\bar{X}, \bar{Y})$. We see that the deviation $Y_i - \bar{Y}$ from the mean is the sum of the deviation of Y_i from the corresponding predicted value $\hat{Y}_i$ and the deviation of $\hat{Y}_i$ from $\bar{Y}$; that is,

$$(Y_i - \bar{Y}) = (Y_i - \hat{Y}_i) + (\hat{Y}_i - \bar{Y}) \qquad 11.24$$

The second element, the deviation of $\hat{Y}_i$ from $\bar{Y}$, is associated with the relationship between Y and X, so that this much of the deviation of Y_i from the mean may be said to be accounted for by the regression of Y on X. If we square and sum each side of Equation (11.24) over all the observations in the sample, we find that, since

$$\Sigma(Y_i - \hat{Y}_i)(\hat{Y}_i - \bar{Y}) = 0$$

we have

$$\Sigma(Y_i - \bar{Y})^2 = \Sigma(Y_i - \hat{Y}_i)^2 + \Sigma(\hat{Y}_i - \bar{Y})^2 \qquad 11.25$$

partitioned   We have **partitioned** the total sum of squares for Y into two parts—one part, $\Sigma(\hat{Y}_i - \bar{Y})^2$, is associated with the regression; the remainder, $\Sigma(Y_i - \hat{Y}_i)^2$, with the random or unexplained variation in the data. These quantities may be found more simply by using the following formulas:

Table 11.1 Partitioned Sum of Squares in Regression

Source of Variation	Degrees of Freedom	Sum of Squares
Regression	1	$b\Sigma(X_i - \bar{X})(Y_i - \bar{Y})$
Deviations from regression	$n - 2$	$\Sigma(Y_i - \bar{Y})^2 - b\Sigma(X_i - \bar{X})(Y_i - \bar{Y})$
Totals	$n - 1$	$\Sigma(Y_i - \bar{Y})^2$

$$\Sigma(\hat{Y}_i - \bar{Y})^2 = b\Sigma(X_i - \bar{X})(Y_i - \bar{Y})$$
$$\Sigma(Y_i - \hat{Y}_i)^2 = \Sigma(Y_i - \bar{Y})^2 - b\Sigma(X_i - \bar{X})(Y_i - \bar{Y})$$

11.26

The partitioning we have discussed is conveniently summarized in Table 11.1. We note that the sum of squares for regression has 1 degree of freedom, the sum of squares of deviations has $n - 2$ degrees of freedom, and the total sum of squares has, as usual, $n - 1$ degrees of freedom. The degrees of freedom, as well as the sum of squares, are partitioned.

For the gasoline-consumption example of Sections 11.1 through 11.4, the total variation of the Ys about their mean $\bar{Y}$ is measured by the total sum of squares (with degrees of freedom equal to $n - 1$, or 4):

$$\Sigma(Y_i - \bar{Y})^2 = 23.80$$

The portion of this variation that is accounted for by the linear relation between the variables is the sum of squares (with 1 degree of freedom) associated with, or due to, regression:

$$b\Sigma(X_i - \bar{X})(Y_i - \bar{Y}) = (-.135)(-135.10) = 18.24$$

The remaining variation—the random variation, or the variation left unexplained by the linear relation—is the variation of the observed Ys about the estimated regression line. This sum of squares (with $n - 2$, or 3, degrees of freedom) is found by subtraction:

$$\Sigma(Y_i - \bar{Y})^2 - b\Sigma(X_i - \bar{X})(Y_i - \bar{Y}) = 23.80 - 18.24 = 5.56$$

The summary table is as follows:

Source	df	Sum of Squares
Regression	1	18.24
Deviations from regression	3	5.56
Totals	4	23.80

SECTION 11.6 PARTITIONING THE SUM OF SQUARES

An informative measure is obtained if we look at the fraction of the total variation in Y that is accounted for by the association between Y and X. If we take the ratio of the sum of squares associated with the regression to the total sum of squares for Y, we have

$$r^2 = \frac{\Sigma(\hat{Y}_i - \bar{Y})^2}{\Sigma(Y_i - \bar{Y})^2} = \frac{b\Sigma(X_i - \bar{X})(Y_i - \bar{Y})}{\Sigma(Y_i - \bar{Y})^2} \qquad 11.27$$

This is called the *coefficient of determination*.

It is obvious that r^2 must always be between zero and one, inclusive, for the sum of squares in the numerator can never be less than zero or greater than the total sum of squares. If all the points are close to the line, the value of r^2 will be close to one; but as the scatter of the points becomes greater, r^2 will become smaller. For this reason it is a useful measure of the strength of the relationship.

Since r^2 is the fraction of the total variation in Y that is accounted for by the regression, $1 - r^2$ must be the fraction of the variation in Y that is unaccounted for—the fraction associated with the errors of prediction. The latter quantity is sometimes called the *coefficient of alienation*.

The sum of squares may be expressed in terms of r^2 as shown in the following table:

Source	df	Sum of Squares
Regression	1	$r^2 \Sigma (Y_i - \bar{Y})^2$
Deviations from regression	$n - 2$	$(1 - r^2)\Sigma(Y_i - \bar{Y})^2$
Totals	$n - 1$	$\Sigma(Y_i - \bar{Y})^2$

In the gasoline-consumption example,

$$r^2 = \frac{18.24}{23.80} = .77$$

Problems

38. Calculate r^2 for Problems 1 and 3.
39. In a regression analysis, $n = 5$, $\Sigma(Y_i - \bar{Y})^2 = 100$, and the sum of squared deviations from the regression is 20.
    a. Derive the table for the partitioned sum of squares in the regression.
    b. Find r^2 and $s^2_{Y \cdot X}$.

40. Give the partitioned sum of squares for the data of Problem 7. Compute the coefficient of determination, and comment on the strength of the linear relationship between the number of production units and the cost per unit.

41. Compute the coefficient of determination and coefficient of alienation for the data of Problem 8.

42. In Problem 9, what fraction of the variance in units sold is unaccounted for by the allocated shelf space?

43. The data given below are measurements made by a physics student to verify that a linear relationship exists between elapsed time (X) and the distance (Y) an object slides along a frictionless rail.
   a. Fit a regression line to the data.
   b. What percentage of the variability in the measurement of distances can be attributed to experimental error, if we assume that without error the student would have observed a perfect linear relationship between elapsed time and distance?

Elapsed time in seconds (X)	15	30	45	60
Distance in centimeters (Y)	32	57	96	115

44. A table of the partitioned sum of squares in a regression is given below. Find r^2 and $s_{Y.X}^2$.

Source	df	Sum of Squares
Regression	1	25.1
Deviation from regression	17	13.8

11.7 Multiple Regression

In the preceding sections we have studied linear regression of a *dependent* variable Y on an *independent* variable X. Many problems involve several independent variables; these are problems in **multiple regression**. Here we deal with the case of two independent variables.

multiple regression

SECTION 11.7 MULTIPLE REGRESSION

Example 1

The fuel-consumption tests in the example in Section 11.1 were carried out on cars of various weights. Here are the data for three of these cars:

Weight of Car (thousands of pounds)	Miles per Gallon at				
	30 mph	40 mph	50 mph	60 mph	70 mph
2.29	21.55	20.07	19.11	17.83	16.72
3.98	18.25	20.00	16.32	15.77	13.61
5.25	18.33	19.28	15.62	14.22	12.74

Here the independent variables are the weight of the automobile and its speed (they are nonrandom), and the dependent variable is, as before, the rate of fuel consumption (it is subject to the influence of chance as well as to the influences of weight and speed). The problem is to express the dependent variable in an approximate way as a linear function of the two independent variables.

Let Y denote the dependent variable, as before, and let X and X' denote the independent variables. (The variables are not independent of each other in the sense of random variables, because they are not random variables at all.) We assume that Y is a random variable and that the mean value of Y for fixed values of X and X' is

$$\mu_{Y|X, X'} = A + BX + B'X' \qquad 11.28$$

This is the analogue of Equation (11.1). We have at hand a sample of n observations:

$$(X_1, X'_1, Y_1), (X_2, X'_2, Y_2), \ldots, (X_n, X'_n, Y_n)$$

The problem is to use these data to estimate the parameters A, B, and B' in (11.28).

In Example 1 there are 15 observations: $n = 15$. Let X stand for weight (in thousands of pounds), let X' stand for speed (in miles per hour), and let Y stand for gasoline consumption (miles per gallon). If the observations are numbered across the rows of the table in the example, then $(X_1, X'_1, Y_1) = (2.29, 30, 21.55)$, $(X_2, X'_2, Y_2) = (2.29, 40, 20.07)$, and so on, and $(X_{15}, X'_{15}, Y_{15}) = (5.25, 70, 12.74)$.

We estimate the values of A, B, and B' by minimizing a sum of squares, namely

$$\Sigma (Y_i - a - bX_i - b'X'_i)^2 \qquad 11.29$$

This corresponds to the sum in (11.4). The procedure is to choose a, b, and b' so as to make this sum of squares as small as possible; the values of a, b, and b' that achieve this minimum are the proper estimates of A, B, and B', respectively.

The means of the observations are

$$\bar{X} = \frac{1}{n}\sum X_i, \quad \bar{X}' = \frac{1}{n}\sum X'_i, \quad \text{and} \quad \bar{Y} = \frac{1}{n}\sum Y_i$$

The equations are most simply expressed in terms of the quantities

$$x_i = X_i - \bar{X}, \quad x'_i = X'_i - \bar{X}', \quad \text{and} \quad y_i = Y_i - \bar{Y}$$

It can be shown that the values of b and b' that minimize the sum of squares in (11.29) are the solution of the following pair of linear equations:

$$b\sum x^2 + b'\sum xx' = \sum xy$$
$$b\sum xx' + b'\sum (x')^2 = \sum x'y$$

11.30

(For simplicity, the subscripts have been omitted; $\sum xx'$ stands for $\sum x_i x'_i$, $\sum (x')^2$ stands for $\sum (x'_i)^2$, and so on.)

The pair of equations in (11.30) have the solutions

$$b = \frac{\sum xy \sum (x')^2 - \sum x'y \sum xx'}{\sum x^2 \sum (x')^2 - (\sum xx')^2}$$

11.31

$$b' = \frac{\sum x'y \sum x^2 - \sum xy \sum xx'}{\sum x^2 \sum (x')^2 - (\sum xx')^2}$$

11.32

And a is determined from b and b' by the equation

$$a = \bar{Y} - b\bar{X} - b'\bar{X}'$$

11.33

The values a, b, and b' derived from the data by means of the equations (11.31), (11.32), and (11.33) are the least-squares estimates of A, B, and B'. The predicted value of Y for given values of X and X' is

$$\hat{Y} = a + bX + b'X'$$

11.34

This is the analogue of Equation (11.3).

Table 11.2 gives the analysis for the data in Example 1. Here $n = 15$ and

$$\bar{X} = \frac{57.60}{15} = 3.84, \quad \bar{X}' = \frac{750}{15} = 50, \quad \bar{Y} = \frac{259.42}{15} = 17.29$$

Since $\sum xx' = 0$, Equations (11.30) take a simple form in this case:

$$22.05b + \quad 0 \quad = -22.87$$
$$0 \quad + 3000b' = -416.5$$

Table 11.2

X	X'	Y	x	x'	y	x^2	$(x')^2$	xx'	xy	x'y
2.29	30	21.55	−1.55	−20	4.26	2.40	400	31.0	−6.60	−85.2
2.29	40	20.07	−1.55	−10	2.78	2.40	100	15.5	−4.31	−27.8
2.29	50	19.11	−1.55	0	1.82	2.40	0	0.0	−2.82	0.0
2.29	60	17.83	−1.55	10	.54	2.40	100	−15.5	−.84	5.4
2.29	70	16.72	−1.55	20	−.57	2.40	400	−31.0	.88	−11.4
3.98	30	18.25	.14	−20	.96	.02	400	−2.8	.13	−19.2
3.98	40	20.00	.14	−10	2.71	.02	100	−1.4	.38	−27.1
3.98	50	16.32	.14	0	−.97	.02	0	0.0	−.14	0.0
3.98	60	15.77	.14	10	−1.52	.02	100	1.4	−.21	−15.2
3.98	70	13.61	.14	20	−3.68	.02	400	2.8	−.52	−73.6
5.25	30	18.33	1.41	−20	1.04	1.99	400	−28.2	1.47	−20.8
5.25	40	19.28	1.41	−10	1.99	1.99	100	−14.1	2.81	−19.9
5.25	50	15.62	1.41	0	−1.67	1.99	0	0.0	−2.35	0.0
5.25	60	14.22	1.41	10	−3.07	1.99	100	14.1	−4.33	−30.7
5.25	70	12.74	1.41	20	−4.55	1.99	400	28.2	−6.42	−91.0
Sums 57.60	750	259.42				22.05	3000	0.0	−22.87	−416.5

Thus

$$b = \frac{-22.87}{22.05} = -1.04$$

$$b' = \frac{-416.5}{3000} = -.14$$

and $a = 17.29 - (-1.04)(3.84) - (-.14)(50) = 28.28$ by (11.33). Finally, the prediction equation is

$$\hat{Y} = 28.28 - 1.04X - .14X'$$

For a car weighing 2500 pounds and traveling at 55 miles per hour, the average fuel consumption should be about $28.28 - 1.04(2.50) - .14(55)$, which comes to 17.98 miles per gallon.

For further information concerning estimation and testing for multiple regression models, see Reference 11.7.

Problems

45. The following table presents the percentage of defective items produced by a machine. This percentage is dependent upon (1) the impurities present in the raw material (a plastic), as measured in grams per cubic centimeter of a solution of this material, and (2) the speeds at which the machine is operated, as represented by the numbers 1 through 4, which are proportional to the four different speeds of the machine. Let X be the amount of impurities in the plastic, X' the speed of the machine, and Y the percentage of defective items in the output.
  a. Fit a multiple linear regression equation expressing the mean of Y in terms of X and X'.
  b. Predict the mean of Y for $X = 3.5$ and $X' = 3$.

	Impurities				
Speed	1	2	3	4	5
1	2.4	2.8	2.8	3.0	3.1
2	3.1	2.9	3.2	3.3	3.5
3	3.4	3.9	3.8	3.9	4.0
4	3.9	3.9	4.4	4.5	4.6

SECTION 11.7 MULTIPLE REGRESSION

46. For the following data:
  a. Fit the regression of Y on X and X'.
  b. Calculate the $\widehat{Y}$s as given by the regression for the values of X and X' in order to see how good the fit is.
  c. The *coefficient of multiple determination* is similar to the coefficient of determination:

$$R^2 = \frac{\Sigma (\widehat{Y}_i - \bar{Y})^2}{\Sigma (Y_i - \bar{Y})^2}$$

Find R^2 using the results in part b.

X	0	1	2	3	4
X'	1	0	1	0	1
Y	3	-1	4	0	4

47. One of the criteria in admitting students to a law school is the student's GPA (undergraduate grade-point average). Another is his LSAT (law-school admission test score). It is important to correlate his first-year average grade (denoted FYA) with the GPA and LSAT. The following data, from actual samples, have been simplified for ease of calculation.
  a. Fit a multiple regression line for the FYA on the GPA and LSAT scores.
  b. Calculate the $\widehat{Y}$s as given by the regression.
  c. Find R^2 (see Problem 46).
  d. Discuss your findings.

GPA	3.2	2.8	3.6	3.1	3.3	3.0	3.3	3.4
LSAT	550	500	550	550	600	500	700	600
FYA	55	60	60	65	65	70	70	70

GPA	3.5	3.6	3.8	3.1	3.6	3.9	3.4	3.8
LSAT	650	550	500	700	650	750	650	700
FYA	70	70	75	75	75	80	80	80

48. The linear relationship $\mu_{Y|X} = A + BX$ may be an inappropriate way to express the relationship between variables X and Y in certain cases. One possible alternative is the quadratic relationship $\mu_{Y|X} = A + BX + B'X^2$. It is possible to fit a quadratic prediction equation $\widehat{Y} = a + bX + b'X^2$ to data by using multiple linear regression. Just let

$X' = X^2$ and apply the formulas of this section. Fit a quadratic prediction equation to the following data. Plot the data and the predicted values on graph paper.

X	0	1	2	3	4
Y	4.5	0.5	0.0	1.5	3.5

49. The heights of 44 college-age males (Y) and the heights of the mother (X) and father (X') of each were obtained as part of a project for a statistics class. The table below summarizes computations that were performed by a computer program for multiple regression. (The results have been rounded.)
   a. Fit a multiple regression equation to the data.
   b. Use the equation to predict the heights of some of the males in your class. Would you consider the equation to be a good predictor?
   c. The coefficient of multiple determination (see Problem 46) may be computed by the following formula:

$$R^2 = \frac{\beta \Sigma xy + \beta' \Sigma x'y}{\Sigma y^2}$$

Compute R^2 for these data and interpret it.

n	$\bar{X}$	$\bar{X}'$	$\bar{Y}$	Σx^2	$\Sigma (x')^2$	Σy^2	Σxy	$\Sigma x'y$	$\Sigma xx'$
44	64	70	70	303	375	372	134	183	74

11.8 Correlation

In the preceding sections of this chapter we have been concerned with the use of regression techniques to estimate the parameters of an assumed linear relation between X and the mean of Y given X. We assumed that the values of X were known and allowed them to be selected and controlled by the experimenter; that is, we did not assume that X was a random variable.

We now consider methods that are appropriate when it is assumed that X and Y are both random variables and have a joint distribution. We want to make inferences about the degree of linear relationship between them without estimating the regression line.

One of the parameters of the joint distribution of X and Y is the *product*

SECTION 11.8 CORRELATION

correlation coefficient

moment correlation coefficient, or simply the **correlation coefficient,** ρ. It is a measure of the *linear* covariation of the variables; that is, it measures the degree of *linear* association between them. It is a dimensionless quantity that may take any value between -1 and 1, inclusive. If ρ is either -1 or 1, the variables have a perfect linear relationship in that all of the points in a sample lie exactly on a line. If ρ is near -1 or 1, there is a high degree of linear association.

A *positive correlation* means that as one variable increases, the other increases. A *negative correlation* means that as one variable increases, the other decreases. Heights and weights of humans are positively correlated, but the age of a car and its trade-in value are negatively correlated. If ρ is equal to zero, we say the variables are *uncorrelated* and that there is no linear association between them. Bear in mind that ρ measures only linear relationship. The variables may be perfectly related in a curvilinear relationship and ρ could still equal zero.

In a correlation analysis our problem is to make inferences about ρ. Given a sample of size n, $(X_1, Y_1), (X_2, Y_2), \ldots, (X_n, Y_n)$, we use the **sample correlation coefficient**, r, as an estimator for ρ:

sample correlation coefficient

$$r = \frac{\Sigma (X_i - \bar{X})(Y_i - \bar{Y})}{\sqrt{\Sigma (X_i - \bar{X})^2 \Sigma (Y_i - \bar{Y})^2}} \qquad 11.35$$

Regression techniques and correlation methods are closely related, for r is the square root of the coefficient of determination. It is primarily in interpretation that the differences lie. In correlation, r is an estimator for the population correlation coefficient ρ. In regression, since X is not a random variable, there is no correlation, and r^2 is simply a measure of closeness of fit.

When we want to make inferences about the population correlation coefficient ρ, we usually assume that the variables X and Y have a joint normal distribution; but even with the assumption of normality, if ρ is not equal to zero, the sampling distribution of r is complicated and not at all easy to use. For this reason, tables and graphs have been made for finding confidence intervals for ρ. See Reference 11.1 or 11.6.

We may also make use of a normal approximation. R. A. Fisher has shown that if X and Y are jointly normally distributed, the quantity

$$z = \frac{1}{2} \log_e \frac{1+r}{1-r} \qquad 11.36$$

is approximately normally distributed with mean

$$\mu_z = \frac{1}{2} \log_e \frac{1+\rho}{1-\rho} \qquad 11.37$$

and variance

$$\sigma_z^2 = \frac{1}{n-3} \qquad 11.38$$

This approximation holds fairly well for sample sizes greater than 50. Closer approximations, some of which hold reasonably well for values of *n* as low as 11, may be found in Reference 11.3.

To use this approximation to find a confidence interval for ρ, we transform the sample correlation coefficient, *r*, to the corresponding *z*-value by Equation (11.36), or by means of Table 7 in the Appendix, which is much easier. We then find the confidence interval for μ_z:

$$L = z - t_{\alpha/2, \infty} \frac{1}{\sqrt{n-3}}$$
$$R = z + t_{\alpha/2, \infty} \frac{1}{\sqrt{n-3}} \qquad 11.39$$

These limits are then transformed via Table 7 to limits for ρ.

For example, suppose that in a sample of size 52 we find $r = .61$. From Table 7 the corresponding *z* is .71, and the 95% confidence limits for μ_z are

$$L = .71 - 1.96 \left(\frac{1}{\sqrt{49}}\right) = .43$$

and

$$R = .71 + 1.96 \left(\frac{1}{\sqrt{49}}\right) = .99$$

Referring again to the table, we find that the corresponding limits for ρ are .4053 and .7574, respectively. We may say that these are approximate 95% confidence limits for the unknown population correlation coefficient.

The *z*-transformation may also be used to test the hypothesis that ρ is equal to a specified value ρ_0, for under this hypothesis

$$t = \frac{z - \mu_{z0}}{\sigma_z} \qquad 11.40$$

has approximately a standard normal distribution (the *t*-distribution with infinite degrees of freedom).

Suppose we want to test the hypothesis $H_0: \rho = .6$ against the alternative hypothesis $H_a: \rho \neq .6$, and that $n = 103$, $r = .5$. Using the conversion table we find

$$z = \frac{1}{2} \log_e \frac{1 + .5}{1 - .5} = .55$$

SECTION 11.9 CORRELATION AND CAUSE

$$\mu_{z0} = \frac{1}{2} \log_e \frac{1 + .6}{1 - .6} = .69$$

The calculated value of the test statistic is

$$t = \frac{.55 - .69}{\frac{1}{10}} = -1.4$$

From the t-table, $Z_{.025}$ is found to be 1.96; therefore, we accept the hypothesis $H_0: \rho = .6$. Our level of significance is approximately 5%.

If we want to test the hypothesis that the variables are not linearly related, that is, that ρ is 0, we may use an ordinary t-test, for when p is 0,

$$t = r\sqrt{\frac{n-2}{1-r^2}} \qquad 11.41$$

has the t-distribution with $n - 2$ degrees of freedom. To test $H_0: \rho = 0$, we merely evaluate t by Equation (11.41) and compare it with the tabular t-value for the given probability of Type I error.

It is worthy of note that the estimated correlation coefficient is unaffected by linear transformations. If we let U be $AY + B$ and let V be $CX + D$, the correlation coefficient r_{UV} for U and V is equal to that for X and Y, r_{XY}. This can be shown by substituting for U and V in terms of X and Y in the equation

$$r_{UV} = \frac{\Sigma (U_i - \bar{U})(V_i - \bar{V})}{\sqrt{\Sigma (U_i - \bar{U})^2 \Sigma (V_i - \bar{V})^2}}$$

as follows:

$$r_{UV} = \frac{\Sigma (AY_i + B - A\bar{Y} - B)(CX_i + D - C\bar{X} - D)}{\sqrt{\Sigma (AY_i + B - A\bar{Y} - B)^2 \Sigma (CX_i + D - C\bar{X} - D)^2}}$$

$$= \frac{AC \Sigma (Y_i - \bar{Y})(X_i - \bar{X})}{\sqrt{A^2 C^2 \Sigma (Y_i - \bar{Y})^2 \Sigma (X_i - \bar{X})^2}}$$

$$= \pm r_{XY}$$

The sign before r_{XY} depends upon the signs of A and C.

11.9 Correlation and Cause

It has been said, and with justification, that among all of the measures treated in this book, the correlation coefficient is the most subject to misinterpretation. One of the main reasons for this is the frequently false assumption that because two variables are related, a change in one *causes* a change in the other. If one has a point to prove, it is extremely easy to succumb to this fallacy and to use a perfectly respectable correlation coefficient to "prove" a **cause-and-effect relationship** that may not exist.

cause-and-effect relationship

It has been shown that there is a negative correlation between smoking and grades. If one objects to smoking, one seizes upon this fact as objective evidence that smoking is harmful—that smoking causes low grades. This may be true, but one could also argue that low grades result in increased nervous tension, which, in turn, causes the individual to smoke more. It could also be true that smoking and low grades are not directly related to each other, but are correlated because both are related to some other factor, such as involvement in extracurricular or social activities.

This last suggestion illustrates another danger in interpretation of the results of a correlation analysis. Frequently, two variables may appear to be highly correlated when, in fact, they are not directly associated with each other but are both highly correlated with a third variable.

Because of sampling variation we can get a significant correlation when the variables are not related, but we are aware of this possibility and control the probability of this kind of error by selecting our level of significance. It is the unwarranted cause-and-effect assumption and the spurious correlation we must most beware of.

Problems

50. Given the following sets of quantities, find the sample correlation coefficient. Test the hypothesis $H_0: \rho = 0$ against the alternative $H_a: \rho \neq 0$ at the indicated level of significance.
    a. $n = 27$, $\Sigma(Y_i - \bar{Y})^2 = 100$, $\Sigma(X_i - \bar{X})^2 = 625$, $\Sigma(X_i - \bar{X})(Y_i - \bar{Y}) = 200$, $\alpha = .01$
    b. $n = 5$, $\Sigma(Y_i - \bar{Y})^2 = 16$, $\Sigma(X_i - \bar{X})^2 = 25$, $\Sigma(X_i - \bar{X})(Y_i - \bar{Y}) = 10$, $\alpha = .05$

51. The high-school grade-point averages (X) and the college grade-point averages (Y) of a random sample of 29 college freshmen showed the following results: $\Sigma(Y_i - \bar{Y})^2 = 64$, $\Sigma(X_i - \bar{X})(Y_i - \bar{Y}) = 40$, $\Sigma(X_i - \bar{X})^2 = 100$. Test the null hypothesis $\rho = 0$ against the alternative hypothesis $\rho > 0$ at the 5% level of significance.

52. The following data are the city and highway mileage ratings of five randomly chosen automobiles. Test the null hypothesis $\rho = 0$ against the alternative $\rho > 0$ at the 5% level of significance.

SECTION 11.9 CORRELATION AND CAUSE   397

Highway mileage	40	22	25	31	29
City mileage	24	16	18	25	22

53. For the data in Problem 46 find the sample correlation coefficient between Y and X, Y and X', and X and X'. Compare these with R.

54. Find the sample correlation coefficient between Y and X, Y and X', and X and X' of Problem 47. Compare these with R.

55. Determinations were made of the amount (ppm) of a soluble compound present at two different depths in a number of soils.
    a. Estimate the correlation between the amounts present at the two depths.
    b. Is there a significant correlation? Let $\alpha = .05$.

12 inches	20 inches	12 inches	20 inches
24	20	66	84
84	103	31	30
13	16	43	62
13	20	19	26
48	86	7	21
61	36	50	73
112	53	72	83

56. Find a 95% confidence interval for ρ when
    a. $n = 84, r = .60$.   b. $n = 67, r = .20$.
    c. $n = 52, r = -.83$.

57. Find a 99% confidence interval for ρ when
    a. $n = 128, r = .14$.   b. $n = 80, r = -.30$.
    c. $n = 75, r = .92$.

58. The correlation between the verbal and quantitative Scholastic Aptitude Test scores of 120 randomly selected high school seniors is $r = .76$. Make a 95% confidence interval for ρ.

59. A manufacturer of a machine for electronically measuring blood pressure claimed that the readings on his machine would correlate at least at the .9 level with readings obtained by the conventional method. A sample of 200 patients gave a correlation $r = .75$. Do these results refute the manufacturer's claim? Let $\alpha = .05$.

60. The correlation between the lifetime of a certain type of light bulb and the voltage applied to it is $\rho = -.50$. After the filament was redesigned, a sample of 62 of the new light bulbs showed a correlation of

$r = -.30$ between lifetime and voltage. Has the correlation changed significantly as a result of the design change? Let $\alpha = .05$.

61. A psychologist has constructed an IQ test much shorter than the standard one. He claims that his short test has a correlation of at least .85 with the standard one. Each of 50 students has been given the two tests. The sample correlation coefficient is .90. Is this result sufficient to reject at the 5% level the null hypothesis $H_0 : \rho = .85$ in favor of the alternative $H_a : \rho > .85$?

62. For each of the following pairs of variables state whether you would expect a positive correlation, a negative correlation, or no correlation at all.
   a. Weight of husband and wife.
   b. IQ of husband and wife.
   c. Education of husband and wife.
   d. IQ and hours per day of television viewing.
   e. IQ and shoe size.
   f. Time of day and time required to perform a manual task.
   g. Price of hamburgers and salaries of teachers at the same time.
   h. Time to run 880 yards and hours of training.
   i. Temperature and inches of snow on the ground.
   j. Education and income.

11.10 The Least-Squares Estimates

The following algebraic manipulations show that the b in (11.5) and the a in (11.6) make the sum in (11.4) as small as possible. It is to be emphasized that it is not necessary to follow through this demonstration in order to understand the least-squares method as applied in the preceding sections.

To simplify the notation we write $x_i = X_i - \bar{X}$ and $y_i = Y_i - \bar{Y}$. Temporarily write a_0 in place of $a - \bar{Y} + b\bar{X}$. The sum in (11.4) is then

$$\Sigma(Y_i - a - bX_i)^2 = \Sigma[(Y_i - \bar{Y}) - (a - \bar{Y} + b\bar{X}) - b(X_i - \bar{X})]^2$$
$$= \Sigma(y_i - a_0 - bx_i)^2 \qquad 11.42$$

By the formula $(r + s + t)^2 = r^2 + s^2 + t^2 + 2rs + 2rt + 2st$, we have

$$\Sigma(y_i - a_0 - bx_i)^2 = \Sigma(y_i^2 + a_0^2 + b^2 x_i^2 - 2y_i a_0 - 2y_i bx_i + 2a_0 bx_i)$$

Since $\Sigma a_0^2 = na_0^2$ and $\Sigma x_i = \Sigma y_i = 0$, this reduces to

$$\Sigma(y_i - a_0 - bx_i)^2 = \Sigma y_i^2 + na_0^2 + b^2 \Sigma x_i^2 - 2b \Sigma x_i y_i \qquad 11.43$$

By the formula $(r + s)^2 = r^2 + 2rs + s^2$, we have

$$\frac{(b \Sigma x_i^2 - \Sigma x_i y_i)^2}{\Sigma x_i^2} = b^2 \Sigma x_i^2 - 2b \Sigma x_i y_i + \frac{(\Sigma x_i y_i)^2}{\Sigma x_i^2}$$

The first two terms on the right here match the last two terms on the right in (11.43), and so a substitution into (11.43) gives

$$\Sigma(y_i - a_0 - bx_i)^2 = \Sigma y_i^2 + na_0^2 + \frac{(b\Sigma x_i^2 - \Sigma x_i y_i)^2}{\Sigma x_i^2} - \frac{(\Sigma x_i y_i)^2}{\Sigma x_i^2}$$

The left side here is, by (11.42), the same thing as the sum in (11.4); replacing the a_0 on the right by $a - \bar{Y} + b\bar{X}$ therefore gives

$$\Sigma(Y_i - a - bX_i)^2 = \Sigma y_i^2 - \frac{(\Sigma x_i y_i)^2}{\Sigma x_i^2} + n(a - \bar{Y} + b\bar{X})^2$$
$$+ \frac{(b\Sigma x_i^2 - \Sigma x_i y_i)^2}{\Sigma x_i^2} \quad 11.44$$

This is an algebraic identity; it holds whatever values a and b may have.

Now the first and second terms on the right in (11.44) do not involve a and b at all. Therefore, to make the right side of (11.44) as small as possible, choose a and b to make the third and fourth terms equal to 0. (Notice that these two terms cannot in any case be negative.) This means choosing b so that $b\Sigma x_i^2 - \Sigma x_i y_i = 0$ (which makes the fourth term 0) and then choosing a so that $a - \bar{Y} + b\bar{X} = 0$ (which makes the third term 0). But these are indeed the b in (11.5) and the a in (11.6).

For this a and b, the right side of (11.4) is $\Sigma y_i^2 - (\Sigma x_i y_i)^2/(\Sigma x_i^2)$. This is the minimum value of $\Sigma(Y_i - a - bX_i)^2$, and it coincides with the sum of squares due to deviations from regression; see Table 11.1.

CHAPTER PROBLEMS

63. In an experiment to determine the effect of the length of heat-treatment time on the resistivity of a type of wire, five samples of wire were treated at each of four times. The observations are of the resistance per unit volume of the wire. Note that there are 20 observations, each value of X occurring with five different Ys.

X = Time (hours)	2	4	6	8
Y = Resistance	21.3	23.2	25.5	25.9
	21.7	23.4	23.6	25.2
	21.4	23.3	24.7	27.6
	22.1	23.7	26.0	27.1
	20.7	23.0	24.3	26.3

a. Find the regression equation.

b. Test the hypothesis that $B = 1.0$ against the alternative that $B \neq 1.0$. Let $\alpha = .05$.
c. What fraction of the variation in Y is accounted for by time?
d. Estimate the mean resistance given that $X = 5$.
e. Find a 95% confidence interval for the mean resistance given $X = 5$.

64. Various known amounts X of a compound were chromatographed and the peak area Y determined, giving the following results. The peak area has been multiplied by a suitable scale factor.
    a. Estimate the relationship between peak area and volume injected.
    b. Find a 95% confidence interval for the slope of the line.
    c. Test the hypothesis that the line passes through the origin. Let $\alpha = .05$.
    d. To estimate the slope of a line through the origin, we use

    $$b = \frac{\Sigma X_i Y_i}{\Sigma X_i^2}$$

    i.e., the uncorrected sums of squares and crossproducts. Find the regression equation $\hat{Y} = bX$ which assumes $A = 0$.

Peak Area	Amount Injected (cc)	Peak Area	Amount Injected (cc)
13	.1	62	.4
25	.2	105	.7
27	.2	88	.6
46	.3	63	.4
18	.1	77	.5
31	.2	109	.7
46	.3	117	.8
57	.4	35	.2
75	.5	98	.6
87	.6	121	.8

65. The following data are observations on the horsepower of an engine at 1800 rpm as a function of the viscosity of the oil.
    a. Plot the points on graph paper. Let X be viscosity and Y be horsepower.
    b. Find and plot the point $(\bar{X}, \bar{Y})$.
    c. Find a and b and, hence, the regression equation $\hat{Y} = a + bX$. Draw this line on the graph.
    d. Find the deviation of each Y from the corresponding $\hat{Y}$. Square and sum these deviations.

hp	Viscosity	hp	Viscosity
16.8	45	18.1	54
18.1	59	19.2	63
18.5	66	16.9	48
17.0	47	18.6	55
18.8	61	21.0	67
19.7	68	16.4	44
17.5	49	17.7	56
19.0	57	18.2	62
20.2	67	16.7	50
16.3	43	20.8	70

e. Find the sum of squares for deviations from regression by the formula

$$\Sigma(Y_i - \hat{Y}_i)^2 = \Sigma(Y_i - \bar{Y})^2 - b\Sigma(X_i - \bar{X})(Y_i - \bar{Y})$$

How do you account for the difference between this figure and that obtained in part (d)?

f. Construct a summary table showing the partitioning of the sum of squares and degrees of freedom.

g. Find the estimated variance $s_{Y.X}^2$ and the variance and standard deviation of b.

h. Find the 95% confidence interval for B.

i. Test the hypothesis that $B = 0$ against the alternative $B \neq 0$. Let $\alpha = .05$.

j. What is the best estimate for the mean horsepower if the viscosity of an oil is 50? What is the standard deviation of this estimate?

k. Find a 99% confidence interval for the mean horsepower, given that the viscosity is 60.

66. In predicting the Y-variable from the X-variable in linear regression, computations may be simplified by changing the scale of the X-variable through a linear transformation. Two sets of data, I and II, are given below. The values of Y in I and II are the same, but the values of X in II were obtained by dividing the values of X in I by 100 and then subtracting 3 from each. That is, $X_{II} = X_I/100 - 3$.

	Set I				
X	100	200	300	400	500
Y	1	3	4	6	7

	Set II				
X	−2	−1	0	1	2
Y	1	3	4	6	7

a. Fit a regression line to both sets of data.
b. Show that the predicted values of Y for corresponding values of X are the same for the two sets of data.
c. Verify that the correlation coefficient between X and Y is unaffected by the linear transformation.

Summary and Keywords

In linear regression, the relation between an **independent** variable and a **dependent** variable (p. 364) is studied. A preliminary idea of the relationship can be obtained from a study of the **scatter diagram** (p. 364). A **regression line** (p. 367) is fitted to the data by the method of **least squares** (p. 367), which estimates the **intercept** and the **slope** (p. 365) and makes possible statistical **inferences** (p. 374) on these parameters and the **prediction** (p. 378) of future values. The variation in the data can be described by the **partitioned** sum of squares (p. 383). Problems involving two or more independent variables come under **multiple regression** (p. 386). If the dependent variable is random, correlation theory applies. The **correlation coefficient** (p. 393) measures the linear association between two variables, and it is estimated by the **sample correlation coefficient** (p. 393). Correlation is *not* to be identified with **cause and effect** (p. 396).

REFERENCES

11.1 David, F. N. *Tables of the Correlation Coefficient*. Cambridge, Mass.: Cambridge University Press, 1954.

11.2 Huff, Darrell. *How to Lie with Statistics*. New York: Norton, 1965. Chapter 8.

11.3 Kendall, M. G., and Stuart, A. *The Advanced Theory of Statistics*. 2d ed. London: Griffin, 1964.

11.4 Mosteller, Frederick; Rourke, Robert E. K.; and Thomas, George B., Jr. *Probability with Statistical Applications*. 2d ed. Reading, Mass.; Addison-Wesley, 1970. Chapter 11.

11.5 Ostle, Bernard. *Statistics in Research*. 2d ed. Ames, Iowa: Iowa State University Press, 1963. Chapters 8 and 9.

REFERENCES

11.6 Richmond, Samuel B. *Statistical Analysis*. 2d ed. New York: Ronald Press, 1964. Chapter 19.

11.7 Snedecor, George W., and Cochran, William G. *Statistical Methods*. 6th ed. Ames, Iowa: Iowa State University Press, 1967. Chapters 6, 7, and 13.

12

Analysis of Variance

12.1 Introduction

In Chapter 9 we studied methods for analyzing and interpreting the results of two simple kinds of experiments for comparing the effect of two treatments. The two-sample techniques are applicable when the experiment is a completely randomized design for two treatments, and the method of paired comparisons is appropriate when the two treatments are randomized *within* each of n pairs of similar experimental units. We now extend these methods to the general case of k samples, or k treatments, in a completely randomized experiment (this parallels the two-sample theory of Sections 9.2, 9.3, and 9.4) and to the randomized complete block experiment (this parallels the paired-comparisons theory of Section 9.6). Here we become acquainted with one of the most powerful of statistical tools, the **analysis of variance.**

The analysis of variance is an arithmetic device for **partitioning the total variation** in a set of data according to the various sources of variation that are present. It results in a summary table similar to Table 11.1, which is, in fact, the analysis-of-variance table for simple linear regression; this table provides a convenient form for summarizing and presenting the information contained in a set of data. Furthermore, study of the complete analysis-of-variance table for a set of experimental or sample survey data will show us whether or not valid tests of certain hypotheses exist, and, if so, how the tests should be performed.

analysis of variance
partitioning the total variation

12.2 The Role of Randomization

When we perform an experiment designed to compare the effects of several treatments, our estimates for these treatment effects may be biased because of known or unknown factors that influence the results and that may favor some treatments more than others. The elimination of the bias due to these factors is the primary purpose of **randomization.**

randomization

Suppose we have four chemicals and want to study their effects on the breaking strength of a ceramic. We divide the slip (clay) into portions that will be cast into cylinders, and then we randomly assign the chemicals to equal numbers of these portions. Randomization at this point guards against nonhomogeneity of the slip. After the cylinders have been cast, they are placed on racks and fired in a kiln that is large enough that all of them can be fired at the same time. We randomize the positions of the cylinders within the kiln, because not all positions will be subjected to the same temperature, and the temperature at which the cylinders are fired will probably affect their breaking strength. If it should happen that all the cylinders containing some one of the four chemicals are subjected to greater heat during firing than those containing the other three, the differences among the mean breaking strengths will estimate the differences in the effects of the chemicals *plus* the effect of the different temperatures. If, however, we randomize the positions of the cylinders within the kiln, then all cylinders have the same chance of being fired under the more favorable temperature, and our estimates will be unbiased. We may argue that any specific randomization may result in an arrangement that favors one or another of the additives. This is true, but it is also allowed for in the procedures we use for testing hypotheses and finding confidence intervals. Furthermore, proper randomization will ensure that our assumption of independent observations is satisfied and will permit us to use the mathematics of probability to attach measures of reliability to the inferences we make on the basis of our experimental results.

12.3 The Completely Randomized Design

In Section 9.2 we studied some of the concepts associated with the completely randomized experiment for comparing the effects of two treatments. We now extend these ideas to the case of k treatments, each of which is assigned at random to n experimental units. The randomization is carried out separately for each of kn experimental units; that is, for each unit we select at random a number from 1 to k to decide which treatment should be applied to that experimental unit. We generally restrict the randomization in such a way that each treatment is applied to the same number of experimental units. If there are no restrictions except this requirement of equal numbers of experimental units per treatment, then the experiment is said to have a **completely randomized design.** In this case all experimental units have the same chance of receiving any one of the treatments, and the experimental units are independent.

completely randomized design

After the experiment has been conducted, we have a set of data consisting of the kn responses of the experimental units, classified into k groups according to the treatments that were applied. We assume (1) that the observed values in any one group constitute a random sample of all possible responses under that treatment of all experimental units, (2) that the variation among units treated alike is the same for all treatments, and (3) that the responses are normally distributed. These assumptions are equivalent to assuming that we have k independent random samples from k normal populations that have common variance.

Example 1

Suppose four teaching methods are to be tested on 12 students. The students (the experimental units) are randomly assigned to the teaching methods (the treatments), three students to each method. The randomization prevents the systematic assignment of the best students to some one method. Suppose at the end of the training period the students take a standardized test, achieving the scores in the following table.

		Method (Treatment)			
		1	2	3	4
	1	110	111	113	118
Student	2	109	116	108	123
	3	105	109	109	125

The analysis of these data is carried through in Section 12.6.

12.4 The Model

When our data consist of k independent random samples of size n, the individual observed values are subject to a single criterion of classification, the sample or treatment to which they belong. Two subscripts will therefore be sufficient to identify completely any observed value; hence we represent an observation by X_{ij}, where the first subscript, i, denotes the sample, and the second, j, the individual observation within the sample. For example, X_{23} is the third observed value in the second sample.

For data of the type considered here, we assume the mathematical

model   **model**

$$X_{ij} = \mu + \tau_i + e_{ij}, \quad i = 1, 2, \ldots, k, \quad j = 1, 2, \ldots, n \quad 12.1$$

which states that any observed value X_{ij} is equal to the overall mean μ for all the populations, plus the deviation τ_i of the ith population mean μ_i from the overall mean, plus a random deviation e_{ij} from the mean of the ith population. In other words, if μ_i is the mean of the ith population, then

$$\mu = \frac{1}{k} \sum \mu_i \quad 12.2$$

SECTION 12.5 CONSTRUCTING THE ANALYSIS-OF-VARIANCE TABLE

$$\tau_i = \mu_i - \mu \qquad \text{12.3}$$

$$e_{ij} = X_{ij} - \mu_i = X_{ij} - \mu - \tau_i \qquad \text{12.4}$$

group effects The τ_i are known as the **group effects**, or **treatment effects**.
treatment For this model, Equation (12.1), we shall make the following three
effects assumptions.

> *Assumption 1* μ is an unknown parameter.
>
> *Assumption 2* The τ_i are unknown constants or parameters.
>
> *Assumption 3* The e_{ij} are normally and independently distributed with mean zero and variance σ^2.

The second assumption is appropriate in cases where the populations from which the samples are obtained constitute the whole set of populations in which we are interested. In cases where the populations from which the samples are drawn are themselves a sample of the populations that might be employed, the τ_i are assumed to be random variables that are normally and independently distributed with mean zero and variance σ_τ^2. In the first case we say we have a *fixed model*, in the latter a *random model*.

In the analyses of variance that we consider here, the only effect of assuming the random model instead of the fixed model lies in the interpretation of the results; the calculations and tests of hypotheses are not affected.

12.5 Constructing the Analysis-of-Variance Table

Consider the set of data represented symbolically in Table 12.1. We have k groups or samples, each consisting of n observed values. In the table a group total is represented by $X_{i.}$, where the dot replacing the subscript j shows that we have summed over j, that we have added within the group. The overall sum for the whole set is denoted by $X_{..}$, the dots showing that we have summed over both i and j; that is, we have added within the groups and then over the groups. We write $\bar{X}_{i.}$ and $\bar{X}_{..}$ for the corresponding means. We have, then,

$$X_{i.} = \sum_{j=1}^{n} X_{ij}, \qquad \bar{X}_{i.} = \frac{X_{i.}}{n}$$

Table 12.1 k Groups, Each of Size n

	Group					
	1	2	3	$\cdots$	k	
	X_{11}	X_{21}	X_{31}	$\cdots$	X_{k1}	
	X_{12}	X_{22}	X_{32}	$\cdots$	X_{k2}	
	X_{13}	X_{23}	X_{33}	$\cdots$	X_{k3}	
	$\vdots$	$\vdots$	$\vdots$		$\vdots$	
	X_{1n}	X_{2n}	X_{3n}	$\cdots$	X_{kn}	
Sum	$X_{1.}$	$X_{2.}$	$X_{3.}$	$\cdots$	$X_{k.}$	$X_{..}$
Mean	$\bar{X}_{1.}$	$\bar{X}_{2.}$	$\bar{X}_{3.}$	$\cdots$	$\bar{X}_{k.}$	$\bar{X}_{..}$

$$X_{..} = \sum_{i=1}^{k} X_{i.} = \sum_{i=1}^{k} \sum_{j=1}^{n} X_{ij} \qquad 12.5$$

$$\bar{X}_{..} = \frac{X_{..}}{kn}$$

analysis-of-variance table   To construct the corresponding **analysis-of-variance table**, we must first consider what sources of variation are present and how the total variation is to be partitioned according to these sources. The *total variation* in the data is measured by the total sum of squares of deviations from the overall mean. One source of variation, *the differences among the group means,* is measured by the sum of squares of the deviations of the group means from the overall mean. The only remaining variation is that among the *observations within each group*—that is, the variation of the elements of each group about the mean for that group. This we measure by the pooled sum of squares of deviations of the individual observations from the group means.

Consider the total sum of squares

$$\sum_{i} \sum_{j} (X_{ij} - \bar{X}_{..})^2$$

If we add and subtract the mean of the ith group within the parentheses, we have

$$\sum_{i} \sum_{j} (X_{ij} - \bar{X}_{i.} + \bar{X}_{i.} - \bar{X}_{..})^2$$

SECTION 12.5 CONSTRUCTING THE ANALYSIS-OF-VARIANCE TABLE

If we now square the quantity contained within the parentheses we get

$$\sum_i \sum_j (X_{ij} - \bar{X}_{i.} + \bar{X}_{i.} - \bar{X}_{..})^2$$

$$= \sum_i \sum_j (X_{ij} - \bar{X}_{i.})^2 + 2 \sum_i \sum_j (X_{ij} - \bar{X}_{i.})(\bar{X}_{i.} - \bar{X}_{..}) + \sum_i \sum_j (\bar{X}_{i.} - \bar{X}_{..})^2$$

The middle term on the right-hand side is zero, since it can be written as

$$2 \sum_i \left[(\bar{X}_{i.} - \bar{X}_{..}) \sum_j (X_{ij} - \bar{X}_{i.}) \right]$$

and here $\sum_j (X_{ij} - \bar{X}_{i.})$ is zero for all i (the sum of the deviations of a set of values about their arithmetic mean is equal to zero).

Thus the total sum of squares can be written

$$\sum_i \sum_j (X_{ij} - \bar{X}_{..})^2 = \sum_i \sum_j (X_{ij} - \bar{X}_{i.})^2 + n \sum_i (\bar{X}_{i.} - \bar{X}_{..})^2 \qquad 12.6$$

The first term of the right-hand member of Equation (12.6) is the pooled sum of squares of deviations of the observations within the groups from the group means. Within each group there are n values, and among these n values there are $n - 1$ degrees of freedom. Since there are k such groups, the pooled sum of squares within groups will have $k(n - 1)$ degrees of freedom.

The second term in the right-hand member of Equation (12.6) is the sum of squares of deviations of the group means from the overall mean. Multiplication by n puts it on a per-observation basis. This is a sum of squares of deviations for k quantities and will therefore have $k - 1$ degrees of freedom.

The analysis-of-variance table corresponding to the identity (12.6) for the partitioning of the sum of squares for Table 12.1 is shown in Table 12.2. The mean squares of Table 12.2 are the sums of squares divided by their degrees of freedom.

In the analysis of variance (Table 12.2), the sources of variation are written in general terms as **among groups** and **within groups**. If our data were the result of a completely randomized experiment for k treatments, each applied to n experimental units, the sources of variation would be *among treatments* and *among experimental units treated alike*. These are usually shortened to **treatments** and **error**, as shown in Table 12.3.

The sums of squares are given in Table 12.2 in terms of deviations from

Table 12.2 Analysis of Variance for the Data of Table 12.1

Source of Variation	Degrees of Freedom	Sum of Squares	Mean Square
Among groups	$k - 1$	$n \sum_i (\bar{X}_{i.} - \bar{X}_{..})^2$	$\dfrac{n}{k-1} \sum_i (\bar{X}_{i.} - \bar{X}_{..})^2$
Within groups	$k(n - 1)$	$\sum_i \sum_j (X_{ij} - \bar{X}_{i.})^2$	$\dfrac{1}{k(n-1)} \sum_i \sum_j (X_{ij} - \bar{X}_{i.})^2$
Totals	$kn - 1$	$\sum_i \sum_j (X_{ij} - \bar{X}_{..})^2$	

Table 12.3 Partitioning of the Degrees of Freedom for a Completely Randomized Experiment with k Treatments and n Experimental Units per Treatment

Source	df
Treatments	$k - 1$
Error	$k(n - 1)$
Total	$kn - 1$

the means. Ordinarily they would be computed by the following machine formulas.

Formula 1: Total sum of squares:

$$\sum_i \sum_j (X_{ij} - \bar{X}_{..})^2 = \sum_i \sum_j X_{ij}^2 - \frac{X_{..}^2}{kn} \qquad 12.7$$

Formula 2: Among-groups sum of squares:

$$n \sum_i (\bar{X}_{i.} - \bar{X}_{..})^2 = \frac{1}{n} \sum_i X_{i.}^2 - \frac{X_{..}^2}{kn} \qquad 12.8$$

SECTION 12.6 A NUMERICAL EXAMPLE

Formula 3: Within-groups sum of squares (total sum of squares minus among-groups sum of squares):

$$\sum_i \sum_j X_{ij}^2 - \frac{1}{n} \sum X_{i.}^2 \qquad 12.9$$

The square of the sum divided by the total number of observed values,

$$\frac{X_{..}^2}{kn}$$

is called the *correction term*.

12.6 A Numerical Example

To illustrate the analysis-of-variance calculations outlined in Section 12.5, we apply the procedures to the data of the example in Section 12.3.

Example 1

As the first step in constructing the analysis of variance we find the totals and means, as shown in Table 12.4. We then find the sums of squares as follows:

Step 1 The total sum of squares is

$$\sum_i \sum_j (X_{ij} - \bar{X}_{..})^2 = (110 - 113)^2 + (109 - 113)^2 + (105 - 113)^2$$
$$+ (111 - 113)^2 + \cdots + (125 - 113)^2$$
$$= 428$$

Step 2 The among-groups sum of squares is

$$n \sum_i (\bar{X}_{i.} - \bar{X}_{..})^2 = 3[(108 - 113)^2 + (112 - 113)^2$$
$$+ (110 - 113)^2 + (122 - 113)^2]$$
$$= 3(116)$$
$$= 348$$

Table 12.4 Numerical Example: Four Groups of Size 3

	Group				
	1	2	3	4	
	110	111	113	118	
	109	116	108	123	
	105	109	109	125	
Sum	324	336	330	366	1356
Mean	108	112	110	122	113

Step 3 The within-groups sum of squares is most easily obtained as the difference between the total sum of squares and the among-groups sum of squares:

$$\sum_i \sum_j (X_{ij} - \bar{X}_{i.})^2 = 428 - 348 = 80$$

We verify this figure by direct calculation.

$$\sum_j (X_{1j} - \bar{X}_{1.})^2 = (110 - 108)^2 + (109 - 108)^2 + (105 - 108)^2 = 14$$

$$\sum_j (X_{2j} - \bar{X}_{2.})^2 = (111 - 112)^2 + (116 - 112)^2 + (109 - 112)^2 = 26$$

$$\sum_j (X_{3j} - \bar{X}_{3.})^2 = (113 - 110)^2 + (108 - 110)^2 + (109 - 110)^2 = 14$$

$$\sum_j (X_{4j} - \bar{X}_{4.})^2 = (118 - 122)^2 + (123 - 122)^2 + (125 - 122)^2 = 26$$

Therefore, the pooled sum of squares within groups is

$$\sum_i \sum_j (X_{ij} - \bar{X}_{i.})^2 = 14 + 26 + 14 + 26 = 80$$

The analysis-of-variance table for these data follows:

SECTION 12.6 A NUMERICAL EXAMPLE

Source	df	SS	MS
Among groups	3	348	116
Within groups	8	80	10
Totals	11	428	

The reader should verify that the use of the machine formulas (12.7), (12.8), and (12.9) will result in the same values for the sum of squares. The use of the machine formulas will, in general, simplify the calculations.

Problems

1. a. Construct the analysis-of-variance table for the data given below. Use deviations from means to calculate sums of squares.
   b. Recalculate the sums of squares by the machine formulas.
   c. Write down the model and list the assumptions.

	Group		
1	2	3	4
15	9	17	13
17	12	20	12
22	15	23	17

2. a. Construct the analysis-of-variance table for the following data. Use deviations from means to calculate the sums of squares.
   b. Calculate the sums of squares by the machine formulas.
   c. Write the model and state the assumptions.

	Group	
1	2	3
50	67	50
48	72	44
53	71	43
48	74	45
51	66	43

3. In a completely randomized experiment, six treatments were each applied to five experimental units. Complete the following analysis-of-variance table for the data.

Source of Variation	df	Sum of Squares	Mean Square
Treatments		200	
Error		___	___
Totals		296	

4. To study the effects of different pressures on the yield of a dye, five lots were produced under each of three pressures, with the results given below. Construct the analysis-of-variance table.

Pressure (mm)	Yield				
200	32.4	32.6	32.1	32.4	32.3
500	37.8	38.2	37.9	38.0	37.8
800	30.3	30.5	30.0	30.1	29.7

12.7 Estimation of Effects

Given the data of Table 12.1 and the analysis of variance in Table 12.2, we may make the following statements concerning the populations from which our data were obtained:

(1) The overall mean $\bar{X}_{..}$ is the best unbiased estimator for μ.

(2) The group mean $\bar{X}_{i.}$ is the best unbiased estimator for the population mean $\mu_{i.}$.

(3) For any two groups, say group i and group m, the difference of the sample group means is an unbiased estimate of the difference of the two population means:

$$E(\bar{X}_{i.} - \bar{X}_{m.}) = \mu_i - \mu_m$$
$$= (\mu + \tau_i) - (\mu + \tau_m)$$
$$= \tau_i - \tau_m$$

SECTION 12.7 ESTIMATION OF EFFECTS

(4) The within-groups mean square is the best unbiased estimator for the common variance σ^2 (see Assumption 3 in Section 12.4).

If we want to find confidence intervals for the means or for differences among the means, we must first estimate the standard deviations for the corresponding estimators. The estimated variance for a group or treatment mean, *as always,* is our estimate for σ^2, divided by the number of observed values that were averaged to obtain the mean in question; therefore

$$s^2_{\bar{X}_{i.}} = \frac{s^2}{n} \qquad 12.10$$

where s^2 is the within-groups, or error, mean square. The estimated standard deviation of $\bar{X}_{i.}$ is

$$s_{\bar{X}_{i.}} = \sqrt{\frac{s^2}{n}}$$

The estimated variance of the difference between two means is the sum of their variances:

$$s^2_{\bar{X}_{i.} - \bar{X}_{m.}} = \frac{s^2}{n} + \frac{s^2}{n} = \frac{2s^2}{n} \qquad 12.11$$

The corresponding standard deviation is

$$s_{\bar{X}_{i.} - \bar{X}_{m.}} = \sqrt{\frac{2s^2}{n}}$$

Since we assume that the random errors, the e_{ij} in our model (see Equation (12.1)), are normally distributed, the group means will be normally distributed. This means that the quantity

$$t = \frac{\bar{X}_{i.} - \mu_i}{s_{\bar{X}_{i.}}} \qquad 12.12$$

has a *t*-distribution with $k(n-1)$ degrees of freedom. Notice that the degrees of freedom are those associated with our estimate of variance, the within-groups mean square. It follows that a $100(1-\alpha)\%$ confidence interval for the group mean (μ_i, or $\mu + \tau_i$) is

$$\begin{aligned} L &= \bar{X}_{i.} - t_{\alpha/2} s_{\bar{X}_{i.}} \\ R &= \bar{X}_{i.} + t_{\alpha/2} s_{\bar{X}_{i.}} \end{aligned} \qquad 12.13$$

Similarly, under our assumptions the difference between two group means is normally distributed; hence

$$\frac{\bar{X}_{i.} - \bar{X}_{m.} - \mu_i + \mu_m}{s_{\bar{X}_{i.} - \bar{X}_{m.}}} \qquad 12.14$$

has a t-distribution with $k(n - 1)$ degrees of freedom. A confidence interval for the difference $(\mu_i - \mu_m, \text{ or } \tau_i - \tau_m)$ is given by

$$L = \bar{X}_{i.} - \bar{X}_{m.} - t_{\alpha/2} s_{\bar{X}_{i.} - \bar{X}_{m.}}$$
$$R = \bar{X}_{i.} - \bar{X}_{m.} + t_{\alpha/2} s_{\bar{X}_{i.} - \bar{X}_{m.}}$$

12.15

In connection with confidence intervals in this context, a word of caution is necessary. If we calculated a 95% confidence interval for the difference between two group means on the basis of a randomly selected pair of groups or upon a pair of groups specified in advance of collection of the data, our confidence level will be equal to .95. If, however, we look at the data and then select the groups, our confidence level will not, in general, be equal to .95 and may, in fact, be much smaller. For k groups, there are a total of $k(k - 1)/2$ pairs of means. If we were to find a 95% confidence interval for every pair, the *average* confidence level would be .95, but for any individual interval its value would depend upon the relative positions of the members of the pair when the means are ranked.

In recent years much has been done to develop procedures for making all possible comparisons among the means through the use of confidence intervals, or by means of t-tests. A discussion of these procedures will be found in References 12.2 and 12.5.

Problems

5. For the data in Problem 1,
   a. Find estimates for the group means and their variances.
   b. Make a 95% confidence interval for the mean of group 1.
6. For the data of Problem 2,
   a. Estimate the variance of a mean.
   b. Estimate the variance of the difference of two means.
   c. Make a 95% confidence interval for the difference between the means of group 1 and group 2.
7. For the data in Problem 4,
   a. Estimate the mean yields and their standard deviations.
   b. Find a 95% confidence interval for the mean yield under each of the three pressures.

8. Four animals were randomly assigned to each of four rations. They were individually fed, and after a period of time a measure of damage to a certain organ was obtained.
   a. Construct the analysis-of-variance table.

b. Estimate the mean for each ration.
c. Find a 95% confidence interval for each mean.

	Ration		
1	2	3	4
7.2	4.5	9.7	7.1
6.8	6.0	8.4	6.1
6.0	4.6	8.8	7.2
6.3	5.3	9.9	6.4

9. Certain types of soft contact lenses gradually accumulate protein from the eye and must be cleaned periodically with a strong chemical solution. In an experiment to compare three brands of lenses, an optometrist randomly assigned each brand to six patients and recorded the time (in months) that the lenses were worn before they had to be chemically cleaned. The data are given below.
  a. Estimate the mean wearing time for each brand, and estimate the standard deviation of the mean.
  b. Estimate the standard deviation of the difference between two means.
  c. Make a 95% confidence interval for the difference between the means of brands C and B.

Brand	Months of Wear
A	4 2 3 2 3 3
B	3 3 4 5 4 3
C	6 7 6 8 7 8

12.8 The Hypothesis of Equal Means

Given the results of an experiment in which each of k treatments is applied to n experimental units, we construct an analysis-of-variance table that may be summarized as follows:

Source	df	MS	EMS
Treatments	$k - 1$	T	$\sigma^2 + n\kappa_\tau^2$
Error	$k(n - 1)$	E	σ^2
Total	$kn - 1$		

T and E are the mean squares for treatments and error, respectively.

A new feature is the column headed EMS, where EMS stands for *expected mean square*. The entries in this column of the table are the expected values of the corresponding mean squares, those functions of the parameters that are estimated unbiasedly by the experimental mean squares T and E.

The error mean square (within-group mean square), E, has an expected value equal to σ^2 and is, therefore, an unbiased estimator for the common variance. The treatment mean square, T, when we assume the fixed model is appropriate (see Section 12.4), estimates

$$\sigma^2 + n\kappa_\tau^2 \qquad 12.16$$

where κ_τ^2 is the mean squared deviation among the treatment effects; that is,

$$\kappa_\tau^2 = \frac{1}{k - 1} \sum_i \tau_i^2 \qquad 12.17$$

If we assume a random model, that is, if the k treatments are considered to be a random sample of size k from an infinite population of possible treatments, then the τ_i are random variables with variance σ_τ^2, and the expected mean square among groups becomes

$$\sigma^2 + n\sigma_\tau^2 \qquad 12.18$$

Having partitioned the variation in our experimental data in an analysis-of-variance table, we turn our attention to testing the hypothesis that there are no differences among the treatment effects—that the mean response to each treatment is the same for all treatments. This hypothesis is

$$H_0: \tau_i = 0, \; i = 1, 2, \ldots, k \qquad 12.19$$

since the τ_i are the deviations of the treatment means from the overall mean and are therefore all equal to zero if and only if the μ_i are all the same. The alternative hypothesis is that not all of the τ_i are equal to zero.

The expected mean square for treatments is $\sigma^2 + n\kappa_\tau^2$, where κ_τ^2, given by Equation (12.17), is a function of the sum of squares of the τ_i with the

SECTION 12.8 THE HYPOTHESIS OF EQUAL MEANS

F-distribution

property that if all the τ_i are equal to zero, then κ_τ^2 vanishes; hence, under the hypothesis of equal treatment means, both mean squares have the same expected value: σ^2. If the data are assumed to consist of independent random samples from normal populations, and if the hypothesis is true, it can be shown that the ratio of the mean squares, T/E, has an **F-distribution** with $k - 1$ and $k(n - 1)$ degrees of freedom.

The F-distributions constitute a two-parameter family of distributions whose parameters are the degrees of freedom ν_1 for the numerator mean square, and the degrees of freedom ν_2 for the denominator mean square. The upper percentage points for F are given in Table 8 of the Appendix and are defined by

$$P(F_{\nu_1, \nu_2} \geq F_{\alpha, \nu_1, \nu_2}) = \alpha \qquad 12.20$$

Only the upper points are given because it is a characteristic of the F-distributions that the lower percentage points may be obtained as reciprocals of the upper ones:

$$F_{1-\alpha, \nu_1, \nu_2} = \frac{1}{F_{\alpha, \nu_2, \nu_1}} \qquad 12.21$$

Under the null hypothesis H_0 that τ_i is zero for all i, the ratio of the *expected* mean squares for treatments and error is $\sigma^2/\sigma^2 = 1$, but if the hypothesis is not true, the ratio is

$$\frac{\sigma^2 + n\kappa_\tau^2}{\sigma^2}$$

Since κ_τ^2 is based on the sum of squares of the τ_i, it will always be greater than zero when not all of the τ_i are equal to zero. Therefore, we would reject the null hypothesis of equal treatment effects only for large values of the ratio; hence the critical region is

$$\frac{T}{E} \geq F_{\alpha, k-1, k(n-1)} \qquad 12.22$$

For purposes of illustration, let us test the hypothesis that the four groups in Table 12.4 are from populations having the same mean. In the analysis of variance for these data we found

Source	df	MS
Among groups	3	116
Within groups	8	10

The ratio of the mean squares is

$$F = \frac{T}{E} = \frac{116}{10} = 11.6$$

From the F-table, $F_{.01, 3, 8}$ is found to be 7.591. Since the calculated value 11.6 exceeds the tabular value, we reject the null hypothesis at the 1% level of significance and conclude that the population means are different.

The F-test tells us whether or not we may conclude that the means are different. It does not tell us where the differences are. The procedures for locating these differences are discussed in References 12.2, 12.4, and 12.5.

If we assume the random model, for which the among-groups mean square has expected value $\sigma^2 + n\sigma_\tau^2$, then the test of the hypothesis that σ_τ^2 is equal to zero is identical with the preceding test of equal group means. For the random model we are generally more interested in estimating σ_τ^2 than in testing this hypothesis.

Problems

10. For the data in Problem 2,
    a. Write the expected mean squares.
    b. Test the hypothesis of equal group means. Let $\alpha = .01$.
11. For the data in Problem 3,
    a. Give the expected mean squares.
    b. Test the hypothesis of equal treatment effects. Let $\alpha = .05$.

12. Test the hypothesis that the mean yields are equal for the three pressures in Problem 4. Let $\alpha = .01$.
13. Test the hypothesis that the means of the 4 groups in Problem 1 are equal. Let $\alpha = .05$.
14. Given an experiment in which each of five treatments was randomly assigned to six experimental units, and given that the pooled estimate for variance is 24,
    a. Complete the following table.

Source of Variation	df	Sum of Squares	Mean Square
Treatments			
Error	___	___	
Totals		1176	

SECTION 12.8 THE HYPOTHESIS OF EQUAL MEANS   423

b. Test the hypothesis of equal treatment effects. Let $\alpha = .05$.
c. Find the standard deviation of a treatment mean.

15. Three packaging materials were tested for moisture retention by storing the same food product in each of them for a fixed period of time and then determining the moisture loss. Each material was used to wrap ten food samples. Given the following results,
   a. Construct the analysis-of-variance table.
   b. Can we reject the hypothesis that the materials are equally effective? Let $\alpha = .05$.
   c. Find a 95% confidence interval for the mean loss that results from using material 2.

	Material		
	1	2	3
Number of packages	10	10	10
Mean loss	224	232	228
$\Sigma(X_{ij} - \bar{X}_{i.})^2$	300	400	380

16. For the data in Problem 8, test the hypothesis of equal damage resulting from the rations. Let $\alpha = .05$.

17. Five treatments were each assigned to ten experimental units. The total sum of squares was 8,550, and the variance of each treatment mean was 15. Reproduce the analysis-of-variance table and test the hypothesis of equal treatment means. Let $\alpha = .05$.

18. The following data are the blood alcohol percentages of 16 people in four weight groups 30 minutes after drinking 2 ounces of 100-proof liquor.
   a. Is there a significant difference among the means of the blood alcohol percentages of the four groups?
   b. Make a 95% confidence interval for the mean blood alcohol percentage of the 170–190 weight group.

	Weight Group		
130–150	150–170	170–190	190–210
.05	.05	.04	.04
.06	.04	.03	.03
.04	.06	.05	.02
.07	.04	.04	.03

12.9 The Effects of Unequal Group Sizes

In our discussion of the analysis of variance we have thus far assumed that we have the same number n of observed values in each group or sample. We will now look at some effects of unequal group sizes on the analysis-of-variance table on the calculations required to complete the table, and on the inferences we may make.

Let n_i be the number of elements in the ith group; that is, we have n_1 observations in the first group, n_2 in the second, and so on. The total number of observed values $N.$ is then the sum of the numbers in the k groups:

$$N. = \sum_{i=1}^{k} n_i \qquad 12.23$$

Table 12.5 Analysis-of-Variance Table for Unequal-Size Groups

Source	df	SS	MS	EMS
Among groups	$k - 1$	$\sum_i n_i (\bar{X}_{i.} - \bar{X}_{..})^2$	G	$\sigma^2 + \dfrac{1}{k-1} \sum_i n_i \tau_i^2$
Within groups	$\sum_i (n_i - 1)$	$\sum_i \sum_j (X_{ij} - \bar{X}_{i.})^2$	W	σ^2
Totals	$N. - 1$	$\sum_i \sum_j (X_{ij} - \bar{X}_{..})^2$		

The analysis-of-variance table for k groups of unequal size is shown in Table 12.5. Study of this table reveals the following.

(1) Among the k groups there are $k - 1$ degrees of freedom, regardless of the sample sizes.

(2) In the among-groups sum of squares, each squared deviation of a group mean from the overall mean is multiplied by the corresponding n_i *before* summation.

(3) There are $n_i - 1$ degrees of freedom among the n_i elements of the ith group, and the pooled degrees of freedom within groups is the sum $\sum_i (n_i - 1)$. This may be written as $N. - k$.

(4) The total degrees of freedom are, as usual, one less than the total number of observations.

SECTION 12.9 THE EFFECTS OF UNEQUAL GROUP SIZES

(5) In the expected mean square for among groups, the second term differs from $n\kappa_\tau^2$ in that the products $n_i\tau_i^2$ are summed, rather than just the τ_i^2.

In finding the sums of squares by the machine formulas, the only difference is in the among-groups sum of squares:

$$\sum_i \frac{X_{i.}^2}{n_i} - \frac{X_{..}^2}{N_.} \qquad 12.24$$

The difference is that in the first term the order of operation is opposite from that in the equal-number formula (12.8). Here we divide the square of each group total by the size of the group and then add, whereas with equal numbers we obtain the sum of squares of the group totals and then divide by n.

The total sum of squares is found in the usual way, and the within-groups sum of squares is the difference between the total sum of squares and the among-groups sum of squares.

Let G denote the among-groups mean square, and W the within-group mean square. Under the hypothesis of equal group means (the hypothesis H_0 that τ_i is zero for all i), both mean squares are estimates of σ^2 and their ratio G/W has an F-distribution with $k - 1$ and $\Sigma(n_i - 1)$ degrees of freedom; hence the test of this hypothesis is performed in the same way as with groups of equal size.

The variances of the group means will be different from one another, because each mean is the average of different numbers of observed values. For the mean of the ith group

$$s_{\bar{X}_{i.}}^2 = \frac{s^2}{n_i} \qquad 12.25$$

where s^2 is the within-group mean square W. For two group means the estimated variance of their difference is

$$s_{\bar{X}_{i.} - \bar{X}_{m.}}^2 = s^2 \left(\frac{1}{n_i} + \frac{1}{n_m} \right) \qquad 12.26$$

If in the unequal-number case we assume the random model, the among-groups expected mean square becomes

$$\sigma^2 + n_0 \sigma_\tau^2 \qquad 12.27$$

where

$$n_0 = \frac{1}{k-1}\left[N_. - \frac{\Sigma n_i^2}{N_.} \right] \qquad 12.28$$

is a sort of average number of observations per group. In the simple analysis of variance under consideration, this quantity has no effect on the F-test for equal group means and would require calculation only if we wanted to obtain an estimate for σ_τ^2.

We illustrate the calculations required for the analysis of variance with unequal numbers by the following simple example. We use the machine formulas for the sums of squares.

Example 1

	Group			
	1	2	3	
	10	6	14	
	8	9	13	
	5	8	10	
	12	13	17	
	14		16	
	11			
Sum	60	36	70	166
Mean	10	9	14	11.1
	$n_1 = 6$	$n_2 = 4$	$n_3 = 5$	$N_. = \sum_i n_i = 15$

The total sum of squares is

$$\sum_i \sum_j X_{ij}^2 - \frac{X_{..}^2}{N_.} = (10)^2 + (8)^2 + (5)^2 + \cdots + (16)^2 - \frac{(166)^2}{15}$$

$$= 2010 - 1837.07 = 172.93$$

The among-groups sum of squares is

$$\sum_i \frac{X_{i.}^2}{n_i} - \frac{X_{..}^2}{N_.} = \frac{(60)^2}{6} + \frac{(36)^2}{4} + \frac{(70)^2}{5} - \frac{(166)^2}{15}$$

$$= 600 + 324 + 980 - 1837.07$$

$$= 1904 - 1837.07 = 66.93$$

The within-groups sum of squares is the total sum of squares minus the group sum of squares:

SECTION 12.9 THE EFFECTS OF UNEQUAL GROUP SIZES

$$172.93 - 66.93 = 106$$

The analysis-of-variance table is as follows:

Source	df	SS	MS
Among groups	2	66.93	33.47
Within groups	12	106.00	8.83
Totals	14	172.93	

To test the hypothesis of equal group means, $H_0: \tau_i = 0$ for all i, the sample F-value is

$$F = \frac{33.47}{8.83} = 3.79$$

From the F-table, $F_{.05, 2, 12}$ is found to be 3.8853; therefore, we cannot reject the hypothesis of equal means at the 5% level of significance.
  The variances for the sample means are

$$s^2_{\bar{X}_{1.}} = \frac{8.83}{6} = 1.47, \quad s^2_{\bar{X}_{2.}} = \frac{8.83}{4} = 2.21, \quad s^2_{\bar{X}_{3.}} = \frac{8.83}{5} = 1.77$$

If we assume the random model, then

$$n_0 = \frac{1}{2}\left[15 - \frac{77}{15}\right] = 4.93$$

and the expected mean square for among groups becomes $\sigma^2 + 4.93\sigma^2_\tau$.

Problems

19. Given the following data,
    a. Construct the analysis-of-variance table. Include the expected mean squares.
    b. Estimate the population means and estimate the standard deviation of each group mean.
    c. Find a 99% confidence interval for the difference between the means of the first and third groups.
    d. Test the hypothesis of equal group means. Let $\alpha = .05$.

	Group		
1	2	3	4
9	17	22	13
6	10	17	14
8	12	21	22
13			16
			10

20. The following are the hourly rates of pay for samples of workers in three different types of firms.
   a. Construct the analysis-of-variance table.
   b. Write the expected mean squares.
   c. Test the hypothesis of equal hourly pay scales.
   d. Estimate the standard deviations of the group means.
   e. Find a 95% confidence interval for $\mu_1 - \mu_3$.
   f. Test the hypothesis $H_0: \mu_1 = \mu_2$. Let $\alpha = .01$.

	Type of Firm	
1	2	3
2.30	2.00	1.65
2.35	1.80	1.90
2.25	2.15	1.85
2.00	2.10	1.80
1.95	1.90	1.75
2.10	1.75	1.95
2.40		2.00
		1.90

21. In a completely randomized experiment three treatments were assigned to $n_1 = 7$, $n_2 = 10$, and $n_3 = 8$ experimental units, respectively.
   a. Complete the following table.

Source	df	SS	MS
Treatments			
Error			10
Total		320	

SECTION 12.9 THE EFFECTS OF UNEQUAL GROUP SIZES

b. Write the model and state the assumptions.
c. Test the hypothesis of equal treatment effects.

22. Given that five treatments were randomly assigned to $n_1 = 5$, $n_2 = 8$, $n_3 = 10$, $n_4 = 13$, and $n_5 = 11$ experimental units, respectively, and the estimated standard deviation of $\bar{X}_{2.}$ was 2,
   a. Complete the following table.
   b. Give the expected mean squares.
   c. Write the model and state the assumptions.
   d. Test the hypothesis of equal treatment effects. Let $\alpha = .05$.
   e. Given that $\bar{X}_{5.} = 50$, find a 99% confidence interval for $\mu + \tau_5$.

Source	df	SS	MS
Treatments			131.2
Error	—	—	
Totals			

23. In the course of a sample survey conducted in one county of a midwestern state, the educational backgrounds of randomly selected respondents were recorded, with the following results.

   Urban: Sample size, $n_1 = 350$; mean number of years of school completed, $\bar{X}_1 = 11.4$.
   Rural-farm: Sample size, $n_2 = 200$; mean number of years of school completed, $\bar{X}_2 = 9.3$.
   Rural-nonfarm: Sample size, $n_3 = 150$; mean number of years of school completed, $\bar{X}_3 = 10.1$.

   The estimated standard deviation was ten years.
   a. Construct the analysis-of-variance table.
   b. Are there real differences in the educational backgrounds of the three groups? Let $\alpha = .01$.
   c. What is the combined mean number of years completed? Find a 95% confidence interval for the mean educational level in this county.

24. To compare the effectiveness of three algebra texts before adopting one, a school board tested them in thirteen classes of approximately the same size and level of ability. The texts were assigned randomly; text A to four classes, B to four classes, and C to five classes. The data given on page 430 are the class averages for the final examination.

a. Is there a difference in the effectiveness of textbooks? Let $\alpha = .01$.
b. Which textbook should the school board adopt?

Textbook	Class Averages
A	75.3 78.7 76.1 74.0
B	79.1 77.3 80.1 82.3
C	73.5 75.7 74.2 72.3 75.8

12.10 Principles of Design

experimental error

For the completely randomized experiment, the pooled estimate of variance—the within-treatment mean square—is an estimate of **experimental error**. Experimental error is defined as the variation among experimental units treated alike. It is clear that in order to have an estimate of experimental error, we must have at least two experimental units assigned to each treatment; otherwise, we would not have units treated alike, and hence we would have no degrees of freedom and no estimate for error.

In designing experiments it is important that the experimental unit be carefully specified. It is defined as the element or group of elements over which the treatments are randomized. For a given experiment, all experimental units have the same chance of receiving any one of the treatments, and they are *independent* of each other.

Consider the following situation. In order to compare the effects of k different rations on the gains in weight of pigs, we conduct a feeding experiment in which n out of a total of kn animals are randomly assigned to each of k pens. After they are replaced in the pens, we use some random device to decide which pen gets which ration. At the completion of the experiment we have n weight gains for the n animals fed on each of the rations, but we do not have an estimate of experimental error, because there was only one experimental unit, the pen, for each ration. The pen is the unit over which the treatments were randomized; the animal in the pen is a unit of observation but *not* the experimental unit. We have no estimate for the experimental error variance and therefore no valid test of the hypothesis of equal ration means.

We can correct this situation in either of two ways: (1) We can put each animal in a separate pen, so that each is individually fed, and randomize the treatments over the animals, or (2) we can have two or more pens that

SECTION 12.10 PRINCIPLES OF DESIGN

receive each ration. The second suggestion is a compromise that may be necessary if our facilities will not permit each animal to be individually fed.

replicated   We say that an experiment is **replicated,** or repeated, when there is more than one experimental unit per treatment. If in a completely randomized experiment we have n experimental units per treatment, we can say that we have n replicates. The reasons for replication are as follows:

1. To obtain an estimate of experimental error.
2. Through increased replication, to obtain an increasingly more precise estimate of error.
3. Through increased replication, to increase the precision with which we estimate a treatment mean.

Increasing the amount of replication *will not* reduce the experimental error, since the variance σ^2 is a function of the experimental design and not of the number of experimental units per treatment.

The precision of an experiment is *inversely proportional* to the variance of the difference between two treatment means, that is, inversely proportional to

$$\sigma^2_{\bar{X}_1 - \bar{X}_2} = \frac{2\sigma^2}{n} \qquad 12.29$$

provided we have equal replication of each treatment.

For our purposes we may think of efficiency as being inversely proportional to the experimental error variance, and the relative efficiency of two different experimental designs as being the inverse ratio of their experimental error variances. If one design has an error variance equal to 10 and a second has an error variance equal to 5, the first requires twice as many experimental units per treatment as the second in order to have the same precision as measured by the variance of the difference between two treatment means given by Equation (12.29). We say that the second design is twice as efficient as the first.

To increase precision and efficiency, then, we must reduce the variance of the difference between two means, $2\sigma^2/n$. If we increase the amount of replication, we achieve a gain in precision but not in efficiency. To get a gain in efficiency, we must increase the precision without increasing the size of the experiment. This can be done by selecting an experimental design that will result in a smaller value of σ^2, the experimental error variance.

blocking   Among the ways in which experimental error can be reduced, one of the most effective makes use of the **blocking** principle. We use prior information about the experimental units to group them into *blocks,* or sets, of units such that under uniform treatment the variation in the responses among the units within a block will be less than the variation among blocks. Just as for the paired comparisons of Section 9.6, if the blocking is effective and if the blocks are large enough that each of our treatments occurs at least once in each block, then estimates for a comparison among the treatment effects are obtained from each block, and the estimate of experimental error is independent of the variation among the blocks.

Since the degrees of freedom for our estimate of error will be reduced (as compared with the corresponding completely randomized design) by the degrees of freedom for differences among the blocks, blocks should be used only when there is a sound basis for blocking. We are reluctant to give up degrees of freedom because that lessens the power and sensitivity of our tests of hypotheses. The more degrees of freedom, the greater the probability of detecting a given difference between two treatment effects. Therefore, we are willing to give up degrees of freedom only if there is a net gain in efficiency—only if the experimental error is reduced enough to make up for the loss. This reduction will not occur if the blocking is not effective. The interested reader should consult one or more of References 12.1, 12.3, 12.4, and 12.5.

12.11 The Randomized Complete Block Design

Given kn experimental units that can be grouped into n blocks of k units each in such a way that the responses of the units within a block may be expected to be more homogeneous than those of units taken from *different* blocks, we may use a *randomized complete block* design for an experiment involving k treatments. The treatments are randomized within each block, the randomization being carried out separately for each block. With k treatments in blocks of size k (k experimental units per block), each treatment occurs once, and only once, in each block. This constitutes the completeness. The paired comparisons of Chapter 9 are randomized complete block experiments with blocks of size 2.

The model associated with the randomized block design is

$$X_{ij} = \mu + \beta_i + \tau_j + e_{ij} \qquad 12.30$$

$$i = 1, 2, \ldots, n, \qquad j = 1, 2, \ldots, k$$

where μ is the overall mean, β_i is the effect of the ith block (the deviation of the block mean from the overall mean), τ_j is the effect of the jth treatment, and e_{ij} is the deviation of the observed value X_{ij} from its expected value. We assume that μ, β_i, and τ_j are unknown parameters and that the e_{ij} are

Table 12.6 Observed Values from a Randomized Block Experiment

Block	Treatment					Block Totals
	1	2	3	...	k	
1	X_{11}	X_{12}	X_{13}	...	X_{1k}	$X_{1.}$
2	X_{21}	X_{22}	X_{23}	...	X_{2k}	$X_{2.}$
3	X_{31}	X_{32}	X_{33}	...	X_{3k}	$X_{3.}$
⋮	⋮	⋮	⋮		⋮	⋮
n	X_{n1}	X_{n2}	X_{n3}	...	X_{nk}	$X_{n.}$
Treatment Totals	$X_{.1}$	$X_{.2}$	$X_{.3}$		$X_{.k}$	$X_{..}$

normally and independently distributed with mean zero and common variance σ^2.

The observed values may be displayed in a two-way table, as shown in Table 12.6, since they are subject to two criteria of classification—blocks and treatments. Notice that a treatment total is represented by $X_{.j}$. We have summed over the blocks, and therefore the subscript i, which designates the block, is replaced by a dot.

The total variation among the kn observed values is measured by the total sum of squares, the sum of squares of the deviations of the X_{ij} from the experiment mean $\bar{X}_{..}$:

$$\sum_i \sum_j (X_{ij} - \bar{X}_{..})^2 = \sum_i \sum_j X_{ij}^2 - \frac{X_{..}^2}{kn} \qquad 12.31$$

Algebraic manipulation leads to the identity

$$\sum_i \sum_j (X_{ij} - \bar{X}_{..})^2 = \sum_i \sum_j (X_{ij} - \bar{X}_{i.} - \bar{X}_{.j} + \bar{X}_{..})^2$$

$$+ k \sum_i (\bar{X}_{i.} - \bar{X}_{..})^2$$

$$+ n \sum_j (\bar{X}_{.j} - \bar{X}_{..})^2 \qquad 12.32$$

This shows us that the total variation is partitioned into *three* parts: (1) the sum of squares of deviations of the responses from the *estimated* expected

Table 12.7 Analysis of Variance for a Randomized Complete Block Experiment

Source	df	SS	MS	EMS
Blocks	$n - 1$	$k \sum_i (\bar{X}_{i.} - \bar{X}_{..})^2$	B	$\sigma^2 + k\kappa_\beta^2$
Treatments	$k - 1$	$n \sum_j (\bar{X}_{.j} - \bar{X}_{..})^2$	T	$\sigma^2 + n\kappa_\tau^2$
Error	$(k - 1)(n - 1)$	$\sum_i \sum_j (X_{ij} - \bar{X}_{i.} - \bar{X}_{.j} + \bar{X}_{..})^2$	E	σ^2
Totals	$kn - 1$	$\sum_i \sum_j (X_{ij} - \bar{X}_{..})^2$		

values, $\bar{X}_{i.} + \bar{X}_{.j} - \bar{X}_{..}$, (2) the sum of squares due to differences among the blocks, and (3) the sum of squares associated with differences among the treatments. This partition is conveniently displayed in the analysis of variance, Table 12.7. In the table *Blocks* stands for variation *among* blocks, and *Treatments* for variation *among* treatments.

To calculate the sums of squares, we use the following machine formulas.

Formula 1: *Block sum of squares:*

$$\frac{1}{k} \sum_i X_{i.}^2 - \frac{X_{..}^2}{kn} \qquad 12.33$$

Formula 2: *Treatment sum of squares:*

$$\frac{1}{n} \sum_j X_{.j}^2 - \frac{X_{..}^2}{kn} \qquad 12.34$$

The error sum of squares is found by subtraction.

Formula 3: *The error sum of squares is equal to the total sum of squares, minus the block sum of squares, minus the treatment sum of squares.* 12.35

The treatment effects are estimated by $\bar{X}_{.j} - \bar{X}_{..}$, and the difference between two treatment effects by the difference between the treatment means, $\bar{X}_{.j} - \bar{X}_{.m}$.

The error mean square E is our unbiased estimate for σ^2, the error variance; therefore, the variance of a treatment mean is estimated by

$$s_{\bar{X}_{.j}}^2 = \frac{s^2}{n} = \frac{E}{n} \qquad 12.36$$

and the variance of $\bar{X}_{.j} - \bar{X}_{.m}$ is estimated by

$$s_{\bar{X}_{.j}-\bar{X}_{.m}}^2 = \frac{2E}{n} \qquad 12.37$$

Under the hypothesis of equal treatment effects (the hypothesis H_0 that τ_j is zero for all j), the treatment mean square T is also an estimate for σ^2; hence, the ratio

$$F_{k-1,\,(k-1)(n-1)} = \frac{T}{E} \qquad 12.38$$

has an F-distribution with $k - 1$ and $(k - 1)(n - 1)$ degrees of freedom.

Because of the way the randomization is carried out, despite the form of the expected mean square for blocks, we do not test the hypothesis of equal block means. We are more concerned with the efficiency of the RCB (randomized complete block) design relative to the efficiency of the CR (completely randomized) design that might have been used for the same kn experimental units.

We can estimate the relative efficiency by

$$\text{Rel. eff. RCB to CR} = \frac{(n-1)B + n(k-1)E}{(kn-1)E} \times 100\% \qquad 12.39$$

where B and E are the block and error mean squares, respectively, from the analysis of variance of the RCB experiment. We see that when the block mean square is greater than the error mean square, the estimated experimental error for the RCB is less than that for the CR design. This will be the case only when we have an effective criterion for blocking.

Example 1

Suppose a randomized complete block experiment is conducted in order to make comparisons among the yields of three varieties of barley. Suppose six blocks are used, each block for a different type of soil. We are interested in the differences, if they exist, among the three varieties. Differences in soil may also affect the yield, and to offset this, each variety is planted in each soil type.

Suppose the yields in bushels per acre are the following:

Block	Variety			Block Totals
	1	2	3	
1	45	40	30	115
2	40	42	37	119
3	46	38	26	110
4	38	42	25	105
5	35	45	27	107
6	43	48	35	126
Variety Totals	247	255	180	682
Means	41.2	42.5	30	37.9

The total sum of squares is

$$(45)^2 + (40)^2 + (30)^2 + \cdots + (35)^2 - \frac{(682)^2}{18} = 843.78$$

the block sum of squares is

$$\frac{1}{3}[(115)^2 + (119)^2 + \cdots + (126)^2] - \frac{(682)^2}{18} = 105.11$$

the variety sum of squares is

$$\frac{1}{6}[(247)^2 + (255)^2 + (180)^2] - \frac{(682)^2}{18} = 565.44$$

and the error sum of squares, by subtraction, is

$$843.78 - (105.11 + 565.44) = 173.23$$

The analysis-of-variance table is as follows:

Source	df	SS	MS
Blocks	5	105.11	21.02
Varieties	2	565.44	282.72
Error	10	173.23	17.32
Total	17		

To test the hypothesis of equal mean yields for the three varieties, we use

$$F_{2,10} = \frac{282.72}{17.32} = 16.32$$

From the *F*-table, $F_{.01, 2, 10}$ is found to equal 7.56; therefore, we reject the hypothesis of equal variety effects. We would now examine the variety means and make comparisons among them. We can see in this case that varieties 1 and 2 are about the same, and that most of the difference among the means is accounted for by the low mean for the third variety. As previously mentioned, one would use techniques for making comparisons among the means that are beyond the scope of this text.

The variance of a variety mean is estimated by

$$s_{\bar{X}_{.j}}^2 = \frac{E}{n} = \frac{17.32}{6} = 2.89$$

Making use of Equation (12.39), we estimate the efficiency of this experiment relative to a CR experiment on the same experimental plots as

$$\frac{5(21.02) + 12(17.32)}{17(17.32)} \times 100\% = 106\%$$

There is a gain of only 6% in efficiency. This could have been anticipated by looking at the block totals. It can be seen that the differences among them are not large, and therefore the blocking was not very effective.

Problems

25. Given the following data for a randomized complete block experiment,
  a. Use the terms in the identity (12.32) to calculate the sums of squares.
  b. Use machine formulas to calculate the sums of squares.
  c. Construct the analysis-of-variance table. Include the expected mean squares.
  d. Write the model and state the assumptions.

	Treatment		
Block	1	2	3
1	6	7	8
2	14	9	16

438 — CHAPTER 12 ANALYSIS OF VARIANCE

26. The following are the yields in kilograms per plot that resulted when four equally spaced levels of nitrogen, N_0, N_4, N_8, and N_{12}, were applied to a variety of a grain in a randomized block experiment. Blocking was based on level of irrigation.
    a. Construct the analysis-of-variance table.
    b. Test the hypothesis that there are no differences among the effects of the nitrogen levels.
    c. Estimate the means and their standard deviations.
    d. Is there a significant difference between the means for the highest and lowest levels of nitrogen?

	Treatment			
Block	N_0	N_4	N_8	N_{12}
1	4.37	4.50	4.41	4.92
2	6.72	8.80	7.82	8.05
3	8.32	8.73	8.91	9.40
4	8.03	8.31	9.62	9.28

27. In a service test of traffic paints four different paints were tested at three different locations. The locations may be considered to be blocks. The following are measures of visibility taken after a period of exposure to weather and traffic.
    a. Construct the analysis-of-variance table.
    b. Test the hypothesis of equal serviceability for the four paints. Let $\alpha = .05$.

	Paint			
Location	1	2	3	4
1	7.4	9.6	11.3	11.8
2	4.2	9.2	9.2	9.9
3	10.0	9.8	9.9	10.4

28. The following data are field weights in pounds of corn for 26-hill plots. The treatments were different methods of application of a fertilizer: (1) check (no fertilizer), (2) 300 pounds per acre plowed under, and (3) 300 pounds per acre broadcast.
    a. Construct the analysis-of variance table.

SECTION 12.11 THE RANDOMIZED COMPLETE BLOCK DESIGN

b. Test the hypothesis of no differences among treatments. Let $\alpha = .05$.

c. Can we conclude that there is a difference between the two methods of fertilizer application? Let $\alpha = .05$.

Treatment	Block					
	1	2	3	4	5	6
1	45.1	46.6	51.2	49.3	52.4	44.2
2	56.7	57.3	54.6	55.0	60.1	58.2
3	53.3	55.0	54.7	58.2	55.2	57.2

29. In order to estimate the efficiency of a randomized complete block design relative to a completely randomized design, uniformity data were obtained—that is, all experimental units were treated in the same fashion and a measure of response under identical conditions obtained—with the following results:

109	142	124	90	135	126	107
102	112	123	148	111	120	130
84	89	111	103	123	152	150
84	69	79	115	110	108	110
96	45	100	140	159	75	100
152	125	133	114	132	160	120

a. Obtain an estimate for the variation among the experimental units ignoring rows and columns. This estimates the error variance for a completely randomized design.

b. Construct the analysis-of-variance table using rows as blocks—the sources will be rows—and experimental units within rows. The mean square for units within rows estimates the error variance for a randomized complete block design.

c. Estimate the gain or loss in efficiency that would result from using the randomized blocks rather than the completely randomized design. Assume there would be either six blocks or six experimental units per treatment.

30. The number of television sets produced by three manufacturing plants during three working shifts are given below.

a. Treating the shifts as the blocks, test for differences among means across plants. Let $\alpha = .05$.

b. Treating the plants as the blocks, test for differences among the means across shifts. Let $\alpha = .05$.

	Plant		
Shift	A	B	C
1	41	52	59
2	42	47	53
3	46	51	56

CHAPTER PROBLEMS

31. To investigate the effects of four different types of plates on the extraction efficiency of a pulse column, five runs were made with each type of plate. The height in theoretical units was determined for each run, with the results given below.
    a. Test the hypothesis of equal extraction efficiency across plates, taking the run number as a blocking variable. Let $\alpha = .01$.
    b. Analyze the data as a completely randomized design, ignoring any block effect due to run number. Let $\alpha = .01$.
    c. Is the run number a significant factor in this experiment?

	Plate			
Run	1	2	3	4
1	4.0	7.2	3.9	7.3
2	3.2	6.9	5.2	6.5
3	3.9	8.3	4.2	6.0
4	4.7	7.5	5.4	5.8
5	3.8	6.8	4.5	5.9

32. In Problem 31 the plate types differed in the number and size of perforations in the plates as follows:

   Type 1: Size of hole, h_1; number of holes, p_1
   Type 2: Size of hole, h_2; number of holes, p_1
   Type 3: Size of hole, h_1; number of holes, p_2
   Type 4: Size of hole, h_2; number of holes, p_2

To test the hypothesis that hole size, averaged over numbers of holes, has no effect, we test $H_0: \mu_2 - \mu_1 + \mu_4 - \mu_3 = 0$. To test the hypothesis that the number of holes has no effect, the appropriate hypothesis is $H_0: \mu_4 - \mu_2 + \mu_3 - \mu_1 = 0$.

Using the data of Problem 31, test these hypotheses. What conclusions do you reach? (HINT: $\bar{X}_{2.} - \bar{X}_{1.} + \bar{X}_{4.} - \bar{X}_{3.}$ estimates $\mu_2 - \mu_1 + \mu_4 - \mu_3$ unbiasedly and has the estimated variance $4s^2/n$; hence,

$$\frac{\bar{X}_{2.} - \bar{X}_{1.} + \bar{X}_{4.} - \bar{X}_{3.}}{\sqrt{4s^2/n}}$$

has a t-distribution if $\mu_2 - \mu_1 + \mu_4 - \mu_3 = 0$.) Analyze as a completely randomized design.

33. In Problem 8 the rations may be described as follows, as regards source and level of fat:

> Ration 1: Vegetable fat, high level
> Ration 2: Vegetable fat, low level
> Ration 3: Animal fat, high level
> Ration 4: Animal fat, low level

   a. Test the hypothesis that there is no difference between the animal and vegetable fats, $H_0: \mu_3 + \mu_4 - \mu_1 - \mu_2 = 0$. Let $\alpha = .05$.
   b. Test the hypothesis that the levels of fat do not differ in their effects, $H_0: \mu_1 + \mu_3 - \mu_2 - \mu_4 = 0$. Let $\alpha = .05$.
   c. Refer to the hint in Problem 32 and solve this problem in a similar way.

34. Refer to Problem 27.
   a. Given that paints numbered 1 and 2 were white, and that those numbered 3 and 4 were yellow, can we conclude that yellow paints have a higher visibility score? Use a t-test, or find a sum of squares for white versus yellow with one degree of freedom as

$$\frac{(X_{1.} + X_{2.})^2}{n_1 + n_2} + \frac{(X_{3.} + X_{4.})^6}{n_3 + n_4} - \frac{X_{..}^2}{\Sigma n_i}$$

   b. Find a 95% confidence interval for the difference between the visibilities of white and yellow paints.

35. Random samples of people currently employed show the number of days lost from work by geographic region. Assume that the estimated standard deviation on a per-person basis is 2.579.
   a. Construct the analysis-of-variance table.
   b. Are there significant differences among regions as regards the number of days lost per person? Let $\alpha = .05$.
   c. Find 95% confidence intervals for the regional means.

	Region			
	Northeast	North Central	South	West
Total days lost	930	1,026	1,257	643
Number in sample	181	200	210	110

36. A type of transceiver was used in three different general kinds of service—fixed base, shipboard, and airborne. A record was kept of the number of tube removals for each socket for each equipment. The following data are the average transformed values for ten transceivers in each type of service. A square-root transformation was applied to the original data in an effort to achieve common variance. Analyze and interpret these results.

Socket Number	Type of Service		
	Fixed Base	Shipboard	Airborne
1	3.33	5.61	9.30
2	3.33	4.85	9.57
3	1.87	2.12	5.15
4	2.12	2.55	7.52
5	1.58	.71	2.55
6	1.87	1.58	3.24
7	.71	1.87	2.92
8	1.22	2.55	2.74
9	2.12	.71	5.61
10	2.92	1.87	2.55
11	3.24	1.58	4.74
12	2.12	2.74	2.74
13	3.24	4.74	8.34
14	3.08	2.34	4.85
15	2.74	1.22	3.81

37. Six treatments were applied to each of eight experimental units. The standard deviation for the difference between two treatment means was equal to 4, and the calculated *F*-value for testing the hypothesis of equal treatment means was equal to 3. Reproduce the analysis-of-variance table.

Summary and Keywords

Analysis of variance is an arithmetic way of **partitioning the total variation** in data according to the sources of variation (p. 406). **Randomization** (p. 406) is introduced into the analysis to eliminate biases, and this is commonly done by a **completely randomized design** (p. 407). The **group** or **treatment effects** (p. 409) in the normal **model** (p. 408) are studied by means of the **analysis-of-variance table** (p. 410), which measures the variation **among groups** (p. 411) by the **treatment** (p. 411) sum of squares, and measures the variation **within groups** (p. 411) by the **error** (p. 411) sum of squares. The ratio of the treatment and error mean squares, important for statistical inference, has an **F-distribution** (p. 421). **Experimental error** (p. 430) is controlled by **replication** (p. 431) and by **blocking** (p. 431).

REFERENCES

12.1 Cochran, W. G., and Cox, Gertrude M. *Experimental Designs*. 2d ed. New York: John Wiley & Sons, 1957. Chapters 2–4.

12.2 Duncan, D. B. "Multiple range and multiple F tests." *Biometrics* 11:1955.

12.3 Kempthorne, Oscar. *The Design and Analysis of Experiments*. New York: John Wiley & Sons, 1952. Chapters 6 and 7.

12.4 Ostle, Bernard. *Statistics in Research*. 2d ed. Ames, Iowa: Iowa State University Press, 1963. Chapters 10–12.

12.5 Snedecor, George W., and Cochran, William G. *Statistical Methods*. 6th ed. Ames, Iowa: Iowa State University Press, 1967. Chapters 10 and 11.

13

Nonparametric Methods

13.1 Introduction

nonparametric methods

Many of the methods we have considered, such as those involving the t-test, apply only to normal populations. The need for techniques that apply more broadly has lead to the development of **nonparametric methods.** These do not require that the underlying populations be normal—or indeed that they have any single mathematical form—and some even apply to nonnumerical data. In place of parameters such as means and variances and their estimators, these methods use ranks and other measures of relative magnitude; hence the term *nonparametric*.

13.2 The Wilcoxon Test

median

The **median** of a continuous population is a number ξ such that an observation X from the population has probability .5 of being less than ξ and probability .5 of being greater than ξ. In symbols,

$$P(X < \xi) = P(X > \xi) = .5 \qquad 13.1$$

The sample quantity corresponding to ξ was considered in Section 3.6. The population is *symmetric* about its median ξ if, for each positive x, the observation X has the same probability of being less than $\xi - x$ as of being greater than $\xi + x$. In symbols,

$$P(X < \xi - x) = P(X > \xi + x) \qquad 13.2$$

Geometrically, this means that the two shaded areas in Figure 13.1 are equal for each x, so that the frequency curve looks the same if viewed through a mirror. A normal curve with mean ξ has median ξ and is symmetric about ξ, but a nonnormal curve like the one in the figure can also have this property.

Consider the null hypothesis that the population median is 0 and the population is symmetric about 0, together with the alternative hypothesis that the population median ξ is positive and the population is symmetric about ξ:

$$H_0: \xi = 0, \text{ symmetric}$$
$$H_a: \xi > 0, \text{ symmetric} \qquad 13.3$$

If the population is normal, the t-test applies to this problem. If the population may be nonnormal, another test is needed.

SECTION 13.2 THE WILCOXON TEST   **447**

Figure 13.1

A frequency curve symmetric about its median ξ

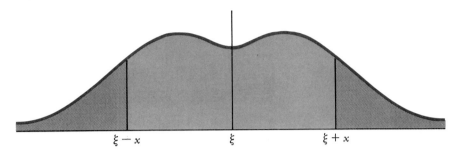

Wilcoxon statistic

Let $X_1, X_2, \ldots, X_n$ be an independent sample from the population. The **Wilcoxon statistic** W is computed in three steps.

(1) Arrange the observations $X_1, X_2, \ldots, X_n$ in order of increasing *absolute value*.

(2) Write down the numbers $1, 2, \ldots, n$ in order, each with the algebraic sign of the observation in the corresponding position in the arrangement of step 1.

(3) Compute the sum W of these numbers with their appropriate signs.

For example, suppose $n = 4$ and the observations are $+5.4$, -7.3, -3.8, and $+9.5$. Arranged in order of increasing absolute value, the observations are

$$-3.8 \quad +5.4 \quad -7.3 \quad +9.5 \qquad\qquad 13.4$$

Here $+5.4$ precedes -7.3 because $|+5.4| < |-7.3|$ (even though $+5.4 > -7.3$). The numbers 1, 2, 3, and 4 with the corresponding signs attached are

$$-1 \quad +2 \quad -3 \quad +4 \qquad\qquad 13.5$$

ranks

The sum W is $-1 + 2 - 3 + 4$, or $+2$. The numbers 1, 2, 3, and 4 are the **ranks** of the quantities 3.8, 5.4, 7.3, and 9.5, respectively.

In the testing problem (13.3) we reject H_0 in favor of H_a if W is excessively large. If the population were normal, so that the t-test applied, we would compute the sample mean $\bar{X}$ and reject H_0 in case of an excessively large value of $\bar{X}$ (large relative to $s/\sqrt{n}$—compare the t-statistic (8.2) for a μ_0 of 0). The statistic W/n is analogous to the mean $\bar{X}$. For the four preceding observations, $\bar{X} = (+5.4 - 7.3 - 3.8 + 9.5)/4$, or $+.95$, and $W/4 = .5$. These are the centers of gravity for the two dispositions of weights in Figure 13.2. To pass from the array (13.4) to the array (13.5) is

Figure 13.2

$\bar{X}$ and W/n compared

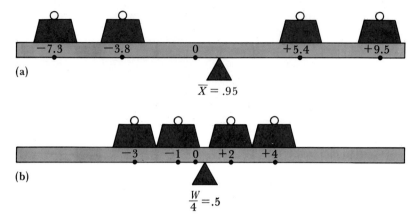

arithmetically to pass from $\bar{X}$ to $W/4$ and is geometrically to pass from (a) to (b) in the figure. In (a), the center of gravity $\bar{X}$ exceeds 0 because the weights on the positive side overbalance those on the negative side. In (b) the center of gravity $W/4$ still exceeds 0 because $+9.5$ (farthest from 0 in (a)) has moved to $+4$ (farthest from 0 in (b)), -7.3 (next farthest from 0 in (a)) has moved to -3 (next farthest from 0 in (b)), and so on. The weights on the positive side still overbalance the others. Thus W/n is a kind of *nonparametric mean*, and it accords with intuition in the testing problem (13.3) to reject H_0 in favor of H_a if W/n is large. Passing from $\bar{X}$ to W/n enables us to dispense with the assumption of normality.

For fixed n the distribution of W is the same for every population satisfying the null hypothesis H_0. There are published tables of this distribution for small values of n (see Reference 13.2). In the examples here we use the normal approximation, which is sufficiently accurate for an n of 10 or more. The mean and variance of W under the null hypothesis are

$$\mu_W = 0 \qquad 13.6$$

and

$$\sigma_W^2 = \frac{n(n+1)(2n+1)}{6} \qquad 13.7$$

Thus the standardized variable is W/σ_W. To test at the level α, we reject H_0 if W/σ_W exceeds the corresponding percentage point $t_{\alpha,\infty}$ of the normal curve. This is the Wilcoxon test. The problem and the rejection region are

$$H_0: \xi = 0, \text{ symmetric}$$
$$H_a: \xi > 0, \text{ symmetric}$$
$$R: W/\sigma_W > t_{\alpha,\infty}$$

SECTION 13.2 THE WILCOXON TEST

This testing problem can arise, for example, from paired comparisons, as in Section 9.6. We have n pairs of experimental units and two competing treatments (or a treatment and a nontreatment). Treatment 1 is given to a randomly selected element of pair i, which results in an observation Y_{i1}, and Treatment 2 is given to the other element of pair i, which results in an observation Y_{i2}. Under the null hypothesis that the two treatments have the same effect, the difference $Y_{i1} - Y_{i2}$, which we denote X_i, has median 0 and is symmetric about 0. Without the assumption of normality required for the t-test, we can use the Wilcoxon test.

Example 1

In Example 1 of Section 9.6, p. 314, we used the t-test on the null hypothesis that a particular heat treatment has no effect on the number of bacteria in skim milk against the alternative that the treatment tends to reduce this number. Here X_i is the count (log DMC) for sample i before treatment minus the count after treatment. The twelve differences are

.03 .14 1.17 .15 −.02 −.04 .44 .22 −.01 .13 .19 .70

Arranged in order of increasing absolute value, they are

−.01 −.02 .03 −.04 .13 .14 .15 .19 .22 .44 .70 1.17

The numbers 1, 2, . . . , 12 with algebraic signs in the same pattern are

−1 −2 +3 −4 +5 +6 +7 +8 +9 +10 +11 +12

The sum W of these comes to $+64$. By the formula (13.7),

$$\sigma_W = \sqrt{\frac{12 \times 13 \times 25}{6}} = \sqrt{650} = 25.5$$

so the standardized W-value, W/σ_W, is $64/25.5$, or 2.51. The upper 5-percent point $t_{.05,\infty}$ on the normal curve being 1.645, we reject the null hypothesis of no effect.

The same method works for the opposite alternative

$$H_0: \xi = 0, \text{ symmetric}$$
$$H_a: \xi < 0, \text{ symmetric}$$
$$R: W/\sigma_W < -t_{\alpha,\infty}$$

and for the two-sided alternative

$$H_0: \xi = 0, \text{ symmetric}$$
$$H_a: \xi \neq 0, \text{ symmetric}$$
$$R: |W/\sigma_W| > t_{\alpha/2,\infty}$$

The median in the null case need not be 0. To test, say, $\xi = 5$ against $\xi > 5$, we subtract 5 from each X_i and proceed as before. Ties (equal observations) can be dealt with by averaging adjacent ranks. If in Example 1 the ordered observations had been $-.01, -.02, .02, -.04$, etc., we would have used ranks 1, 2.5, 2.5, 4, etc. (instead of 1, 2, 3, 4, etc.) and signed ranks $-1, -2.5, +2.5, -4$, etc.

Problems

1. Assume that the following measurements come from a population symmetric about its median ξ:

-12.6	-20.5	8.1	-17.3	-21.2	-29.4	5.5	-25.3
15.9	$.3$	-11.7	-9.3	-28.8		2.7	19.2	-27.1

Test $H_0: \xi = 0$ against $H_a: \xi < 0$ at the 2.5% level.

2. Do part a of Problem 31, Chapter 9, by the Wilcoxon test. Let $\alpha = .05$. Use the normal approximation for the distribution of W/σ_W.

3. To test a new sleeping pill, 12 people were observed in a sleep laboratory for two nights. On one of the nights the subjects were given the sleeping pill, and on the other night they were not. The amount of time each subject was awake each night is given below. Does it appear from the differences of amounts of time awake that the sleeping pill was effective in reducing the amount of time the subjects remained awake at night? Let $\alpha = .05$.

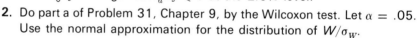
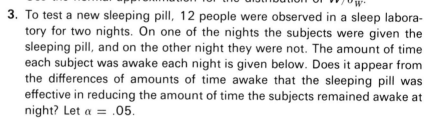

	Time Awake (minutes)					
	1	2	3	4	5	6
Subject with sleeping pill	12.5	7.1	12.3	19.2	16.9	22.1
Subject without sleeping pill	17.4	9.0	14.4	23.5	15.9	20.4
	7	8	9	10	11	12
Subject with sleeping pill	15.8	11.3	9.7	8.6	9.7	13.2
Subject without sleeping pill	18.9	17.8	14.4	10.1	15.3	13.1

4. Use the Wilcoxon test to do part a of Problem 32, Chapter 9. Let $\alpha = .05$.

13.3 The Sign Test

sign test

Unless the population is symmetric about its median, the Wilcoxon test cannot be used to test the hypothesis that the median is 0. In the absence of symmetry, we can use the **sign test** instead. The sign test, being less sensitive than the Wilcoxon test, should be used only if symmetry cannot be assumed.

On the basis of a random sample $X_1, X_2, \ldots, X_n$, we are to test the null hypothesis that the population median ξ is 0 against the alternative that it exceeds 0:

$$H_0: \xi = 0$$
$$H_a: \xi > 0$$
13.8

The sign test is simple. We count the number Y of positive values among $X_1, X_2, \ldots, X_n$. Because of the definition (13.1) of the median, the distribution of Y under H_0 is binomial with $p = .5$, while under H_a it is binomial with $p > .5$. Thus we can use the methods of Section 8.3.

Example 1

Twenty patients on a certain diet made the following weight gains (in pounds):

| +7 | −6 | +3 | +1 | +6 | +4 | +9 | −5 | +9 | −7 |
| −3 | +7 | −9 | +8 | +6 | −4 | +4 | +9 | −6 | +1 |

We are to test at the 5% level the null hypothesis that the median weight gain is 0 against the alternative that it is positive. There are 13 positive observations among the 20: $Y = 13$. The mean and standard deviation of Y are

$$n \times .5 = 10$$

and

$$\sqrt{n \times .5 \times .5} = \sqrt{5} = 2.24$$

so the standardized variable is $(Y - 10)/2.24$, which has the value 1.34. The upper 5 percentage point $t_{.05, \infty}$ is 1.645, and so we do not reject the null hypothesis.

By an adaptation of the methods of Section 8.3, it is possible to test against a two-sided alternative ($\xi \neq 0$) as well. To test for a median of ξ_0 (not necessarily 0), we subtract ξ_0 from each X_i and then carry through the test as before.

Problems

5. Use the sign test to test at the 5% level whether the following measurements come from a population with median 10.

13.6	8.0	15.8	12.2	3.5	17.3	9.5	11.4	2.8	10.7
7.1	12.7	13.9	11.4	7.5	16.3	13.2	17.0	12.1	6.3

6. Do Problem 1 without the assumption of symmetry.

7. Two laboratories were asked to determine the bacterial count of 14 samples of water. Does it appear that the laboratories differ in their determination of bacterial counts? Let $\alpha = .05$. Use the sign test.

Water sample	1	2	3	4	5	6	7
Laboratory A	151	75	129	180	232	166	53
Laboratory B	148	77	118	171	239	155	50

Water sample	8	9	10	11	12	13	14
Laboratory A	41	99	162	185	112	117	60
Laboratory B	33	89	161	170	108	130	54

8. The nationwide median hourly wage of workers in a certain trade is $7.80 per hour. A random sample of 65 workers in this trade in Boston showed 40 who earned above the nationwide median. Does this indicate a higher median wage for this trade in Boston? Let $\alpha = .05$.

13.4 The Rank Sum Test

The rank sum test stands to the Wilcoxon test as the two-sample t-test of Section 9.4 stands to the one-sample t-test of Section 8.2. The rank sum

SECTION 13.4 THE RANK SUM TEST

Figure 13.3

The alternative hypothesis H_a

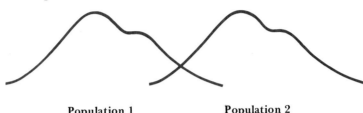

Population 1   Population 2

test can be used to check whether two populations have the same median, and it does not require the assumption of normality.

Let $X_{11}, X_{12}, \ldots, X_{1n_1}$ be a sample of size n_1 from population 1, and let $X_{21}, X_{22}, \ldots, X_{2n_2}$ be a sample of size n_2 from population 2. Let H_0 be the null hypothesis that the two populations are the same, and let H_a be the hypothesis that the frequency curve for population 2 has the same shape as the curve for population 1 but is shifted to the right as shown in Figure 13.3.

rank sum

The statistic for this problem, the **rank sum,** is computed in four steps.

(1) Combine the two samples into one, underlining the observations in the second sample to keep track of them.

(2) Arrange the observations in the combined sample in order of increasing size, carrying along the lines under the observations in the second sample.

(3) Write down the numbers $1, 2, \ldots, n_1 + n_2$ in order, underlining those for which the corresponding position in the arrangement of step 2 is occupied by an element of the second sample (that is, by an underlined number).

(4) From the average of the underlined numbers in step 3 subtract the average of the others. Call the difference V.

For example, suppose $n_1 = 3$, the observations being 3.1, -2.4, and 0.6; and suppose $n_2 = 2$, the observations being 7.7 and 2.8. The combined sample is

3.1 -2.4 0.6   <u>7.7</u>   <u>2.8</u>

Arranged in increasing order, this is

-2.4 0.6   <u>2.8</u>   3.1   <u>7.7</u>   13.9

The numbers 1, 2, 3, 4, and 5 ($n_1 + n_2 = 5$) with underlines in the same pattern are

$$\underline{1} \quad 2 \quad \underline{3} \quad 4 \quad \underline{5} \qquad\qquad 13.10$$

The underlined numbers average to $(3 + 5)/2$, or 4; the others, to $(1 + 2 + 4)/3$, or 2.33. The difference V of these averages is $4 - 2.33$, or 1.67.

If we were to compute the two-sample t-statistic (see (9.12) on p. 305), we would first form the average $\bar{X}_2$ of the underlined numbers in (13.9), then the average $\bar{X}_1$ of the others, and then the difference $\bar{X}_2 - \bar{X}_1$ (which we would then normalize). To compute V, we apply the same procedure to (13.10). Although the numbers in (13.10) are different from those in (13.9), they come in the same order. Therefore, in the same way as does a large value of $\bar{X}_2 - \bar{X}_1$, a large value of V indicates that observations from population 2 tend to exceed those of population 1—indicates, that is, that the data in the pair of samples favor H_a over H_0. The statistic V can be thought of as a *nonparametric difference of means*.

For each n_1 and n_2, the distribution of V under H_0 has the desirable property that it does not depend on the shape of the frequency curve common to the two populations. Tables are available for small values of n_1 and n_2 (see Reference 13.2). In the examples here, the normal approximation to this distribution can be used. The mean and variance of V are

$$\mu_V = 0 \qquad\qquad 13.11$$

and

$$\sigma_V^2 = \frac{(n_1 + n_2)^2(n_1 + n_2 + 1)}{12 n_1 n_2} \qquad\qquad 13.12$$

We reject H_0 in favor of H_a if V/σ_V exceeds $t_{\alpha, \infty}$.

Example 1

Healthy and infected plants of a common strain achieved the following heights (in inches):

Infected	40.1	37.5	46.3	44.2	35.1	48.2	41.7	37.3	50.3	43.9
Healthy	49.4	41.3	45.8	39.7	44.4	43.1	48.8	47.5		

We are to test at the 5% level the null hypothesis of no difference in heights against the alternative that healthy plants tend to be taller, and we are to do this without assuming normality. Here $n_1 = 10$ and $n_2 = 8$. The two samples merged into one and ordered are

35.1	37.3	37.5	<u>39.7</u>	40.1	<u>41.3</u>	41.7	<u>43.1</u>	43.9
44.2	<u>44.4</u>	<u>45.8</u>	46.3	<u>47.5</u>	48.2	<u>48.8</u>	<u>49.4</u>	50.3

SECTION 13.4 THE RANK SUM TEST

The numbers 1, 2, ..., 18 ($n_1 + n_2 = 18$) underlined in the same pattern are

---

1 2 3 <u>4</u> 5 <u>6</u> 7 <u>8</u> 9 10 <u>11</u> <u>12</u> 13 <u>14</u> 15 <u>16</u> <u>17</u> 18

The ranks for the second sample (the underlined numbers) average to 88/8, or 11.0, and the ranks for the first sample (the remaining numbers) average to 83/10, or 8.3. The difference V is $11.0 - 8.3$, or 2.7. The standard deviation according to the formula (13.12) is

$$\sigma_V = \sqrt{\frac{18^2 \times 19}{12 \times 10 \times 8}} = 2.53$$

Thus the standardized statistic V/σ_V has the value 2.7/2.53, or 1.07. Since this is less than 1.645, the upper 5% point for the normal curve, the data favor the null hypothesis.

As in Section 13.2, ties are dealt with by averaging adjacent ranks. If in Example 1 the observation 37.5 were 39.7 instead, creating a tie for third from the bottom, the ranks 3 and 4 would each be replaced by 3.5.

Paired comparisons, as in Example 1 of Section 13.2, are preferable to the group method of this section if variation among pairs exceeds variation between units within the pairs, but not otherwise. The situation is just as in the parameteric case; see Section 9.7.

Problems

9. Use the rank sum test to check whether the following two samples come from the same population. Let $\alpha = .05$.

Sample 1	120	136	107	109	129	117	115	110	124
Sample 2	131	144	146	111	103	122	121	139	130
	133	132	135	128					

10. Do Problem 22, Chapter 9, by the rank sum test. Let $\alpha = .05$.
11. Do Problem 24, Chapter 9, by the rank sum test. Let $\alpha = .05$.

12. Two groups of children, thought to be different in their abilities to

manipulate geometric images mentally, were timed in assembling a puzzle. Test the null hypothesis that the distributions of times for the two groups are the same against the alternative that group A is faster than group B. Let $\alpha = .05$.

	Time (seconds)							
Group A	90	195	77	150	124	134	100	
Group B	177	135	166	211	209	206	181	199

13. The nicotine content of 16 cigarettes selected randomly from two brands is given below. Does it appear that the brands differ significantly in nicotine content? Let $\alpha = .05$.

| Brand I | .29 | .57 | .36 | .46 | .16 | .82 | .66 | .34 |
| Brand II | .42 | .65 | .51 | .40 | .63 | .47 | 1.24 | .43 |

13.5 Association

Suppose four students, A, B, C, and D, are ranked in two subjects, mathematics and history, with the following results.

	Student			
	A	B	C	D
Mathematics	4	2	3	1
History	3	1	4	2

That is to say, in mathematics student D is worst (rank 1), B is next worst (rank 2), C is next (rank 3), and A is best (rank 4); while in history the ranking (worst to best) is B, D, A, C. To check whether there is a connection between performance in the two subjects, we want a measure of the extent to which these two rankings agree.

We can compute such a measure as follows. For each of the six possible pairs of students we check whether the ranks in the two subjects go in the same direction or in opposite directions and score a $+1$ or a -1 accordingly. Since student A did better than student B both in mathematics and in

SECTION 13.5 ASSOCIATION

history (the ranks go in the same direction), the score for the pair AB is $+1$. Since A did better than C in mathematics but worse in history (the ranks go in opposite directions), the score for the pair AC is -1. The six pairs and the corresponding scores are

Pair	AB	AC	AD	BC	BD	CD
Score	+1	−1	+1	+1	−1	+1

measure of association

The sum of the six scores is $+2$. This sum provides a **measure of association** for the two rankings. That it is positive in this case indicates that high standing in mathematics tends to go with high standing in history. (The sum would be the same if the students were ranked best to worst instead of worst to best.)

For the rankings

	Student			
	A	B	C	D
Mathematics	1	2	3	4
History	1	2	3	4

the score is $+6$, there being a $+1$ for each of the six pairs of students. This, the highest possible value of the measure, indicates perfect agreement between the two rankings. For the rankings

	Student			
	A	B	C	D
Mathematics	1	2	3	4
History	4	3	2	1

the score is -6, there being a -1 for each pair. This, the lowest possible value of the measure, indicates perfect disagreement. The measure always lies between $+6$ and -6. If we divide it by 6, the new measure lies between $+1$ (the value for perfect agreement) and -1 (the value for perfect disagreement). For our original set of rankings it is $+2/6$, or $+.33$. This is Kendall's **rank correlation coefficient**, denoted by τ.

rank correlation coefficient

In the general case we have n subjects (four students in the example) ranked in two ways (by standing in two subjects in the example). Thus we have two permutations of the numbers 1, 2, ..., n. There are $\binom{n}{2}$ pairs of subjects, and for each pair we score $+1$ or -1 according as the two rankings go in the same or in opposite directions. We then compute τ, the sum of these scores divided by $\binom{n}{2}$. A positive τ indicates *positive association,* a tendency for the rankings to agree, with a τ of $+1$ for perfect agreement. A negative τ indicates *negative association,* a tendency for the rankings to disagree, with a τ of -1 for complete disagreement. A τ of 0 indicates that the two rankings are independent of one another.

Example 1

Ten students are tested before a course of study and again afterward. Their ranks in the tests are as follows:

	\multicolumn{10}{c	}{Student}
	A	B	C	D	E	F	G	H	I	J
Test 1	8	9	4	1	5	10	2	3	7	6
Test 2	9	1	8	3	4	2	10	5	6	7

The computation of τ is facilitated by rearranging the students so the ranks for the first test increase:

	\multicolumn{10}{c	}{Student}
	D	G	H	C	E	J	I	A	B	F
Test 1	1	2	3	4	5	6	7	8	9	10
Test 2	3	10	5	8	4	7	6	9	1	2

(The identifications D, G, H, etc., are irrelevant to the computation.) The number of pairs is $\binom{10}{2}$, or 45. For each pair of numbers in the bottom row, we score $+1$ or -1 according as the right element of the pair does or does not exceed the left element, and we sum the scores. To put it another way, for each entry in the bottom row we compute the number of entries to the right of it that are larger minus the number of entries to the right of it that are smaller, and we add up these differences. Here the sum is -9, so that τ is $-9/45$, or $-.2$.

As usual, the question arises whether a τ far removed from 0 is really significant—whether it may not be an accident of sampling. If the two rankings are independent in the sense that each is merely a random permutation of the other, then τ has mean 0 and variance

$$\sigma_\tau^2 = \frac{2(2n + 5)}{9n(n - 1)}$$

If n is 10 or more, τ has approximately a normal distribution.
In Example 1, $n = 10$, so that

$$\sigma_\tau = \sqrt{\frac{2 \times 25}{9 \times 10 \times 9}} = .25$$

The standardized statistic τ/σ_τ is $-.2/.25$, or $-.8$. This is not significant even at the 10% level ($-t_{.1,\,\infty}$ is -1.282).

Problems

14. Ten children given two reading tests ranked as follows:

	\multicolumn{10}{c}{Child}
	A	B	C	D	E	F	G	H	I	J
Test 1	5	6	3	9	4	8	1	7	10	2
Test 2	3	4	1	8	5	10	6	7	9	2

Compute the rank correlation coefficient. Is it significantly greater than 0 at the 2.5% level?

15. Ten students received the following grades in each of two tests they took.
    a. Find r and test the hypothesis $H_0: \rho = 0$ (see the end of Section 11.8). Let $\alpha = .05$.
    b. Find τ and test the hypothesis of no association between the two groups. Let $\alpha = .05$.

	\multicolumn{10}{c}{Student}
	1	2	3	4	5	6	7	8	9	10
Test 1	86	79	67	92	65	75	61	77	91	68
Test 2	80	81	60	87	71	72	56	75	93	67

16. To see if a relationship exists between the size of boulders in a stream and the distance from a source, samples of boulders were measured every half mile downstream beginning at one mile.
   a. Find r and test the hypothesis $H_0: \rho = 0$. Let $\alpha = .05$.
   b. Rank the data and compute τ. Is it significant? Let $\alpha = .05$.

Distance Downstream (miles)	Average Size (inches)
1.0	41.3
1.5	34.0
2.0	32.5
2.5	35.2
3.0	28.8
3.5	25.6
4.0	30.1
4.5	25.7
5.0	22.6
5.5	20.3
6.0	18.8
6.5	17.1

17. Two judges ranked the contestants in a diving competition. Test the null hypothesis that there is no association between the rankings of the two judges against the alternative hypothesis that there is a positive association. Let $\alpha = .05$.

Contestant	1	2	3	4	5	6	7	8	9	10	11
Ranking by Judge 1	2	1	8	9	6	7	3	11	10	4	5
Ranking by Judge 2	3	2	7	10	4	8	6	9	11	1	5

18. The following data are the high and low temperatures of 12 randomly selected cities for an April day. Do the data indicate a positive association between the high and low temperatures, as one might expect? Let $\alpha = .01$.

City	1	2	3	4	5	6	7	8	9	10	11	12
High	69	61	62	51	54	70	64	71	75	80	43	42
Low	42	33	41	35	31	38	46	51	54	53	32	36

CHAPTER PROBLEMS

19. If, in applying the rank sum test, it happens that two or more of the observations in the combined sample have the same numerical value, then the values of V and σ_V^2 must be adjusted to account for ties in the ranks. The value of V is adjusted by the way in which ranks are assigned to the observations. If two or more of the observations have the same numerical value, then the rank assigned to each is just the average of their ranks in the combined sample. The variance is adjusted as follows. Let d_1 denote the number of observations having the smallest value, d_2 the number having the next smallest value, and so on. Define the correction factor by

$$\text{C.F.} = \frac{(n_1 + n_2) \sum (d_i^3 - d_i)}{12 n_1 n_2 (n_1 + n_2 - 1)}$$

The adjusted variance is given by

$$\sigma_V^2(\text{adj.}) = \sigma_V^2 - \text{C.F.}$$

The computations given below illustrate the procedure.

	Data			
Group 1	12	13	13	15
Group 2	13	15	16	

	Adjusted Ranks						
Observation	12	13	13	13	15	15	16
Av. Rank	1	3	3	3	5.5	5.5	7

$$V = \frac{3 + 5.5 + 7}{3} - \frac{1 + 3 + 3 + 5.5}{4} = 2.042$$

$$d_1 = 1, \quad d_2 = 3, \quad d_3 = 2, \quad d_4 = 1$$

$$\text{C.F.} = \frac{7(1^3 - 1 + 3^3 - 3 + 2^3 - 2 + 1^3 - 1)}{12(4)(3)(6)}$$

$$= .243$$

$$\sigma_V^2(\text{adj.}) = 2.722 - .243 = 2.479$$

$$V/\sigma_V(\text{adj.}) = 1.297$$

$$t_{.025, \infty} = 1.96: \quad \text{do not reject } H_0.$$

Work Problem 23, Chapter 9, by applying the rank sum test with adjustments for ties.

20. Thirty consumers who volunteered to evaluate a food product were randomly divided into two groups of 15 each. One group rated the taste of pudding A, while the other group rated the taste of pudding B. The rating scale was 1, 2, and 3 for poor, fair, and good, respectively. On the basis of the data given below, does it appear that one pudding is preferred to the other? Use the rank sum statistic with adjustments for ties. Let $\alpha = .05$. (See Problem 19.)

	Rating														
Pudding A	2	2	3	2	3	2	1	2	1	2	2	2	2	1	1
Pudding B	1	1	3	3	3	3	3	3	3	1	2	3	3	3	2

21. A nonparametric counterpart of the F-test for equality of means (see Sections 12.8 and 12.9) is the *Kruskal-Wallis test*. The procedure is outlined below.

   (1) Determine the number of groups, k; the number of observations for each group, n_i; and the total number of observations, N.
   (2) Arrange the observations in the combined sample in order of increasing size, and assign ranks 1 to N as in the rank sum test.
   (3) Compute the average rank, $\bar{R}_i$, for each group.
   (4) Compute the Kruskal-Wallis statistic

$$K\text{-}W = \frac{12}{N(N+1)} \sum_{i=1}^{k} n_i \left(\bar{R}_i - \frac{N+1}{2} \right)^2$$

   (5) Refer to the chi-square table (Table 6) with $k - 1$ degrees of freedom for critical values. Reject the null hypothesis of equality of distributions if $K\text{-}W > \chi^2_{\alpha,\, k-1}$ (upper-tail test).

The following computations illustrate the procedure.

$$k = 3, \quad N = 8$$

$$K\text{-}W = \frac{12}{(8)(9)} [3(6.33 - 4.50)^2 + 2(5.50 - 4.50)^2 + 3(2 - 4.50)^2]$$

$$= 5.13$$

$$\chi^2_{.05,\, 2} = 5.99: \quad \text{do not reject } H_0.$$

	Data and Ranks		
	\multicolumn{3}{c}{Group}		
	1	2	3
	12(4)	14(6)	11(3)
	15(7)	13(5)	10(2)
	16(8)		9(1)
n_i	3	2	3
$\bar{R}_i$	6.33	5.50	2

Rework part a of Problem 24, Chapter 12, using the Kruskal-Wallis test. Compare with the F-test for equality of means.

Summary and Keywords

Nonparametric methods (p. 446) apply without the assumption that the underlying population has some prescribed form, such as normal. Nonparametric statistics involve **medians** (p. 446) and **ranks** (p. 447) and other measures of relative magnitude. The **Wilcoxon statistic** (p. 447), a kind of nonparametric mean, can be used to test hypotheses on the population median, provided the population is symmetric about the median; otherwise the **sign test** (p. 451) applies. The **rank sum** (p. 453), a kind of nonparametric difference of means, can be used to test whether two populations have the same location. The **rank correlation coefficient** (p. 457) measures the **association** (p. 457) between rankings, and can be used to test for independence of rankings.

REFERENCES

13.1 Hogg, Robert V., and Craig, Allen T. *Introduction to Mathematical Statistics*. 3d ed. New York: Macmillan, 1970. Chapter 11.

13.2 Kraft, Charles H., and van Eeden, Constance. *A Nonparametric Introduction to Statistics*. New York: Macmillan, 1968.

13.3 Lehmann, E. L. *Nonparametrics, Statistical Methods Based on Ranks*. San Francisco: Holden-Day, Inc., 1975.

13.4 Mosteller, Frederick, and Rourke, Robert E. *Sturdy Statistics*. Reading, Mass.: Addison-Wesley, 1973.

Answers to Odd-Numbered Problems

Chapter 2

1. **a.** 6 **b.** 9 **c.** 513–1024.
3. Suggestion: Use nine intervals of length 10 each.
5. Class limits: −4–20, 21–45, 46–70, 71–95, 96–120
   Class boundaries: −4.5, 20.5, 45.5, 70.5, 95.5, 120.5
9. 33.5, 34.1, 67.1 are outliers.
15.

Class	Frequency
0.0–0.3	3
0.4–0.7	10
0.8–1.1	16
1.2–1.5	9
1.6–1.9	5

19. **a.** Boundaries: 0–10, 10–20, 20–30, 30–40, 40–50
    Class marks: 5, 15, 25, 35, 45
25. **b.** 35.7%

Chapter 3

1. **a.** 22 **b.** 2 **c.** 20 **d.** 1 **e.** 15 **f.** 41 **g.** 2 **h.** −6 **i.** 12
3. **a.** $\sum_{i=1}^{4} P_i X_i$ **b.** $a_1 \sum_{i=1}^{5} q^{i-1}$ **c.** $\sum_{i=1}^{4} \frac{X_i}{i}$ **d.** $\frac{\sum_{i=1}^{3} i^2 Y_i}{\sum_{i=1}^{3} i^3 Y_i}$ **e.** $\sum_{i=1}^{4} (-1)^{i+1} U_i^i$

f. $\left(\sum_{i=1}^{3} a_i\right)^4$ g. $\sum_{i=1}^{3} m_i X_i - \sum_{i=1}^{3} n_i Y_i$

7. **a.** -1.4 **b.** 186.2 **c.** .0015
9. Team B
11. Store A
13. 338.1
15. 22.13
17. 22.36
19. 6.9
21. **a.** 3 **b.** 17.1 **c.** 1
23. 22
25. The median
27. **a.** Country A
    b. m is between 1250 and 4000, and $\bar{X} = 4250$
    c. Country B
29. $M = 0$
33. Major mode at 1.0, minor mode at 1.4.
35. 3.36, 4.9
37. 3.143×10^{-4}, .018
39. $s = 11.6$
41. Team B
43. **a.** All data values are the same.
    b. A mistake was made in the computation.
45. **a.** 11, 10 **b.** 18, 90 **c.** 23, 90
47. 16, 1.6
49. **a.** 103, 36, 6 **b.** 534, 4, 2
51. $\bar{U} = 1.9$, $s_U^2 = 24.8$, $s_U = 4.98$, $\bar{X} = 2.4$, $s_X^2 = 24.8 \times 10^{-18}$, $s_X = 4.98 \times 10^{-9}$
53. .41, .12
55. 64.4, 65.5
57. 23.4, 34.4
59. **a.** 15 **b.** 19 **c.** 15.75 **d.** 6, 30.5
61. **a.** -1, 5, 3, 2.5, 6
    b. 8, 24, 16, 16, 16
    c. 2.5, 5, 3.5, 3.625, 2.5
63. The mean, the variance, the range.

65.

	1977	1978	1979	1980
1977 base	100.0	98.2	102.0	94.7
1978 base	101.8	100.0	103.9	96.4

67. 114.7, 112.1

69.

	a.	b.	c.	d.	e.	f.
(i)	7	7	—	8	10	3.16
(ii)	2.5	2.5	1, 4	3	1.90	1.38
(iii)	0	0	0	6	4	2
(iv)	5	4	4, 8	9	9.25	3.04
(v)	2.33	2	2	4	1.75	1.32
(vi)	-2	-2	3	11	16.67	4.08
(vii)	21	22	23	10	11	3.32
(viii)	1	1.5	3	6	4.86	2.20

71. 2.53, 2, 1, 4.89, 2.21
75. **a.** 5131, 5500 **b.** 5021, 1846

ANSWERS TO ODD-NUMBERED PROBLEMS

Chapter 4

1. **a.** 9 **b.** 27 **c.** 18 **d.** 27 **e.** 9
3. **a.** 4 **b.** 6 **c.** 4 **d.** 4 **e.** 2
5. **b.** 10
7. **a.** 15 **b.** $\frac{4}{15}$
9. **a.** $\frac{12}{52}$ **b.** $\frac{13}{52}$ **c.** $\frac{8}{52}$ **d.** 0 **e.** $\frac{16}{52}$ **f.** $\frac{16}{52}$ **g.** $\frac{28}{52}$
11. **a.** $\frac{1}{2}$ **b.** $\frac{1}{2}$ **c.** $\frac{1}{4}$ **d.** $\frac{3}{4}$ **e.** $\frac{1}{4}$
13. $\frac{1}{3}$
15. **a.** $\frac{5}{95}$ **b.** $\frac{2}{8}$
17. $\frac{100}{190}$
19. **a.** $2 \times \frac{26}{52} \times \frac{25}{51} = \frac{25}{51}$ **b.** $2 \times \frac{2}{52} \times \frac{1}{51}/\frac{25}{51} = 1/(13 \times 25)$ **c.** No **d.** No
    e. $1, 4 \times \frac{13}{52} \times \frac{12}{51}/\frac{25}{51} = \frac{12}{25}$
21. $1 - .8^3 = .488$
23. **a.** No **b.** $\frac{2}{3}$ **c.** No
25. **a.** 2 **b.** 4 **c.** 8 **d.** 14
27. **a.** .6 **b.** .7 **c.** .1 **d.** .1 **e.** .5 **f.** .9 **g.** $\frac{1}{6}$ **h.** $\frac{5}{6}$
29. **a.** $10^3 = 1000$ **b.** $10 \times 9 \times 8 = 720$ **c.** $1000 - 10 = 990$
31. **a.** $20 \times 19 \times 18 = 6840$ **b.** $\binom{20}{3} = 1140$
33. **a.** $26^5 = 11{,}881{,}376$ **b.** $32^5 = 33{,}554{,}432$
35. $\binom{5}{3} = 10$
37. $2 \times 25!$
39. **a.** $\frac{10}{10^3} = \frac{1}{100}$ **b.** $1 - \frac{10 \times 9 \times 8}{10^3} = .28$
41. **a.** $\binom{52}{13} = 635{,}013{,}559{,}600$ **b.** $\binom{52}{13}\binom{39}{13}\binom{26}{13}\binom{13}{13}$
43. **a.** $\binom{13}{2}\binom{4}{2}\binom{4}{2} \times 44 = 123{,}552$ **b.** $13\binom{4}{3} 12 \binom{4}{2} = 3744$
    c. $4 \times 10 = 40$ **d.** $13 \times 48 = 624$
45. $40 \times 39^3 = 2{,}372{,}760$
47. $6!/6 = 120$
49. $5 \times 6 \times 5 \times 3 = 450$
51. $\binom{7}{4} = 35$
53. **a.** $4/\binom{52}{13}$ **b.** $\binom{48}{13}/\binom{52}{13}$ **c.** $1 - \left[\binom{39}{13}/\binom{52}{13}\right]$ **d.** $4\binom{48}{12}/\binom{52}{13}$

55. **a.** $\binom{5}{2} 6 \times \frac{5}{6^5} = \frac{300}{7776}$ **b.** $2 \times \frac{5!}{6^5} = \frac{240}{7776}$
    c. $6 \binom{5}{2} 5 \times 4 \times \frac{3}{6^5} = \frac{3600}{7776}$ **d.** $\binom{4}{2} \frac{1}{2} \times 5 \times 6 \times \frac{4}{6^5} = \frac{1800}{7776}$
    e. $\binom{5}{2} 6 \times 5 \times \frac{4}{6^5} = \frac{1200}{7776}$

57. $1 - .683 = .317$ (binomial, $n = 5$, $p = .4$)

59. $.667$ (binomial, $n = 20$, $p = .1$)

61. $\frac{1}{64}, \frac{63}{64}, \frac{20}{64}$

63. **a.** $\binom{5}{4}(\frac{1}{3})^4(\frac{2}{3})^1 + \binom{5}{5}(\frac{1}{3})^5(\frac{2}{3})^0 = .04$ **b.** $(\frac{2}{3})^5 = .13$ **c.** $.87$

65. $\frac{1}{8}, \frac{1}{4}, \frac{5}{16}, \frac{5}{16}$

67. $1 - .648 = .352$ (binomial, $n = 15$, $p = .2$)

69. $1 - .6^n > .90$, $n = 5$

71. **a.** $.8521$ **b.** $.1420$ **c.** $.0059$

73. **a.** $.8521$ **b.** $.1420$ **c.** $.0059$

75. **a.** $\frac{1}{210}$ **b.** $\frac{8}{21}$ **c.** $\frac{25}{210}$

77. **a.** $\binom{4}{3}\binom{3}{0}/\binom{7}{3}$ **b.** $\binom{4}{1}\binom{3}{2}/\binom{7}{3} + \binom{4}{0}\binom{3}{3}/\binom{7}{3}$

79. $(\frac{3}{5} \times 1)/(\frac{3}{5} \times 1 + \frac{2}{5} \times \frac{1}{2}) = \frac{3}{4}$

81. **a.** $.009/.037 = .24$ **b.** $.291/.963 = .30$

83. **a.** $.30$ **b.** $9/60$

85. $1 - .879 = .121$ (binomial, $n = 10$, $p = .2$)

87. $(.8)(.7)(.1)/[(.8)(.7)(.1) + (.8)(.3)(.9) + (.2)(.7)(.9)] = .14$

Chapter 5

1. (i) discrete, integers from 0 to 100
   (ii) continuous, positive real numbers
   (iii) discrete, integers from 0 to 12
   (iv) discrete, nonnegative integers
   (v) discrete, nonnegative integers
   (vi) continuous, nonnegative real numbers
   (vii) discrete, nonnegative integers
   (viii) continuous, positive real numbers

3. **a.** $.30$ **b.** $.14$

5. **a.** $.6$ **b.** $.7$ **c.**

x	-2	-1	0	1	2	3
$P(x)$	$.1$	$.2$	$.1$	$.3$	$.2$	$.1$

7. **a.** 8

x	0	1	2
b. $P(x)$	$\frac{1}{4}$	$\frac{1}{2}$	$\frac{1}{4}$
c. $F(x)$	$\frac{1}{4}$	$\frac{3}{4}$	1

ANSWERS TO ODD-NUMBERED PROBLEMS

9. **a.** .50 **b.** .25 **c.** .50 **d.** .10
11. **a.** 4 **b.** .25 **c.** .50 **d.** .50 **e.** 1.6
13. **a.** $\frac{1}{16}$ **b.** $F(x) = x^2, 0 \leq x \leq 1$
15. **a.** .1470 **b.** .0206
17. .115
19. 2.4, 1.5
21. 50 cents, 5 cents
23. $\frac{3}{4}$
25. 2.4, 1.84
27.

x	0	1	2	3	4	5
a. $5005 P(x)$	210	1260	2100	1200	225	10
b. $5005 F(x)$	210	1470	3570	4770	4995	5005

    c. $E(X) = 2$ **d.** $\text{Var}(X) = .8571$

29. **a.** 2, $\sqrt{2}$ **b.** $p = .1$
31. $-\frac{1}{3}, \frac{4}{9}$
33. 35 cents, 15 cents
35. **a.**

X	1	2	3
$P(X)$	$\frac{2}{8}$	$\frac{5}{8}$	$\frac{1}{8}$

    b.

Y	1	2	3
$P(Y)$	$\frac{3}{8}$	$\frac{3}{8}$	$\frac{2}{8}$

    c. $E(X) = \frac{15}{8}, E(Y) = \frac{15}{8}$ **d.**

$X + Y$	2	3	4	5	6
$P(X + Y)$	$\frac{1}{8}$	$\frac{2}{8}$	$\frac{4}{8}$	0	$\frac{1}{8}$

    e. $E(X + Y) = \frac{30}{8}$ **f.** $\text{Var}(X) = \frac{23}{64}, \text{Var}(Y) = \frac{39}{64}$ **g.** $\frac{76}{64}$ **h.** $-\frac{15}{8}$

37. **a.** .0228 **b.** .0013 **c.** .4332 **d.** .8186
39. **a.** .6103 **b.** .8907 **c.** .0087 **d.** .0104 **e.** .2481 **f.** .9086
41. **a.** 1.43 **b.** −.80 **c.** 1.04 **d.** −.60 **e.** 1.70 **f.** 1.35
43. **a.** .8413 **b.** .9987 **c.** .6295 **d.** .2638
45. **a.** .5948 **b.** .8490
47. **a.** 48 **b.** 28.8, 71.2
49. 5
51. .0026
53. **a.** .2024 **b.** .0260
55. 8856, 4650
57. **a.** .2743 **b.** .5138
59. 19,459
61. **a.** $P(r) = \binom{3}{r}(\frac{1}{3})^r(\frac{2}{3})^{3-r}$, $r = 0, 1, 2, 3$ **b.** 1, $\frac{2}{3}$
63. **a.** .292 **b.** .656 **c.** .344
65. **a.** $(\frac{2}{3})^{10} = .017$
    b. $\sum_{r=5}^{10}\binom{10}{r}(\frac{1}{3})^r(\frac{2}{3})^{10-r} \cong .21$
67. (i) **a.** .451 **b.** .110 **c.** .207
    (ii) **a.** .575 **b.** .154 **c.** .270
69. .387, .402
71. **a.** .7667 **b.** .0044
73. .11
75. .16
77. **a.** .677 **b.** .023 **c.** .191
79. **a.** .135
    b. Yes: $P(X \geq 4) = .019$
81. .217, .699

83.

	−30	−20	−10	0	10	20	30
Normal probability	.04	.19	.50	.80	.96	1.0	1.0
Observed proportion	.04	.15	.49	.80	.95	.98	.99

85. **a.** .932 .912 **b.** .334 .329
87. .556 by the binomial approximation. .600 exact probability.
89. $\sum_{r=3}^{10} \binom{10}{r} (.2119)^r (.7881)^{10-r} \cong .36$

Chapter 6

1. **a.** 25, 2.5, 1.58 **b.** 25, .25, .50 **c.** 25, .025, .158
3. **a.** 2 **b.** 1.9990 **c.** 1.9019 **d.** 1.0101
5. **a.** Samples may be listed as ordered pairs (i, j), $i = 1, \ldots, 5$, $j = 1, \ldots, 5$, with probability $\frac{1}{25}$ each.
   b.

$\bar{X}$	1	1.5	2	2.5	3	3.5	4	4.5	5
$P(\bar{X})$	$\frac{1}{25}$	$\frac{2}{25}$	$\frac{3}{25}$	$\frac{4}{25}$	$\frac{5}{25}$	$\frac{4}{25}$	$\frac{3}{25}$	$\frac{2}{25}$	$\frac{1}{25}$

   c. $\mu = 3$, $\sigma^2 = 2$, $\sigma^2/n = 1$
7. **a.** Samples may be listed as ordered pairs (i, j), $i \neq j$, with probability $\frac{1}{20}$ each.
   b.

$\bar{X}$	1.5	2	2.5	3	3.5	4	4.5
$P(\bar{X})$	$\frac{1}{10}$	$\frac{1}{10}$	$\frac{2}{10}$	$\frac{2}{10}$	$\frac{2}{10}$	$\frac{1}{10}$	$\frac{1}{10}$

   c. $\mu = 3$, $\sigma^2 = 2$, $\text{Var}(\bar{X}) = .75$
9. **a.** .007 **b.** .033 **c.** .960
11. 16
13. $n = 28$
15. .9876
17. .0188
19. **a.** 1.25, 1.5208 **b.** 1.25, .0152 **c.** 6.25, 6.17 **d.** .021
21. .0668, .0013, .9756

Chapter 7

1. **a.** $\bar{X} = 5$, $s^2 = 16$, $s^2/n = 1$, $s = 4$, $s/\sqrt{n} = 1$
   b. $\bar{X} = 25$, $s^2 = 121$, $s^2/n = 12.1$, $s = 11$, $s/\sqrt{n} = 3.48$
   c. $\bar{X} = 4$, $s^2 = 25$, $s^2/n = 1$, $s = 5$, $s/\sqrt{n} = 1$
3. **a.** $\bar{X} = 3.25$, $s^2 = 10$ **b.** $s/\sqrt{n} = .32$
5. **a.** $\bar{X} = 0$, $s^2 = 1$ **b.** $s^2/n = \frac{1}{9}$
7. **a.** 1.81, 6.19 **b.** 27.63, 30.37 **c.** 101.78, 108.23
9. Decreased by a factor of $1/\sqrt{2}$, $\frac{1}{2}$
11. 1.95, 2.25

ANSWERS TO ODD-NUMBERED PROBLEMS

13. **a.** 2.086 **b.** 2.787 **c.** 1.671
15. **a.** 2.160 **b.** 1.321 **c.** 2.750 **d.** 2.947
17. **a.** 2.87, 7.13 **b.** 17.13, 32.87 **c.** 1.94, 6.06
19. 1896, 2064
21. 117.976, 146.624
23. 12.247, 12.493
25. -1.7794, 4.9794
27. 397.6102, 448.3898
29. 200.64, 259.92
31. 29.653, 34.347
33. 21.802, 23.948
35. 42048.55, 44951.45
37. 14
39. 97
41. $n = 63$
43. **a.** $f = .20$, $nf(1 - f) = 4$, $\sqrt{nf(1 - f)} = 2$, $f(1 - f)/n = .0064$, $\sqrt{f(1 - f)/n} = .08$ **b.** .25, 9, 3, .0039, .0625
    c. .80, 16, 4, .0016, .04 **d.** .533, 74.67, 8.64, .000830, .0288
45. **a.** .19, .64 **b.** 0, .22 **c.** .67, .84 **d.** .15, .26 **e.** .47, .53
    f. .67, .73
47. **a.** .335, .665 **b.** .07, .415 **c.** .25, .42 **d.** .16, .25
49. .05, .18
51. **a.** 5, 8 **b.** 4.5, 7.5
53. .186, .314
55. .487, .553
57. (.16, .21)(.156, .204)
61. $n = 2401$
63. $f = .75$, $w = .06$, $n = 801$

Chapter 8

1. **a.** $t = 2$, $t_{.01, 15} = 2.602$; do not reject H_0
   b. $t = .3$, $-t_{.05, 8} = -1.860$; do not reject H_0
   c. $t = 2.75$, $t_{.005, 24} = 2.797$; do not reject H_0
3. **a.** $t = .75$, $t_{.025, 8} = 2.306$, do not reject H_0
   b. $t = 1.5$, $t_{.10, 24} = 1.318$, reject H_0
5. **a.** $t = 2.7$, $-t_{.10, 15} = -1.341$, yes **b.** $-t_{.05, 15} = -1.753$, yes
   c. $-t_{.01, 15} = -2.602$, yes
7. $t = -1.6$, $-t_{.025, 9} = -2.262$, do not reject $H_0: \mu = 150$
9. $t = 2.30$, $t_{.01, 19} = 2.539$, do not reject $H_0: \mu = 80$
11. $t = 1.652$, $t_{.01, 14} = 2.624$, do not reject $H_0: \mu = 12$
13. $t = 4.583$, $t_{.05, 20} = 1.725$, yes (reject $H_0: \mu = 150$)
15. $t = 8.3$, $t_{.01, 34} \leq 2.457$, yes (reject $H_0: \mu = 75.5$)
17. $t = -1.677$, $-t_{.05, 4} = -2.132$, no (do not reject $H_0: \mu = 20\%$)
19. $Z = 1.69$, $Z_{.05} = 1.64$, reject $H_0: p = .05$
21. $Z = 3.00$, $Z_{.005} = 2.576$, reject $H_0: p = .50$
23. 23 correct identifications

25. $Z = 1.25$, $Z_{.05} = 1.645$, do not reject $H_0: p = \frac{1}{3}$
27. $Z = 1.10$, $Z_{.10} = 1.282$, no
29. a. $p = .8414$ b. $p = .0456$
31. a has the smaller p-value
33. $1, .6651, .2632, .0623, .0087, .0007, .00002$
35. $\frac{1}{2}$
37. a. $\frac{1}{2}$ b. $\frac{1}{4}$ c. $\frac{3}{4}$ d. $\frac{3}{4}$ e. $\frac{9}{16}$ f. $\frac{15}{16}$ g. $75 \times 74 / 100 \times 99$
39. a. $.383$ b. $.322, .070$
41. a. $.224$ b. $.474$ c. $.756$ d. $.994$
43. For Problem 9: $Z = 2.30$, $Z_{.01} = 2.326$, do not reject $H_0: \mu = 80$.
    For Problem 17: $Z = -1.677$, $-Z_{.05} = -1.645$, reject $H_0: \mu = 20\%$.
    Greater probability of rejecting a false null hypothesis for a given α.
45. $\beta = -Z_{.755} = .2251$
49. $Z = 3.81$, $Z_{.05} = 1.645$, reject H_0
51. $9.03 < \mu < 13.57$

Chapter 9

1. $14, 28, 30, 3, \frac{39}{8}$
3. $\frac{31}{5}$
5. a. $(23.433, 32.557)$ b. $(-6.016, .016)$
7. $-9.34, 89.34$
9. a. $(110.384, 113.616), (116.788, 119.212), (3.981, 8.019)$
   b. Yes, because the confidence interval does not contain 0.
11. $(10.285, 13.715), (5.792, 12.208), (-.638, 6.638)$
13. $(2.0, 4.4)$
15. $(4, 11)$
17. $t = 2$, $t_{.05, 25} = 1.708$, yes
19. a. $t = -1.225$, $-t_{.05, 22} = -1.717$, do not reject H_0
    b. $t = .949$, $t_{.01, 22} = 2.508$, do not reject H_0
    c. $t = .75$, $t_{.025, 22} = 2.074$, do not reject H_0
21. $t_{.025, 69} = 1.99$; age: $t = 4.464$, reject H_0;
    years at school: $t = 4.465$, reject H_0;
    years on bench: $t = -1.324$, do not reject H_0;
    visits to prisons: $t = 3.977$, reject H_0
23. $t = 1.841$, $t_{.025, 18} = 2.101$, no
25. $t = 2.2366$, $t_{.025, 14} = 2.145$, there is a difference
27. $t = 4.56$, $t_{.025, 56} \cong 2$, yes
29. a. $(.594, 4.746)$ b. $t = 1.797$, $t_{.01, 8} = 2.896$, no

ANSWERS TO ODD-NUMBERED PROBLEMS

31. **a.** $t = 2.00$, $t_{.05, 8} = 1.860$, A produces the larger gain
    b. $(-1.792, 9.792)$
33. **a.** $t = 7.48$, $t_{.05, 4} = 2.132$, yes **b.** $(.020, .036)$
35. **a.** $(-8.64, 13.98)$ **b.** $t = .316$, $t_{.01, 16} = 2.538$, no
37. **a.** $t = 2.924$ **b.** $t = 3.0216$ **c.** Group-test rejects H_0, paired-test does not reject H_0
39. **a.** $f_1 = .32$, $f_2 = .44$, $f_2 - f_1 = .12$ **b.** $.058$ **c.** $.00558, .23442$
    d. $Z = 2.00$, $t_{.005, \infty} = 2.576$; do not reject H_0
41. **a.** $Z = 2.345$, $Z_{.05} = 1.645$, yes **b.** $Z_{.025} = 1.96$, yes
    c. $Z_{.01} = 2.326$, yes
43. $Z = -1.365$, $-Z_{.05} = -1.645$, no
45. $Z = -1.724$, $-Z_{.05} = -1.645$, yes
47. $Z = 1.34$, $Z_{.05} = 1.645$, no
49. $t = -2$, $-t_{.05, 48} \leq -1.684$, yes
51. **a.** $(145.6, 156.8)$ **b.** $(156.5, 166.9)$ **c.** $(9.0, 12.0)$ **d.** No
53. $p = P(Z > 3.146) = .0008$
55. $Z = 2.06$, $Z_{.025} = 1.96$, yes

Chapter 10

1. $\chi^2 = 1.02$, $\chi^2_{.05, 2} = 5.99$, yes
3. $\chi^2 = 4.15$, $\chi^2_{.05, 6} = 12.59$, do not reject $H_0: p_i = \frac{1}{7}$
5. **a.** $.1833, .0400$ **b.** $\chi^2 = 35.02$, $\chi^2_{.01, 10} = 23.21$, no
7. $\chi^2 = 3.375$, $\chi^2_{.05, 10} = 18.31$, yes
9. $\chi^2 = 10$, $\chi^2_{.05, 3} = 7.81$, reject H_0
11. $\chi^2 = 4.12$, $\chi^2_{.05, 2} = 5.99$, yes
13. $\chi^2 = 10.8$, $\chi^2_{.01, 1} = 6.63$, yes
15. $\chi^2 = 8.80$, $\chi^2_{.05, 9} = 16.92$, yes
17. $\chi^2 = 4.40$, $\chi^2_{.05, 1} = 3.04$, no
19. $\chi^2 = 8.80$, $\chi^2_{.05, 9} = 16.92$, yes
21. $\chi^2 = .748$, $\chi^2_{.05, 2} = 5.99$, accept normality
23. $\chi^2 = 5.29$, $\chi^2_{.05, 6} = 12.59$, accept normality
25. On the basis of 6 intervals for observed frequency, we have $\chi^2 = 2.68$, $\chi^2_{.05, 3} = 7.81$, accept normality
27. **a.** $\chi^2 = 1.77$, $\chi^2_{.05, 5} = 11.07$, accept normality
    b. $\chi^2 = 1.66$, $\chi^2_{.05, 3} = 7.81$, accept normality
29. $\chi^2 = 4.81$, $\chi^2_{.05, 3} = 7.81$, accept the hypothesis of a Poisson distribution, 5 classes used in computing χ^2
31. $\chi^2 = 10.79$, $\chi^2_{.05, 2} = 5.99$, math background affects course preference

33. $\chi^2 = .606$, $\chi^2_{.05, 4} = 9.49$, no
35. $\chi^2 = 193.9$, $\chi^2_{.005, 6} = 18.5$, yes
37. $\chi^2 = 2.47$, $\chi^2_{.05, 3} = 7.81$, no
39. $\chi^2 = 2.27$, $\chi^2_{.05, 2} = 5.99$, yes
45. $\chi^2 = 7.033$, $\chi^2_{.05, 2} = 5.99$, reject independence
47. a. $\chi^2 = 19.00$, $\chi^2_{.05, 1} = 3.84$, yes
    b. $\chi^2 = 2.82$, no
    c. $\chi^2 = 1.797$, no

Chapter 11

1. b. $\hat{Y} = -.6 + 3.2X$
3. $\hat{Y} = -58 + 4X$
5. a. $\hat{Y} = 18 - 6X$ b. 36
7. a. $\hat{Y} = 7.3138 - .009X$ c. 6.6838
9. a. $\hat{Y} = 265.8 + 38.7X$
11. .038, .017, 1.32×10^{-6}
13. The estimated standard deviation of daughter's height for a given value of mother's height is 3.12 inches.
15. 227, 204, 11.4
17. $(-9.5436, -4.4564)$
19. a. $Y = -1 + 3X$ b. $t = -.344$, $-t_{.01, 36} = -2.437$, do not reject H_0
    c. (2.3105, 3.6895)
21. $t = 2.2$, $t_{.025, 16} = 2.120$, conclude $B > 2$, (2.04, 4.16)
23. $(-6.290, 2.290)$
25. (6.95, 7.67), $(-.012, -.006)$
27. $t = 18.60$, $t_{.025, 8} = 2.306$, reject H_0
29. a. $\hat{Y} = 4 + 4X$ b. 28 c. (20.906, 35.094) d. (8.565, 47.435)
31. (6.45, 6.91)
33. Confidence intervals: (324, 362) (368, 395) (410, 432) (446, 473) (479, 517)
    Prediction intervals: (304, 383) (345, 419) (384, 457) (422, 497) (458, 538)
35. $Y = 6.662 + .863X$, $(-2.01, -1.93)$, $(-2.07, -1.86)$
37. 95% prediction interval (6.286, 7.714). 8000 an unusual deduction.
39. a.

Source	df	Sum of Squares
Regression	1	80
Deviations from regression	3	20
Totals	4	100

ANSWERS TO ODD-NUMBERED PROBLEMS

     b. $r^2 = .8$, $s_{Y \cdot X}^2 = 2013$

41. $r^2 = .94$, $1 - r^2 = .06$

43. $\widehat{Y} = 3.00 + 1.92X$, $1 - R^2 = .016$, 1.6%

45. a. $\widehat{Y} = 1.84 + .15X + .492X'$ **b.** 3.841

47. a. $\widehat{Y} = 7.8049 + 9.6063X + .0482X'$
    b. 65.05, 58.81, 68.90, 64.10, 68.43, 60.73, 73.25, 69.39, 72.76, 68.90, 68.41, 71.33, 73.72, 81.42, 71.79, 78.05
    c. $R^2 = .5825$

49. a. $\widehat{Y} = 18.83 + .339X + .421X'$ **c.** $.33$

51. $t = 2.8$, $t_{.05, 27} = 1.703$, reject H_0

53. $r_{XY} = .202$, $r_{X'Y} = .973$, $r_{XX'} = 0$, $R = .9947$

55. $r = .70255$, $t = 3.42$, $t_{.025, 12} = 2.179$, reject $\rho = 0$

57. a. $(-.0878, .3540)$ **b.** $(-.5393, -.0259)$ **c.** $(.8580, .9556)$

59. $t = -7.018$, $-t_{.05, \infty} = -1.645$, yes

61. $t = 1.488$, $t_{.05, \infty} = 1.645$, do not reject H_0

63. a. $\widehat{Y} = 19.89 + .822X$
    b. $t = -2.5538$, $-t_{.025, 18} = -2.101$, reject H_0
    c. 88.5% **d.** 24 **e.** $(23.6724, 24.3276)$

65. c. $\widehat{Y} = 9.906 + .148X$ **d.** $\Sigma(Y_i - \widehat{Y}_i)^2 = 5.936$
    e. Same as d, except for roundoff error
    f.

Source	df	Sum of Squares
Regression	1	32.038
Deviations from regression	18	5.902
Totals	19	37.940

    g. .3279, .015   **h.** (.1165, .1795)
    i. $t = 9.87$, $t_{.025, 18} = 2.101$, reject H_0 **j.** 17.306, .1612
    k. (18.389, 19.183)

Chapter 12

1. a.

Source	df	SS	MS
Among groups	3	120	40
Within groups	8	76	9.5
Totals	11	196	

3.

Source	df	SS	MS
Treatments	5	200	40
Error	24	96	4
Totals	29	296	

5. a. $\bar{X}_1 = 18$, $\bar{X}_2 = 12$, $\bar{X}_3 = 20$, $\bar{X}_4 = 14$, $9.5/3$
    b. $18 \mp 2.306\sqrt{9.5/3}$

7. a. $\bar{X}_1 = 32.36$, $\bar{X}_2 = 37.94$, $\bar{X}_3 = 30.12$, $s_{\bar{X}_1} = .101$
   b. (32.14, 32.58), (37.72, 38.16), (29.90, 30.34)
9. a. 2.8, 3.7, 7, .34 b. .48 c. (2.3, 4.3)
11. b. $F = 10$, $F_{.05, 5, 24} = 2.6207$, means are not equal
13. $F = 4.21$, $F_{.05, 3, 8} = 4.0662$, means are not equal
15. a.

Source	df	SS	MS
Treatments	2	320	160
Error	27	1080	40
Totals	29	1400	

17.

Source	df	SS	MS
Treatments	4	1800	450
Error	45	6750	150
Totals	49	8550	

$F = 3$, $F_{.05, 4, 45} = 2.58$, reject hypothesis of equal means

19. a.

Source	df	SS	MS	EMS
Groups	3	216	72	$\sigma^2 + \frac{1}{3}\Sigma n_i \tau_i^2$
Error	11	146	13.27	σ^2
Totals	14	362		

b. Means: 9, 13, 20, 15
Standard deviations: 1.82, 2.103, 2.10, 1.63
c. $(-19.63, -2.37)$
d. $F = 5.426$, $F_{.05, 3, 11} = 3.59$, reject hypothesis of equal means

21. a.

Source	df	SS	MS
Treatments	2	100	50
Error	22	220	10
Totals	24	320	

c. $F = 5$, $F_{.05, 2, 22} = 3.44$; reject H_0

23. a.

Source	df	SS	MS
Groups	2	595	297
Error	697	69,700	100
Totals	699	70,295	

b. $F = 2.97$, $F_{.01, 2, \infty} = 4.61$; do not reject H_0
c. 10.52; 9.78; 11.26

25.

Source	df	SS	MS	EMS
Blocks	1	54	54	$\sigma^2 + 3\kappa_B^2$
Treatments	2	16	8	$\sigma^2 + 2\kappa_T^2$
Error	2	12	6	σ^2
Totals	5	82		

27. a.

Source	df	SS	MS
Locations	2	9.63	4.82
Paints	3	21.26	7.09
Error	6	11.66	1.94
Totals	11	42.55	

b. $F = 3.65$, $F_{.05, 3, 6} = 4.76$; do not reject H_0

29. a. For CR design $s^2 = 648.3$

b.

Source	df	SS	MS
Blocks	5	6,378	
Error	36	20,203	561.2
Totals	41	26,581	

c. Rel. eff. RCB to CR = 115.5%. RCB 15.5% more efficient

31. a. $F = 28.9357$, $F_{.01, 3, 12} = 5.9526$, efficiencies differ across plates
b. $F = 33.3862$, $F_{.01, 3, 16} = 5.2922$, efficiencies differ across plates
c. The mean square of the blocking variable is negligible, so run number is not a significant factor.

33. a. $t = 6.749$, $t_{.025, 12} = 2.179$, reject H_0
b. $t = 6.3498$, $t_{.025, 12} = 2.179$, reject H_0

35. a.

Source	df	SS	MS
Treatments	3	113.792	37.931
Error	697	4632.321	6.646
Totals	700	4746.113	

b. $F = 5.707$, $F_{.05, 3, \infty} = 3.1161$, yes
c. (4.762, 5.514), (4.773, 5.487), (5.637, 6.335), (5.363, 6.327)

37.

Source	df	SS	MS
Treatments	5	960	192
Error	42	2688	64
Totals	47	3648	

Chapter 13

1. $W/\sigma_W = -2.067$, $-t_{.025, \infty} = -1.96$, reject H_0
3. $W/\sigma_W = 2.51$, $t_{.05, \infty} = 1.645$, yes
5. $Z = 1.339$, $t_{.025, \infty} = 1.96$, do not reject the hypothesis that the median is 10
7. $Z = 2.14$, $t_{.025, \infty} = 1.96$, yes
9. $V/\sigma_V = 2.035$, $t_{.025, \infty} = 1.96$, conclude that the populations are different
11. $V/\sigma_V = -.7912$, $-t_{.025, \infty} = -1.96$, do not reject hypothesis of no difference

13. $V/\sigma_V = 1.05$, $t_{.025, \infty} = 1.96$, no
15. a. $r = .9324$, $t = 7.298$, $t_{.025, 8} = 2.306$, reject the hypothesis $\rho = 0$
    b. $\tau/\sigma_\tau = 3.315$, $t_{.025, \infty} = 1.96$, reject the hypothesis of no association
17. $\tau/\sigma_\tau = 2.86$, $t_{.05, \infty} = 1.645$, do not reject the hypothesis of positive association
19. $V(\text{adj.})/\sigma_V(\text{adj.}) = 1.55$, $t_{.025, \infty} = 1.96$, no
21. K-W $= 8.01$, $\chi^2_{.01, 2} = 9.21$, do not reject H_0

Appendix

Table 1. Cumulative binomial probabilities

Table 2. Areas of the standard normal distribution

Table 3. Values of $e^{-\mu}$

Table 4. Values of t for given probability levels

Table 5. 95% confidence intervals (percent) for binomial distributions

Table 6. Percentage points for the chi-square distribution

Table 7. Relationship between z and r (or μ_z and ρ)

Table 8. Percentage points of the F-distribution

Table 9. Random numbers

Table 1 Cumulative Binomial Probabilities

n = 5

x \ p	.1	.2	.3	.4	.5
0	590	328	168	078	031
1	919	737	528	337	188
2	991	942	837	683	500
3	1	993	969	913	812
4	1	1	998	990	969
5	1	1	1	1	1

n = 10

x \ p	.1	.2	.3	.4	.5
0	349	107	028	006	001
1	736	376	149	046	011
2	930	678	383	167	055
3	987	879	650	382	172
4	998	967	850	633	377
5	1	994	953	834	623
6	1	999	989	945	828
7	1	1	998	988	945
8	1	1	1	998	989
9	1	1	1	1	999
10	1	1	1	1	1

n = 25

x \ p	.1	.2	.3	.4	.5
0	072	004	000	000	000
1	271	027	002	000	000
2	537	098	009	000	000
3	764	234	033	002	000
4	902	421	090	009	000
5	967	617	193	029	002
6	991	780	341	074	007
7	998	891	512	154	022
8	1	953	677	274	054
9	1	983	811	425	115
10	1	994	902	586	212
11	1	998	956	732	345
12	1	1	983	846	500
13	1	1	994	922	655
14	1	1	998	966	788
15	1	1	1	987	885
16	1	1	1	996	946
17	1	1	1	999	978
18	1	1	1	1	993
19	1	1	1	1	998
20	1	1	1	1	1
21	1	1	1	1	1
22	1	1	1	1	1
23	1	1	1	1	1
24	1	1	1	1	1
25	1	1	1	1	1

n = 15

x \ p	.1	.2	.3	.4	.5
0	206	035	005	000	000
1	549	167	035	005	000
2	816	398	127	027	004
3	944	648	297	091	018
4	987	836	515	217	059
5	998	939	722	403	151
6	1	982	869	610	304
7	1	996	950	787	500
8	1	999	985	905	696
9	1	1	996	966	849
10	1	1	999	991	941
11	1	1	1	998	982
12	1	1	1	1	996
13	1	1	1	1	1
14	1	1	1	1	1
15	1	1	1	1	1

n = 20

x \ p	.1	.2	.3	.4	.5
0	122	012	001	000	000
1	392	069	008	001	000
2	677	206	035	004	000
3	867	411	107	016	001
4	957	630	238	051	006
5	989	804	416	126	021
6	998	913	608	250	058
7	1	968	772	416	132
8	1	990	887	596	252
9	1	997	952	755	412
10	1	999	983	872	588
11	1	1	995	943	748
12	1	1	999	979	868
13	1	1	1	994	942
14	1	1	1	998	979
15	1	1	1	1	994
16	1	1	1	1	999
17	1	1	1	1	1
18	1	1	1	1	1
19	1	1	1	1	1
20	1	1	1	1	1

The entry gives the probability $\sum_{r=0}^{x} \binom{n}{r} p^r (1-p)^{n-r}$ of getting x or fewer successes in n independent trials with probability p of success at each. The entry has a decimal point to the left (for $n = 5$, $x = 2$, $p = .1$, the probability of two or fewer successes is .991 to three decimal places), except for the centered 1s (for $n = 5$, $x = 3$, $p = .1$, the probability of three or fewer successes is 1.000 to three decimal places).

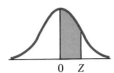

Table 2 Areas of the Standard Normal Distribution

Z	.00	.01	.02	.03	.04	.05	.06	.07	.08	.09
0.0	.0000	.0040	.0080	.0120	.0160	.0199	.0239	.0279	.0319	.0359
0.1	.0398	.0438	.0478	.0517	.0557	.0596	.0636	.0675	.0714	.0753
0.2	.0793	.0832	.0871	.0910	.0948	.0987	.1026	.1064	.1103	.1141
0.3	.1179	.1217	.1255	.1293	.1331	.1368	.1406	.1443	.1480	.1517
0.4	.1554	.1591	.1628	.1664	.1700	.1736	.1772	.1808	.1844	.1879
0.5	.1915	.1950	.1985	.2019	.2054	.2088	.2123	.2157	.2190	.2224
0.6	.2257	.2291	.2324	.2357	.2389	.2422	.2454	.2486	.2517	.2549
0.7	.2580	.2611	.2642	.2673	.2704	.2734	.2764	.2794	.2823	.2852
0.8	.2881	.2910	.2939	.2967	.2995	.3023	.3051	.3078	.3106	.3133
0.9	.3159	.3186	.3212	.3238	.3264	.3289	.3315	.3340	.3365	.3389
1.0	.3413	.3438	.3461	.3485	.3508	.3531	.3554	.3577	.3599	.3621
1.1	.3643	.3665	.3686	.3708	.3729	.3749	.3770	.3790	.3810	.3830
1.2	.3849	.3869	.3888	.3907	.3925	.3944	.3962	.3980	.3997	.4015
1.3	.4032	.4049	.4066	.4082	.4099	.4115	.4131	.4147	.4162	.4177
1.4	.4192	.4207	.4222	.4236	.4251	.4265	.4279	.4292	.4306	.4319
1.5	.4332	.4345	.4357	.4370	.4382	.4394	.4406	.4418	.4429	.4441
1.6	.4452	.4463	.4474	.4484	.4495	.4505	.4515	.4525	.4535	.4545
1.7	.4554	.4564	.4573	.4582	.4591	.4599	.4608	.4616	.4625	.4633
1.8	.4641	.4649	.4656	.4664	.4671	.4678	.4686	.4693	.4699	.4706
1.9	4713	.4719	.4726	.4732	.4738	.4744	.4750	.4756	.4761	.4767
2.0	.4772	.4778	.4783	.4788	.4793	.4798	.4803	.4808	.4812	.4817
2.1	.4821	.4826	.4830	.4834	.4838	.4842	.4846	.4850	.4854	.4857
2.2	.4861	.4864	.4868	.4871	.4875	.4878	.4881	.4884	.4887	.4890
2.3	.4893	.4896	.4898	.4901	.4904	.4906	.4909	.4911	.4913	.4916
2.4	.4918	.4920	.4922	.4925	.4927	.4929	.4931	.4932	.4934	.4936
2.5	.4938	.4940	.4941	.4943	.4945	.4946	.4948	.4949	.4951	.4952
2.6	.4953	.4955	.4956	.4957	.4959	.4960	.4961	.4962	.4963	.4964
2.7	.4965	.4966	.4967	.4968	.4969	.4970	.4971	.4972	.4973	.4974
2.8	.4974	.4975	.4976	.4977	.4977	.4978	.4979	.4979	.4980	.4981
2.9	.4981	.4982	.4982	.4983	.4984	.4984	.4985	.4985	.4986	.4986
3.0	.4987	.4987	.4987	.4988	.4988	.4989	.4989	.4989	.4990	.4990
3.1	.4990	.4991	.4991	.4991	.4992	.4992	.4992	.4992	.4993	.4993
3.2	.4993	.4993	.4994	.4994	.4994	.4994	.4994	.4995	.4995	.4995
3.3	.4995	.4995	.4995	.4996	.4996	.4996	.4996	.4996	.4996	.4997
3.4	.4997	.4997	.4997	.4997	.4997	.4997	.4997	.4997	.4997	.4998
3.5	.4998	.4998	.4998	.4998	.4998	.4998	.4998	.4998	.4998	.4998

To find the area under the standard normal curve between 0 and a Z of 1.12, say, find 1.1 at the left of the table and .02 at the top (1.12 = 1.1 + .02); in the body of the table, the entry in the row for 1.1 and the column for .02 is .3686, and this is the required area (to four decimal places).

Abridged from *Table of Probability Functions*, V. II, of the Federal Works Agency, Work Project Administration for the City of New York, New York, 1942.

Table 3 Values of $e^{-\mu}$

μ	$e^{-\mu}$	μ	$e^{-\mu}$
0.0	1.000	3.1	.045
0.1	.905	3.2	.041
0.2	.819	3.3	.037
0.3	.741	3.4	.033
0.4	.670	3.5	.030
0.5	.607	3.6	.027
0.6	.549	3.7	.025
0.7	.497	3.8	.022
0.8	.449	3.9	.020
0.9	.407	4.0	.018
1.0	.368	4.1	.017
1.1	.333	4.2	.015
1.2	.301	4.3	.014
1.3	.272	4.4	.012
1.4	.247	4.5	.011
1.5	.223	4.6	.010
1.6	.202	4.7	.009
1.7	.183	4.8	.008
1.8	.165	4.9	.007
1.9	.150	5.0	.007
2.0	.135	5.1	.006
2.1	.122	5.2	.006
2.2	.111	5.3	.005
2.3	.100	5.4	.005
2.4	.091	5.5	.004
2.5	.082	5.6	.004
2.6	.074	5.7	.003
2.7	.067	5.8	.003
2.8	.061	5.9	.003
2.9	.055	6.0	.002
3.0	.050		

Table 4 Values of t for Given Probability Levels

Area = h

$t_{h,d}$

Degrees of Freedom	Probability of a Larger Value				
	.1	.05	.025	.01	.005
1	3.078	6.314	12.706	31.821	63.657
2	1.886	2.920	4.303	6.965	9.925
3	1.638	2.353	3.182	4.541	5.841
4	1.533	2.132	2.776	3.747	4.604
5	1.476	2.015	2.571	3.365	4.032
6	1.440	1.943	2.447	3.143	3.707
7	1.415	1.895	2.365	2.998	3.499
8	1.397	1.860	2.306	2.896	3.355
9	1.383	1.833	2.262	2.821	3.250
10	1.372	1.812	2.228	2.764	3.169
11	1.363	1.796	2.201	2.718	3.106
12	1.356	1.782	2.179	2.681	3.055
13	1.350	1.771	2.160	2.650	3.012
14	1.345	1.761	2.145	2.624	2.977
15	1.341	1.753	2.131	2.602	2.947
16	1.337	1.746	2.120	2.583	2.921
17	1.333	1.740	2.110	2.567	2.898
18	1.330	1.734	2.101	2.552	2.878
19	1.328	1.729	2.093	2.539	2.861
20	1.325	1.725	2.086	2.528	2.845
21	1.323	1.721	2.080	2.518	2.831
22	1.321	1.717	2.074	2.508	2.819
23	1.319	1.714	2.069	2.500	2.807
24	1.318	1.711	2.064	2.492	2.797
25	1.316	1.708	2.060	2.485	2.787
26	1.315	1.706	2.056	2.479	2.779
27	1.314	1.703	2.052	2.473	2.771
28	1.313	1.701	2.048	2.467	2.763
29	1.311	1.699	2.045	2.462	2.756
30	1.310	1.697	2.042	2.457	2.750
40	1.303	1.684	2.021	2.423	2.704
60	1.296	1.671	2.000	2.390	2.660
120	1.290	1.661	1.984	2.358	2.626
∞	1.282	1.645	1.960	2.326	2.576

In the row for $d = 11$ degrees of freedom and the column for probability $h = .05$, the entry is $t_{h,d} = t_{.05,11} = 1.796$; if t has the t-distribution with 11 degrees of freedom, then $P(t > 1.796) = .05$.

Abridged from Table III of Fisher and Yates, *Statistical Tables for Biological, Agricultural, and Medical Research*, published by Longman Group Ltd., London (previously published by Oliver and Boyd Ltd., Edinburgh), and by permission of the authors and publishers.

Table 5 95% Confidence Intervals (Percent) for Binomial Distributions

Number Observed (X)	Size of Sample, n						Fraction Observed (X/n)	Size of Sample	
	10	15	20	30	50	100		250	1000
0	0– 31	0– 22	0– 17	0– 12	0– 07	0– 4	.00	0– 1	0– 0
1	0– 45	0– 32	0– 25	0– 17	0– 11	0– 5	.01	0– 4	0– 2
2	3– 56	2– 40	1– 31	1– 22	0– 14	0– 7	.02	1– 5	1– 3
3	7– 65	4– 48	3– 38	2– 27	1– 17	1– 8	.03	1– 6	2– 4
4	12– 74	8– 55	6– 44	4– 31	2– 19	1–10	.04	2– 7	3– 5
5	19– 81	12– 62	9– 49	6– 35	3– 22	2–11	.05	3– 9	4– 7
6	26– 88	16– 68	12– 54	8– 39	5– 24	2–12	.06	3–10	5– 8
7	35– 93	21– 73	15– 59	10– 43	6– 27	3–14	.07	4–11	6– 9
8	44– 97	27– 79	19– 64	12– 46	7– 29	4–15	.08	5–12	6–10
9	55–100	32– 84	23– 68	15– 50	9– 31	4–16	.09	6–13	7–11
10	69–100	38– 88	27– 73	17– 53	10– 34	5–18	.10	7–14	8–12
11		45– 92	32– 77	20– 56	12– 36	5–19	.11	7–16	9–13
12		52– 96	36– 81	23– 60	13– 38	6–20	.12	8–17	10–14
13		60– 98	41– 85	25– 63	15– 41	7–21	.13	9–18	11–15
14		68–100	46– 88	28– 66	16– 43	8–22	.14	10–19	12–16
15		78–100	51– 91	31– 69	18– 44	9–24	.15	10–20	13–17
16			56– 94	34– 72	20– 46	9–25	.16	11–21	14–18
17			62– 97	37– 75	21– 48	10–26	.17	12–22	15–19
18			69– 99	40– 77	23– 50	11–27	.18	13–23	16–21
19			75–100	44– 80	25– 53	12–28	.19	14–24	17–22
20			83–100	47– 83	27– 55	13–29	.20	15–26	18–23
21				50– 85	28– 57	14–30	.21	16–27	19–24
22				54– 88	30– 59	14–31	.22	17–28	19–25
23				57– 90	32– 61	15–32	.23	18–29	20–26
24				61– 92	34– 63	16–33	.24	19–30	21–27
25				65– 94	36– 64	17–35	.25	20–31	22–28
26				69– 96	37– 66	18–36	.26	20–32	23–29
27				73– 98	39– 68	19–37	.27	21–33	24–30
28				78– 99	41– 70	19–38	.28	22–34	25–31
29				83–100	43– 72	20–39	.29	23–35	26–32
30				88–100	45– 73	21–40	.30	24–36	27–33
31					47– 75	22–41	.31	25–37	28–34
32					50– 77	23–42	.32	26–38	29–35
33					52– 79	24–43	.33	27–39	30–36
34					54– 80	25–44	.34	28–40	31–37
35					56– 82	26–45	.35	29–41	32–38
36					57– 84	27–46	.36	30–42	33–39
37					59– 85	28–47	.37	31–43	34–40
38					62– 87	28–48	.38	32–44	35–41
39					64– 88	29–49	.39	33–45	36–42
40					66– 90	30–50	.40	34–46	37–43
41					69– 91	31–51	.41	35–47	38–44
42					71– 93	32–52	.42	36–48	39–45
43					73– 94	33–53	.43	37–49	40–46
44					76– 95	34–54	.44	38–50	41–47
45					78– 97	35–55	.45	39–51	42–48
46					81– 98	36–56	.46	40–52	43–49
47					83– 99	37–57	.47	41–53	44–50
48					86–100	38–58	.48	42–54	45–51
49					89–100	39–59	.49	43–55	46–52
50					93–100	40–60	.50	44–56	47–53
						*		**	**

Reprinted by permission from *Statistical Methods* by George W. Snedecor and William G. Cochran, Sixth Edition © 1967 by The Iowa State University Press, Ames, Iowa 50010.

*If X exceeds 50, read $100 - X$ = number observed and subtract each confidence limit from 100.

**If X/n exceeds .50, read $1.00 - X/n$ = fraction observed and subtract each confidence limit from 100.

Table 6 Percentage Points of the Chi-Square Distribution

$\chi^2_{h,d}$ Area = h

df	.995	.990	.975	.950	.050	.025	.010	.005
1	—	—	—	.004	3.84	5.02	6.63	7.88
2	.01	.02	.05	.10	5.99	7.38	9.21	10.60
3	.07	.11	.22	.35	7.81	9.35	11.34	12.84
4	.21	.30	.48	.71	9.49	11.14	13.28	14.86
5	.41	.55	.83	1.15	11.07	12.83	15.09	16.75
6	.68	.87	1.24	1.64	12.59	14.45	16.81	18.55
7	.99	1.24	1.69	2.17	14.07	16.01	18.48	20.28
8	1.34	1.65	2.18	2.73	15.51	17.53	20.09	21.96
9	1.73	2.09	2.70	3.33	16.92	19.02	21.67	23.59
10	2.16	2.56	3.25	3.94	18.31	20.48	23.21	25.19
11	2.60	3.05	3.82	4.57	19.68	21.92	24.72	26.76
12	3.07	3.57	4.40	5.23	21.03	23.34	26.22	28.30
13	3.57	4.11	5.01	5.89	22.36	24.74	27.69	29.82
14	4.07	4.66	5.63	6.57	23.68	26.12	29.14	31.32
15	4.60	5.23	6.26	7.26	25.00	27.49	30.58	32.80
16	5.14	5.81	6.91	7.96	26.30	28.85	32.00	34.27
17	5.70	6.41	7.56	8.67	27.59	30.19	33.41	35.72
18	6.26	7.01	8.23	9.39	28.87	31.53	34.81	37.16
19	6.84	7.63	8.91	10.12	30.14	32.85	36.19	38.58
20	7.43	8.26	9.59	10.85	31.41	34.17	37.57	40.00
21	8.03	8.90	10.28	11.59	32.67	35.48	38.93	41.40
22	8.64	9.54	10.98	12.34	33.92	36.78	40.29	42.80
23	9.26	10.20	11.69	13.09	35.17	38.08	41.64	44.18
24	9.89	10.86	12.40	13.85	36.42	39.36	42.98	45.56
25	10.52	11.52	13.12	14.61	37.65	40.65	44.31	46.93
26	11.16	12.20	13.84	15.38	38.89	41.92	45.64	48.29
27	11.81	12.88	14.57	16.15	40.11	43.19	46.96	49.64
28	12.46	13.56	15.31	16.93	41.34	44.46	48.28	50.99
29	13.12	14.26	16.05	17.71	42.56	45.72	49.59	52.34
30	13.79	14.95	16.79	18.49	43.77	46.98	50.89	53.67
40	20.71	22.16	24.43	26.51	55.76	59.34	63.69	66.77
50	27.99	29.71	32.36	34.76	67.50	71.42	76.15	79.49
60	35.53	37.48	40.48	43.19	79.08	83.30	88.38	91.95
70	43.28	45.44	48.76	51.74	90.53	95.02	100.43	104.22
80	51.17	53.54	57.15	60.39	101.88	106.63	112.33	116.32
90	59.20	61.75	65.65	69.13	113.14	118.14	124.12	128.30
100	67.33	70.06	74.22	77.93	124.34	129.56	135.81	140.17

Probability of a Larger Value

In the row for $d = 6$ degrees of freedom and the column for probability $h = .050$, the entry is $\chi^2_{h,d} = \chi^2_{.05,6} = 12.59$; if χ^2 has the chi-square distribution with 6 degrees of freedom, then $P(\chi^2 > 12.59) = .05$.

Abridged from Thompson, Catherine M., "Table of percentage points of the χ^2 distribution," *Biometrika*, Vol. 32 (1942), p. 187, by permission.

Table 7 Relationship between z and r (or μ_z and ρ)

z	.00	.01	.02	.03	.04	.05	.06	.07	.08	.09
.0	.0000	.0100	.0200	.0300	.0400	.0500	.0599	.0699	.0798	.0898
.1	.0997	.1096	.1194	.1293	.1391	.1489	.1587	.1684	.1781	.1878
.2	.1974	.2070	.2165	.2260	.2355	.2449	.2543	.2636	.2729	.2821
.3	.2913	.3004	.3095	.3185	.3275	.3364	.3452	.3540	.3627	.3714
.4	.3800	.3885	.3969	.4053	.4136	.4219	.4301	.4382	.4462	.4542
.5	.4621	.4700	.4777	.4854	.4930	.5005	.5080	.5154	.5227	.5299
.6	.5370	.5441	.5511	.5581	.5649	.5717	.5784	.5850	.5915	.5980
.7	.6044	.6107	.6169	.6231	.6291	.6352	.6411	.6469	.6527	.6584
.8	.6640	.6696	.6751	.6805	.6858	.6911	.6963	.7014	.7064	.7114
.9	.7163	.7211	.7259	.7306	.7352	.7398	.7443	.7487	.7531	.7574
1.0	.7616	.7658	.7699	.7739	.7779	.7818	.7857	.7895	.7932	.7969
1.1	.8005	.8041	.8076	.8110	.8144	.8178	.8210	.8243	.8275	.8306
1.2	.8337	.8367	.8397	.8426	.8455	.8483	.8511	.8538	.8565	.8591
1.3	.8617	.8643	.8668	.8693	.8717	.8741	.8764	.8787	.8810	.8832
1.4	.8854	.8875	.8896	.8917	.8937	.8957	.8977	.8996	.9015	.9033
1.5	.9052	.9069	.9087	.9104	.9121	.9138	.9154	.9170	.9186	.9202
1.6	.9217	.9232	.9246	.9261	.9275	.9289	.9302	.9316	.9329	.9342
1.7	.9354	.9367	.9379	.9391	.9402	.9414	.9425	.9436	.9447	.9458
1.8	.9468	.9478	.9488	.9498	.9508	.9518	.9527	.9536	.9545	.9554
1.9	.9562	.9571	.9579	.9587	.9595	.9603	.9611	.9619	.9626	.9633
2.0	.9640	.9647	.9654	.9661	.9668	.9674	.9680	.9687	.9693	.9699
2.1	.9705	.9710	.9716	.9722	.9727	.9732	.9738	.9743	.9748	.0753
2.2	.9757	.9762	.9767	.9771	.9776	.9780	.9785	.9789	.9793	.9797
2.3	.9801	.9805	.9809	.9812	.9816	.9820	.9823	.9827	.9830	.9834
2.4	.9837	.9840	.9843	.9846	.9849	.9852	.9855	.9858	.9861	.9863
2.5	.9866	.9869	.9871	.9874	.9876	.9879	.9881	.9884	.9886	.9888
2.6	.9890	.9892	.9895	.9897	.9899	.9901	.9903	.9905	.9906	.9908
2.7	.9910	.9912	.9914	.9915	.9917	.9919	.9920	.9922	.9923	.9925
2.8	.9926	.9928	.9929	.9931	.9932	.9933	.9935	.9936	.9937	.9938
2.9	.9940	.9941	.9942	.9943	.9944	.9945	.9946	.9947	.9949	.9950
3.0	.9951									
4.0	.9993									
5.0	.9999									

The r-values appear in the body of the table, the z-values in scales at the left and above the table.

Table 8 Percentage Points of the F-Distribution

F-Distribution: 5% Points

v_2 \ v_1	1	2	3	4	5	6	7	8	9
1	161.45	199.50	215.71	224.58	230.16	233.99	236.77	238.88	240.54
2	18.513	19.000	19.164	19.247	19.296	19.330	19.353	19.371	19.385
3	10.128	9.5521	9.2766	9.1172	9.0135	8.9406	8.8868	8.8452	8.8123
4	7.7086	6.9443	6.5914	6.3883	6.2560	6.1631	6.0942	6.0410	5.9988
5	6.6079	5.7861	5.4095	5.1922	5.0503	4.9503	4.8759	4.8183	4.7725
6	5.9874	5.1433	4.7571	4.5337	4.3874	4.2839	4.2066	4.1468	4.0990
7	5.5914	4.7374	4.3468	4.1203	3.9715	3.8660	3.7870	3.7257	3.6767
8	5.3177	4.4590	4.0662	3.8378	3.6875	3.5806	3.5005	3.4381	3.3881
9	5.1174	4.2565	3.8626	3.6331	3.4817	3.3738	3.2927	3.2296	3.1789
10	4.9646	4.1028	3.7083	3.4780	3.3258	3.2172	3.1355	3.0717	3.0204
11	4.8443	3.9823	3.5874	3.3567	3.2039	3.0946	3.0123	2.9480	2.8962
12	4.7472	3.8853	3.4903	3.2592	3.1059	2.9961	2.9134	2.8486	2.7964
13	4.6672	3.8056	3.4105	3.1791	3.0254	2.9153	2.8321	2.7669	2.7144
14	4.6001	3.7389	3.3439	3.1122	2.9582	2.8477	2.7642	2.6987	2.6458
15	4.5431	3.6823	3.2874	3.0556	2.9013	2.7905	2.7066	2.6408	2.5876
16	4.4940	3.6337	3.2389	3.0069	2.8524	2.7413	2.6572	2.5911	2.5377
17	4.4513	3.5915	3.1968	2.9647	2.8100	2.6987	2.6143	2.5480	2.4943
18	4.4139	3.5546	3.1599	2.9277	2.7729	2.6613	2.5767	2.5102	2.4563
19	4.3808	3.5219	3.1274	2.8951	2.7401	2.6283	2.5435	2.4768	2.4227
20	4.3513	3.4928	3.0984	2.8661	2.7109	2.5990	2.5140	2.4471	2.3928
21	4.3248	3.4668	3.0725	2.8401	2.6848	2.5757	2.4876	2.4205	2.3661
22	4.3009	3.4434	3.0491	2.8167	2.6613	2.5491	2.4638	2.3965	2.3419
23	4.2793	3.4221	3.0280	2.7955	2.6400	2.5277	2.4422	2.3748	2.3201
24	4.2597	3.4028	3.0088	2.7763	2.6207	2.5082	2.4226	2.3551	2.3002
25	4.2417	3.3852	2.9912	2.7587	2.6030	2.4904	2.4047	2.3371	2.2821
26	4.2252	3.3690	2.9751	2.7426	2.5868	2.4741	2.3883	2.3205	2.2655
27	4.2100	3.3541	2.9604	2.7278	2.5719	2.4591	2.3732	2.3053	2.2501
28	4.1960	3.3404	2.9467	2.7141	2.5581	2.4453	2.3593	2.2913	2.2360
29	4.1830	3.3277	2.9340	2.7014	2.5454	2.4324	2.3463	2.2782	2.2229
30	4.1709	3.3158	2.9223	2.6896	2.5336	2.4205	2.3343	2.2662	2.2107
40	4.0848	3.2317	2.8387	2.6060	2.4495	2.3359	2.2490	2.1802	2.1240
60	4.0012	3.1504	2.7581	2.5252	2.3683	2.2540	2.1665	2.0970	2.0401
120	3.9201	3.0718	2.6802	2.4472	2.2900	2.1750	2.0867	2.0164	1.9588
∞	3.8415	2.9957	2.6049	2.3719	2.2141	2.0986	2.0096	1.9384	1.8799

(Continued)

Reproduced from Merrington, Maxine, and Thompson, Catherine M., "Tables of percentage points of the inverted Beta (F) distribution," *Biometrika*, Vol. 33 (1943), pp. 73–88, by permission.

Table 8 (*Continued*)

F-Distribution: 5% Points (*Continued*)

ν_2 \ ν_1	10	12	15	20	24	30	40	60	120	∞
1	241.88	243.91	245.95	248.01	249.05	250.09	251.14	252.20	253.25	254.32
2	19.396	19.413	19.429	19.446	19.454	19.462	19.471	19.479	19.487	19.496
3	8.7855	8.7446	8.7029	8.6602	8.6385	8.6166	8.5944	8.5720	8.5494	8.5265
4	5.9644	5.9117	5.8578	5.8025	5.7744	5.7459	5.7170	5.6878	5.6581	5.6281
5	4.7351	4.6777	4.6188	4.5581	4.5272	4.4957	4.4638	4.4314	4.3984	4.3650
6	4.0600	3.9999	3.9381	3.8742	3.8415	3.8082	3.7743	3.7398	3.7047	3.6688
7	3.6365	3.5747	3.5108	3.4445	3.4105	3.3758	3.3404	3.3043	3.2674	3.2298
8	3.3472	3.2840	3.2184	3.1503	3.1152	3.0794	3.0428	3.0053	2.9669	2.9276
9	3.1373	3.0729	3.0061	2.9365	2.9005	2.8637	2.8259	2.7872	2.7475	2.7067
10	2.9782	2.9130	2.8450	2.7740	2.7372	2.6996	2.6609	2.6211	2.5801	2.5379
11	2.8536	2.7876	2.7186	2.6464	2.6090	2.5705	2.5309	2.4901	2.4480	2.4045
12	2.7534	2.6866	2.6169	2.5436	2.5055	2.4663	2.4259	2.3842	2.3410	2.2962
13	2.6710	2.6037	2.5331	2.4589	2.4202	2.3803	2.3392	2.2966	2.2524	2.2064
14	2.6021	2.5342	2.4630	2.3879	2.3487	2.3082	2.2664	2.2230	2.1778	2.1307
15	2.5437	2.4753	2.4035	2.3275	2.2878	2.2468	2.2043	2.1601	2.1141	2.0658
16	2.4935	2.4247	2.3522	2.2756	2.2354	2.1938	2.1507	2.1058	2.0589	2.0096
17	2.4499	2.3807	2.3077	2.2304	2.1898	2.1477	2.1040	2.0584	2.0107	1.9604
18	2.4117	2.3421	2.2686	2.1906	2.1497	2.1071	2.0629	2.0166	1.9681	1.9168
19	2.3779	2.3080	2.2341	2.1555	2.1141	2.0712	2.0264	1.9796	1.9302	1.8780
20	2.3497	2.2776	2.2033	2.1242	2.0825	2.0391	1.9938	1.9464	1.8963	1.8432
21	2.3210	2.2504	2.1757	2.0960	2.0540	2.0102	1.9645	1.9165	1.8657	1.8117
22	2.2967	2.2258	2.1508	2.0707	2.0283	1.9842	1.9380	1.8895	1.8380	1.7831
23	2.2747	2.2036	2.1282	2.0476	2.0050	1.9605	1.9139	1.8649	1.8128	1.7570
24	2.2547	2.1834	2.1077	2.0267	1.9838	1.9390	1.8920	1.8424	1.7897	1.7331
25	2.2365	2.1649	2.0889	2.0075	1.9643	1.9192	1.8718	1.8217	1.7684	1.7110
26	2.2197	2.1479	2.0716	1.9898	1.9464	1.9010	1.8533	1.8027	1.7488	1.6906
27	2.2043	2.1323	2.0558	1.9736	1.9299	1.8842	1.8361	1.7851	1.7307	1.6717
28	2.1900	2.1179	2.0411	1.9586	1.9147	1.8687	1.8203	1.7689	1.7138	1.6541
29	2.1768	2.1045	2.0275	1.9446	1.9005	1.8543	1.8055	1.7537	1.6981	1.6377
30	2.1646	2.0921	2.0148	1.9317	1.8874	1.8409	1.7918	1.7396	1.6835	1.6223
40	2.0772	2.0035	1.9245	1.8389	1.7929	1.7444	1.6928	1.6373	1.5766	1.5089
60	1.9926	1.9174	1.8364	1.7480	1.7001	1.6491	1.5943	1.5343	1.4673	1.3893
120	1.9105	1.8337	1.7505	1.6587	1.6074	1.5543	1.4952	1.4290	1.3519	1.2539
∞	1.8307	1.7522	1.6664	1.5705	1.5173	1.4591	1.3940	1.3180	1.2214	1.0000

(*Continued*)

Table 8 (*Continued*)

F-Distribution: 2.5% Points

v_2 \ v_1	1	2	3	4	5	6	7	8	9
1	647.79	799.50	864.16	899.58	921.85	937.11	948.22	956.66	963.28
2	38.506	39.000	39.165	39.248	29.298	39.331	39.355	39.373	39.387
3	17.443	16.044	15.439	15.101	14.885	14.735	14.624	14.540	14.473
4	12.218	10.649	9.9792	9.6045	9.3645	9.1973	9.0741	8.9796	8.9047
5	10.007	8.4336	7.7636	7.3879	7.1464	6.9777	6.8531	6.7572	6.6810
6	8.8131	7.2598	6.5988	6.2272	5.9876	5.8197	5.6955	5.5996	5.5234
7	8.0727	6.5415	5.8898	5.5226	5.2852	5.1186	4.9949	4.8994	4.8232
8	7.5709	6.0595	5.4160	5.0526	4.8173	4.6517	4.5286	4.4332	4.3572
9	7.2093	5.7147	5.0781	4.7181	4.4844	4.3197	4.1971	4.1020	4.0260
10	6.9367	5.4564	4.8256	4.4683	4.2361	4.0721	3.9498	3.8549	3.7790
11	6.7241	5.2559	4.6300	4.2751	4.0440	3.8807	3.7586	3.6638	3.5879
12	6.5538	5.0959	4.4742	4.1212	3.8911	3.7283	3.6065	3.5118	3.4358
13	6.4143	4.9653	4.3472	3.9959	3.7667	3.6043	3.4827	3.3880	3.3120
14	6.2979	4.8567	4.2417	3.8919	3.6634	3.5014	3.3799	3.2853	3.2093
15	6.1995	4.7650	4.1528	3.8043	3.5764	3.4147	3.2934	3.1987	3.1227
16	6.1151	4.6867	4.0768	3.7294	3.5021	3.3406	3.2194	3.1248	3.0488
17	6.0420	4.6189	4.0112	3.6648	3.4379	3.2767	3.1556	3.0610	2.9849
18	5.9781	4.5597	3.9539	3.6083	3.3820	3.2209	3.0999	3.0053	2.9291
19	5.9216	4.5075	3.9034	3.5587	3.3327	3.1718	3.0509	2.9563	2.8800
20	5.8715	4.4613	3.8587	3.5147	3.2891	3.1283	3.0074	2.9128	2.8365
21	5.8266	4.4199	3.8188	3.4754	3.2501	3.0895	2.9686	2.8740	2.7977
22	5.7863	4.3828	3.7829	3.4401	3.2151	3.0546	2.9338	2.8392	2.7628
23	5.7498	4.3492	3.7505	3.4083	3.1835	3.0232	2.9024	2.8077	2.7313
24	5.7167	4.3187	3.7211	3.3794	3.1548	2.9946	2.8738	2.7791	2.7027
25	5.6864	4.2909	3.6943	3.3530	3.1287	2.9685	2.8478	2.7531	2.6766
26	5.6586	4.2655	3.6697	3.3289	3.1048	2.9447	2.8240	2.7293	2.6528
27	5.6331	4.2421	3.6472	3.3067	3.0828	2.9228	2.8021	2.7074	2.6309
28	5.6096	4.2205	3.6264	3.2863	3.0625	2.9027	2.7820	2.6872	2.6106
29	5.5878	4.2006	3.6072	3.2674	3.0438	2.8840	2.7633	2.6686	2.5919
30	5.5675	4.1821	3.5894	3.2499	3.0265	2.8667	2.7460	2.6513	2.5746
40	5.4239	4.0510	3.4633	3.1261	2.9037	2.7444	2.6238	2.5289	2.4519
60	5.2857	3.9253	3.3425	3.0077	2.7863	2.6274	2.5068	2.4117	2.3344
120	5.1524	3.8046	3.2270	2.8943	2.6740	2.5154	2.3948	2.2994	2.2217
∞	5.0239	3.6889	3.1161	2.7858	2.5665	2.4082	2.2875	2.1918	2.1136

(*Continued*)

Table 8 *(Continued)*

F-Distribution: 2.5% Points *(Continued)*

ν_2 \ ν_1	10	12	15	20	24	30	40	60	120	∞
1	968.63	976.71	984.87	993.10	997.25	1001.4	1005.6	1009.8	1014.0	1018.3
2	39.398	39.415	39.431	39.448	39.456	39.465	39.473	39.481	39.490	39.498
3	14.419	14.337	14.253	14.167	14.124	14.081	14.037	13.992	13.947	13.902
4	8.8439	8.7512	8.6565	8.5599	8.5109	8.4613	8.4111	8.3604	8.3092	8.2573
5	6.6192	6.5246	6.4277	6.3285	6.2780	6.2269	6.1751	6.1225	6.0693	6.0153
6	5.4613	5.3662	5.2687	5.1684	5.1172	5.0652	5.0125	4.9589	4.9045	4.8491
7	4.7611	4.6658	4.5678	4.4667	4.4150	4.3624	4.3089	4.2544	4.1989	4.1423
8	4.2951	4.1997	4.1012	3.9995	3.9472	3.8940	3.8398	3.7844	3.7279	3.6702
9	3.9639	3.8682	3.7694	3.6669	3.6142	3.5604	3.5055	3.4493	3.3918	3.3329
10	3.7168	3.6209	3.5217	3.4186	3.3654	3.3110	3.2554	3.1984	3.1399	3.0798
11	3.5257	3.4296	3.3299	3.2261	3.1725	3.1176	3.0613	3.0035	2.9441	2.8828
12	3.3736	3.2773	3.1772	3.0728	3.0187	2.9633	2.9063	2.8478	2.7874	2.7249
13	3.2497	3.1532	3.0527	2.9477	2.8932	2.8373	2.7797	2.7204	2.6590	2.5955
14	3.1469	3.0501	2.9493	2.8437	2.7888	2.7324	2.6742	2.6142	2.5519	2.4872
15	3.0602	2.9633	2.8621	2.7559	2.7006	2.6437	2.5850	2.5242	2.4611	2.3953
16	2.9862	2.8890	2.7875	2.6808	2.6252	2.5678	2.5085	2.4471	2.3831	2.3163
17	2.9222	2.8249	2.7230	2.6158	2.5598	2.5021	2.4422	2.3801	2.3153	2.2474
18	2.8664	2.7689	2.6667	2.5590	2.5027	2.4445	2.3842	2.3214	2.2558	2.1869
19	2.8173	2.7196	2.6171	2.5089	2.4523	2.3937	2.3329	2.2695	2.2032	2.1333
20	2.7737	2.6758	2.5731	2.4645	2.4076	2.3486	2.2873	2.2234	2.1562	2.0853
21	2.7348	2.6368	2.5338	2.4247	2.3675	2.3082	2.2465	2.1819	2.1141	2.0422
22	2.6998	2.6017	2.4984	2.3890	2.3315	2.2718	2.2097	2.1446	2.0760	2.0032
23	2.6682	2.5699	2.4665	2.3567	2.2989	2.2389	2.1763	2.1107	2.0415	1.9677
24	2.6396	2.5412	2.4374	2.3273	2.2693	2.2090	2.1460	2.0799	2.0099	1.9353
25	2.6135	2.5149	2.4110	2.3005	2.2422	2.1816	2.1183	2.0517	1.9811	1.9055
26	2.5895	2.4909	2.3867	2.2759	2.2174	2.1565	2.0928	2.0257	1.9545	1.8781
27	2.5676	2.4688	2.3644	2.2533	2.1946	2.1334	2.0693	2.0018	1.9299	1.8527
28	2.5473	2.4484	2.3438	2.2324	2.1735	2.1121	2.0477	1.9796	1.9072	1.8291
29	2.5286	2.4295	2.3248	2.2131	2.1540	2.0923	2.0276	1.9591	1.8861	1.8072
30	2.5112	2.4120	2.3072	2.1952	2.1359	2.0739	2.0089	1.9400	1.8664	1.7867
40	2.3882	2.2882	2.1819	2.0677	2.0069	1.9429	1.8752	1.8028	1.7242	1.6371
60	2.2702	2.1692	2.0613	1.9445	1.8817	1.8152	1.7440	1.6668	1.5810	1.4822
120	2.1570	2.0548	1.9450	1.8249	1.7597	1.6899	1.6141	1.5299	1.4327	1.3104
∞	2.0483	1.9447	1.8326	1.7085	1.6402	1.5660	1.4835	1.3883	1.2684	1.0000

(Continued)

Table 8 (*Continued*)

F-Distribution: 1% Points

v_2 \ v_1	1	2	3	4	5	6	7	8	9
1	4052.2	4999.5	5403.3	5624.6	5763.7	5859.0	5928.3	5981.6	6022.5
2	98.503	99.000	99.166	99.249	99.299	99.332	99.356	99.374	99.388
3	34.116	30.817	29.457	28.710	28.237	27.911	27.672	27.489	27.345
4	21.198	18.000	16.694	15.977	15.522	15.207	14.976	14.799	14.659
5	16.258	13.274	12.060	11.392	10.967	10.672	10.456	10.289	10.158
6	13.745	10.925	9.7795	9.1483	8.7459	8.4661	8.2600	8.1016	7.9761
7	12.246	9.5466	8.4513	7.8467	7.4604	7.1914	6.9928	6.8401	6.7188
8	11.259	8.6491	7.5910	7.0060	6.6318	6.3707	6.1776	6.0289	5.9106
9	10.561	8.0215	6.9919	6.4221	6.0569	5.8018	5.6129	5.4671	5.3511
10	10.044	7.5594	6.5523	5.9943	5.6363	5.3858	5.2001	5.0567	4.9424
11	9.6460	7.2057	6.2167	5.6683	5.3160	5.0692	4.8861	4.7445	4.6315
12	9.3302	6.9266	5.9526	5.4119	5.0643	4.8206	4.6395	4.4994	4.3875
13	9.0738	6.7010	5.7394	5.2053	4.8616	4.6204	4.4410	4.3021	4.1911
14	8.8616	6.5149	5.5639	5.0354	4.6950	4.4558	4.2779	4.1399	4.0297
15	8.6831	6.3589	5.4170	4.8932	4.5556	4.3183	4.1415	4.0045	3.8948
16	8.5310	6.2262	5.2922	4.7726	4.4374	4.2016	4.0259	3.8896	3.7804
17	8.3997	6.1121	5.1850	4.6690	4.3359	4.1015	3.9267	3.7910	3.6822
18	8.2854	6.0129	5.0919	4.5790	4.2479	4.0146	3.8406	3.7054	3.5971
19	8.1850	5.9259	5.0103	4.5003	4.1708	3.9386	3.7653	3.6305	3.5225
20	8.0960	5.8489	4.9382	4.4307	4.1027	3.8714	3.6987	3.5644	3.4567
21	8.0166	5.7804	4.8740	4.3688	4.0421	3.8117	3.6396	3.5056	3.3981
22	7.9454	5.7190	4.8166	4.3134	3.9880	3.7583	3.5867	3.4530	3.3458
23	7.8811	5.6637	4.7649	4.2635	3.9392	3.7102	3.5390	3.4057	3.2986
24	7.8229	5.6136	4.7181	4.2184	3.8951	3.6667	3.4959	3.3629	3.2560
25	7.7698	5.5680	4.6755	4.1774	3.8550	3.6272	3.4568	3.3239	3.2172
26	7.7213	5.5263	4.6366	4.1400	3.8183	3.5911	3.4210	3.2884	3.1818
27	7.6767	5.4881	4.6009	4.1056	3.7848	3.5580	3.3882	3.2558	3.1494
28	7.6356	5.4529	4.5681	4.0740	3.7539	3.5276	3.3581	3.2259	3.1195
29	7.5976	5.4205	4.5378	4.0449	3.7254	3.4995	3.3302	3.1982	3.0920
30	7.5625	5.3904	4.5097	4.0179	3.6990	3.4735	3.3045	3.1726	3.0665
40	7.3141	5.1785	4.3126	3.8283	3.5138	3.2910	3.1238	2.9930	2.8876
60	7.0771	4.9774	4.1259	3.6491	3.3389	3.1187	2.9530	2.8233	2.7185
120	6.8510	4.7865	3.9493	3.4796	3.1735	2.9559	2.7918	2.6629	2.5586
∞	6.6349	4.6052	3.7816	3.3192	3.0173	2.8020	2.6393	2.5113	2.4073

(*Continued*)

Table 8 (*Continued*)

F-Distribution: 1% Points (*Continued*)

ν_2 \ ν_1	10	12	15	20	24	30	40	60	120	∞
1	6055.8	6106.3	6157.3	6208.7	6234.6	6260.7	6286.8	6313.0	6339.4	6366.0
2	99.399	99.416	99.432	99.449	99.458	99.466	99.474	99.483	99.491	99.501
3	27.229	27.052	26.872	26.690	26.598	26.505	26.411	26.316	26.221	26.125
4	14.546	14.374	14.198	14.020	13.929	13.838	13.745	13.652	13.558	13.463
5	10.051	9.8883	9.7222	9.5527	9.4665	9.3793	9.2912	9.2020	9.1118	9.0204
6	7.8741	7.7183	7.5590	7.3958	7.3127	7.2285	7.1432	7.0568	6.9690	6.8801
7	6.6201	6.4691	6.3143	6.1554	6.0743	5.9921	5.9084	5.8236	5.7372	5.6495
8	5.8143	5.6668	5.5151	5.3591	5.2793	5.1981	5.1156	5.0316	4.9460	4.8588
9	5.2565	5.1114	4.9621	4.8080	4.7290	4.6486	4.5667	4.4831	4.3978	4.3105
10	4.8492	4.7059	4.5582	4.4054	4.3269	4.2469	4.1653	4.0819	3.9965	3.9090
11	4.5393	4.3974	4.2509	4.0990	4.0209	3.9411	3.8596	3.7761	3.6904	3.6025
12	4.2961	4.1553	4.0096	3.8584	3.7805	3.7008	3.6192	3.5355	3.4494	3.3608
13	4.1003	3.9603	3.8154	3.6646	3.5868	3.5070	3.4253	3.3413	3.2548	3.1654
14	3.9394	3.8001	3.6557	3.5052	3.4274	3.3476	3.2656	3.1813	3.0942	3.0040
15	3.8049	3.6662	3.5222	3.3719	3.2940	3.2141	3.1319	3.0471	2.9595	2.8684
16	3.6909	3.5527	3.4089	3.2588	3.1808	3.1007	3.0182	2.9330	2.8447	2.7528
17	3.5931	3.4552	3.3117	3.1615	3.0835	3.0032	2.9205	2.8348	2.7459	2.6530
18	3.5082	3.3706	3.2273	3.0771	2.9990	2.9185	2.8354	2.7493	2.6597	2.5660
19	3.4338	3.2965	3.1533	3.0031	2.9249	2.8422	2.7608	2.6742	2.5839	2.4893
20	3.3682	3.2311	3.0880	2.9377	2.8594	2.7785	2.6947	2.6077	2.5168	2.4212
21	3.3098	3.1729	3.0299	2.8796	2.8011	2.7200	2.6359	2.5484	2.4568	2.3603
22	3.2576	3.1209	2.9780	2.8274	2.7488	2.6675	2.5831	2.4951	2.4029	2.3055
23	3.2106	3.0740	2.9311	2.7805	2.7017	2.6202	2.5355	2.4471	2.3542	2.2559
24	3.1681	3.0316	2.8887	2.7380	2.6591	2.5773	2.4923	2.4035	2.3099	2.2107
25	3.1294	2.9931	2.8502	2.6993	2.6203	2.5383	2.4530	2.3637	2.2695	2.1694
26	3.0941	2.9579	2.8150	2.6640	2.5848	2.5026	2.4170	2.3273	2.2325	2.1315
27	3.0618	2.9256	2.7827	2.6316	2.5522	2.4699	2.3840	2.2938	2.1984	2.0965
28	3.0320	2.8959	2.7530	2.6017	2.5223	2.4397	2.3535	2.2629	2.1670	2.0642
29	3.0045	2.8685	2.7256	2.5742	2.4946	2.4118	2.3253	2.2344	2.1378	2.0342
30	2.9791	2.8431	2.7002	2.5487	2.4689	2.3860	2.2992	2.2079	2.1107	2.0062
40	2.8005	2.6648	2.5216	2.3689	2.2880	2.2034	2.1142	2.0194	1.9172	1.8047
60	2.6318	2.4961	2.3523	2.1978	2.1154	2.0285	1.9360	1.8363	1.7263	1.6006
120	2.4721	2.3363	2.1915	2.0346	1.9500	1.8600	1.7628	1.6557	1.5330	1.3805
∞	2.3209	2.1848	2.0385	1.8783	1.7908	1.6964	1.5923	1.4730	1.3246	1.0000

(*Continued*)

Table 8 (*Continued*)

F-Distribution: .5% Points

v_2 \ v_1	1	2	3	4	5	6	7	8	9
1	16211	20000	21615	22500	23056	23437	23715	23925	24091
2	198.50	199.00	199.17	199.25	199.30	199.33	199.36	199.37	199.39
3	55.552	49.799	47.467	46.195	45.392	44.838	44.434	44.126	43.882
4	31.333	26.284	24.259	23.155	22.456	21.975	21.622	21.352	21.139
5	22.785	18.314	16.530	15.556	14.940	14.513	14.200	13.961	13.772
6	18.635	14.544	12.917	12.028	11.464	11.073	10.786	10.566	10.391
7	16.236	12.404	10.882	10.050	9.5221	9.1554	8.8854	8.6781	8.5138
8	14.688	11.042	9.5965	8.8051	8.3018	7.9520	7.6942	7.4960	7.3386
9	13.614	10.107	8.7171	7.9559	7.4711	7.1338	6.8849	6.6933	6.5411
10	12.826	9.4270	8.0807	7.3428	6.8723	6.5446	6.3025	6.1159	5.9676
11	12.226	8.9122	7.6004	6.8809	6.4217	6.1015	5.8648	5.6821	5.5368
12	11.754	8.5096	7.2258	6.5211	6.0711	5.7570	5.5245	5.3451	5.2021
13	11.374	8.1865	6.9257	6.2335	5.7910	5.4819	5.2529	5.0761	4.9351
14	11.060	7.9217	6.6803	5.9984	5.5623	5.2574	5.0313	4.8566	4.7173
15	10.798	7.7008	6.4760	5.8029	5.3721	5.0708	4.8473	4.6743	4.5364
16	10.575	7.5138	6.3034	5.6378	5.2117	4.9134	4.6920	4.5207	4.3838
17	10.384	7.3536	6.1556	5.4967	5.0746	4.7789	4.5594	4.3893	4.2535
18	10.218	7.2148	6.0277	5.3746	4.9560	4.6627	4.4448	4.2759	4.1410
19	10.073	7.0935	5.9161	5.2681	4.8526	4.5614	4.3448	4.1770	4.0428
20	9.9439	6.9865	5.8177	5.1743	4.7616	4.4721	4.2569	4.0900	3.9564
21	9.8295	6.8914	5.7304	5.0911	4.6808	4.3931	4.1789	4.0128	3.8799
22	9.7271	6.8064	5.6524	5.0168	4.6088	4.3225	4.1094	3.9440	3.8116
23	9.6348	6.7300	5.5823	4.9500	4.5441	4.2591	4.0469	3.8822	3.7502
24	9.5513	6.6610	5.5190	4.8898	4.4857	4.2019	3.9905	3.8264	3.6949
25	9.4753	6.5982	5.4615	4.8351	4.4327	4.1500	3.9394	3.7758	3.6447
26	9.4059	6.5409	5.4091	4.7852	4.3844	4.1027	3.8928	3.7297	3.5989
27	9.3423	6.4885	5.3611	4.7396	4.3402	4.0594	3.8501	3.6875	3.5571
28	9.2838	6.4403	5.3170	4.6977	4.2996	4.0197	3.8110	3.6487	3.5186
29	9.2297	6.3958	5.2764	4.6591	4.2622	3.9830	3.7749	3.6130	3.4832
30	9.1797	6.3547	5.2388	4.6233	4.2276	3.9492	3.7416	3.5801	3.4505
40	8.8278	6.0664	4.9759	4.3738	3.9860	3.7129	3.5088	3.3498	3.2220
60	8.4946	5.7950	4.7290	4.1399	3.7600	3.4918	3.2911	3.1344	3.0083
120	8.1790	5.5393	4.4973	3.9207	3.5482	3.2849	3.0874	2.9330	2.8083
∞	7.8794	5.2983	4.2794	3.7151	3.3499	3.0913	2.8968	2.7444	2.6210

(*Continued*)

Table 8 (*Continued*)

F-Distribution: .5% Points (*Continued*)

ν_2 \ ν_1	10	12	15	20	24	30	40	60	120	∞
1	24224	24426	24630	24836	24940	25044	25148	25253	25359	25465
2	199.40	199.42	199.43	199.45	199.46	199.47	199.47	199.48	199.49	199.51
3	43.686	43.387	43.085	42.778	42.622	42.466	42.308	42.149	41.989	41.829
4	20.967	20.705	20.438	20.167	20.030	19.892	19.752	19.611	19.468	19.325
5	13.618	13.384	13.146	12.903	12.780	12.656	12.530	12.402	12.274	12.144
6	10.250	10.034	9.8140	9.5888	9.4741	9.3583	9.2408	9.1219	9.0015	8.8793
7	8.3803	8.1764	7.9678	7.7540	7.6450	7.5345	7.4225	7.3088	7.1933	7.0760
8	7.2107	7.0149	6.8143	6.6082	6.5029	6.3961	6.2875	6.1772	6.0649	5.9505
9	6.4171	6.2274	6.0325	5.8318	5.7292	5.6248	5.5186	5.4104	5.3001	5.1875
10	5.8467	5.6613	5.4707	5.2740	5.1732	5.0705	4.9659	4.8592	4.7501	4.6385
11	5.4182	5.2363	5.0489	4.8552	4.7557	4.6543	4.5508	4.4450	4.3367	4.2256
12	5.0855	4.9063	4.7214	4.5299	4.4315	4.3309	4.2282	4.1229	4.0149	3.9039
13	4.8199	4.6429	4.4600	4.2703	4.1726	4.0727	3.9704	3.8655	3.7577	3.6465
14	4.6034	4.4281	4.2468	4.0585	3.9614	3.8619	3.7600	3.6553	3.5473	3.4359
15	4.4236	4.2498	4.0698	3.8826	3.7859	3.6867	3.5850	3.4803	3.3722	3.2602
16	4.2719	4.0994	3.9205	3.7342	3.6378	3.5388	3.4372	3.3324	3.2240	3.1115
17	4.1423	3.9709	3.7929	3.6073	3.5112	3.4124	3.3107	3.2058	3.0971	2.9839
18	4.0305	3.8599	3.6827	3.4977	3.4017	3.3030	3.2014	3.0962	2.9871	2.8732
19	3.9329	3.7631	3.5866	3.4020	3.3062	3.2075	3.1058	3.0004	2.8908	2.7762
20	3.8470	3.6779	3.5020	3.3178	3.2220	3.1234	3.0215	2.9159	2.8058	2.6904
21	3.7709	3.6024	3.4270	3.2431	3.1474	3.0488	2.9467	2.8408	2.7302	2.6140
22	3.7030	3.5350	3.3600	3.1764	3.0807	2.9821	2.8799	2.7736	2.6625	2.5455
23	3.6420	3.4745	3.2999	3.1165	3.0208	2.9221	2.8198	2.7132	2.6016	2.4837
24	3.5870	3.4199	3.2456	3.0624	2.9667	2.8679	2.7654	2.6585	2.5463	2.4276
25	3.5370	3.3704	3.1963	3.0133	2.9176	2.8187	2.7160	2.6088	2.4960	2.3765
26	3.4916	3.3252	3.1515	2.9685	2.8728	2.7738	2.6709	2.5633	2.4501	2.3297
27	3.4499	3.2839	3.1104	2.9275	2.8318	2.7327	2.6296	2.5217	2.4078	2.2867
28	3.4117	3.2460	3.0727	2.8899	2.7941	2.6949	2.5916	2.4834	2.3689	2.2469
29	3.3765	3.2111	3.0379	2.8551	2.7594	2.6601	2.5565	2.4479	2.3330	2.2102
30	3.3440	3.1787	3.0057	2.8230	2.7272	2.6278	2.5241	2.4151	2.2997	2.1760
40	3.1167	2.9531	2.7811	2.5984	2.5020	2.4015	2.2958	2.1838	2.0635	1.9318
60	2.9042	2.7419	2.5705	2.3872	2.2898	2.1874	2.0789	1.9622	1.8341	1.6885
120	2.7052	2.5439	2.3727	2.1881	2.0890	1.9839	1.8709	1.7469	1.6055	1.4311
∞	2.5188	2.3583	2.1868	1.9998	1.8983	1.7891	1.6691	1.5325	1.3637	1.0000

Table 9 Random Numbers

	00 04	05 09	10 14	15 19	20 24	25 29	30 34	35 39	40 44	45 49
00	68229	08998	86193	71307	23918	15007	05652	14729	33615	49324
01	67232	50746	98262	69386	36565	90714	28469	65634	71367	05183
02	06065	26307	51532	35490	90578	87874	17211	22335	26452	73616
03	42880	32240	45422	23910	35440	81252	43224	08303	18579	54828
04	18234	71762	77571	48693	64240	37654	86799	82851	42137	13571
05	40013	21962	28509	18452	63858	62376	49896	05761	18509	35900
06	98262	44042	65063	85998	06634	57981	44240	25840	85498	94414
07	89610	02030	88376	56127	05906	66872	90616	94147	84270	03202
08	40461	32082	53311	97046	78443	42070	60277	32590	73717	38658
09	91804	03047	82495	12382	40074	42578	07082	76760	54521	47275
10	58671	10180	87184	40342	20908	11477	98569	08212	87716	73083
11	94247	64833	26516	85193	73917	94664	43665	11657	43098	91295
12	15686	44790	28320	21253	00654	30964	53126	92245	09397	13182
13	88880	20807	95456	49139	48137	49527	51817	23447	82440	30145
14	84763	40564	21984	78880	84867	15927	55377	79590	45714	13119
15	59758	84745	16600	55997	40190	27753	74270	64619	62954	94793
16	62232	10205	88867	94699	12943	30192	30723	95907	34665	22728
17	08499	26818	27026	50751	28900	95339	57406	99476	58436	27082
18	51343	60528	56060	12804	57461	30276	68753	48307	57723	05294
19	93496	35287	75714	43706	45551	64593	95638	53042	30511	32147
20	33575	62869	12285	03652	45157	32253	71347	87869	85371	63429
21	29342	33444	49059	87355	81796	42695	26367	21109	74259	10305
22	08664	65565	56609	39014	57041	14870	27535	97339	72882	97037
23	41145	12766	63409	06061	50999	17519	76902	32303	23637	68157
24	71680	08843	14602	30941	19265	00435	20357	93917	81399	60230
25	94145	74528	54109	83619	28247	58659	08099	38772	62627	73023
26	02035	30805	61982	16812	83532	54932	17137	58312	66804	69348
27	81096	19599	48288	35928	03790	95981	96859	03439	50101	38080
28	24820	51463	35250	34029	16197	22205	83961	63607	04505	94207
29	98834	64842	65381	53679	81008	39411	04232	70081	00313	56998
30	71163	81649	88195	00030	42378	11554	44461	35731	27261	63799
31	02720	47806	69858	29738	96658	88667	71124	90451	10516	07672
32	96816	22586	41639	48811	07737	05318	06832	63962	71193	06997
33	23224	06157	59494	80321	85513	59627	29207	57618	36384	44230
34	35516	41589	63715	64144	94055	54907	08046	72167	98848	95625
35	48863	13992	10936	32041	16325	87121	44992	33185	42393	34069
36	97638	40220	55444	11156	54966	69564	86883	29969	58881	80134
37	37587	34049	43124	70667	07812	76084	64856	64545	84704	85650
38	53545	93561	78487	00329	41493	47627	69016	30150	38946	46043
39	09917	16646	93963	70280	76646	44280	17876	61714	51400	15567
40	16064	97674	17001	66299	22585	81589	85862	80992	93901	31752
41	50861	48240	64988	52370	99082	68486	06361	10755	83475	80760
42	40407	51269	98199	53741	14811	36587	00352	19430	80609	88731
43	68136	58422	18981	93782	10757	96648	08635	90170	69669	98405
44	57193	74393	41359	39630	56701	02509	82582	18022	79156	90751
45	37407	91572	63117	01994	77916	05970	47143	28564	47496	75676
46	95213	12432	41682	85747	01216	87158	12422	74845	66795	25446
47	93432	31205	45210	83507	19495	06972	08837	04636	26365	08770
48	66919	80926	47584	99241	59864	99639	31889	72827	37397	69602
49	63644	59862	57733	69506	97059	06279	06218	24554	98734	23213

(Continued)

Table 9 (Continued)

	50 54	55 59	60 64	65 69	70 74	75 79	80 84	85 89	90 94	95 99
00	78732	18312	52903	62757	88599	34661	30461	67325	28811	90689
01	10994	35478	82290	55820	90996	86108	60581	49336	68732	66248
02	93019	59363	94827	19766	66449	63825	39403	96792	81098	79981
03	21997	96881	40624	67908	93058	87431	13401	37430	51171	77501
04	83236	17699	21456	05955	94839	57123	46685	39332	12938	12890
05	77642	55094	27974	28812	84582	26096	87176	84397	29346	78445
06	80686	35166	80409	15451	11229	15451	75164	55392	47124	86712
07	99476	47017	14018	63550	63702	11234	20116	75220	10985	67390
08	64657	32193	13654	55536	91991	97331	98656	66564	68885	07558
09	36330	23444	46671	17362	41985	14481	36070	18079	05510	85742
10	58812	76602	09948	13670	15360	82809	59179	47708	13437	77647
11	49610	13757	07636	82178	22666	57588	80251	47220	23699	63887
12	77682	17803	56038	33672	49930	33097	03212	43971	01668	59218
13	98398	48112	12595	33487	22942	42917	95100	37650	85678	38535
14	94650	67741	52491	61473	17095	45630	31155	39340	79071	33405
15	60330	02983	86561	76410	88351	30555	81784	64257	24645	03310
16	65407	19178	94797	19020	00995	60721	45367	40787	27764	18471
17	00737	13431	82136	33344	19178	58796	27014	20732	04580	11127
18	82321	17397	57259	85307	33810	72539	10771	99980	48023	97617
19	27052	45856	82093	39215	60639	21212	46465	36311	01025	41180
20	70416	77623	43151	74525	90056	87966	56141	65080	09455	38433
21	05314	14546	92597	85473	44145	24826	53097	12814	22999	84527
22	58779	27806	79115	65079	24657	27296	64749	37886	86099	09630
23	10182	79043	91485	65440	68839	78053	70764	45766	79583	78523
24	98241	78299	97000	63365	38453	21293	54635	41264	53661	06865
25	14564	44708	46448	82225	41869	16671	77654	46660	08046	04977
26	37773	57523	11693	30106	95421	90526	57168	83044	16445	68171
27	80749	69743	42172	37965	27844	54761	39589	60312	90784	44703
28	44318	23974	19201	30557	17741	30277	96493	51328	70167	90578
29	49919	15939	51979	25474	07224	11151	99466	20192	72306	28873
30	43971	35349	86495	54780	89827	73873	74485	91104	04945	02558
31	86020	24836	97385	77020	83982	82619	97388	00344	02203	73133
32	43133	87886	29399	04890	97693	58128	72217	12943	68215	17855
33	26610	39929	02298	96321	97064	31302	29283	97898	95903	81018
34	15190	82397	80678	24623	52449	78915	58430	37639	86335	12416
35	19168	74692	06658	79183	23405	47723	96823	61999	76769	13933
36	60366	05860	35445	92210	48257	92167	06388	12327	19595	99390
37	26774	32108	68192	75434	65252	20082	08205	06300	76767	19887
38	52355	21502	92372	26111	34366	77653	47487	90545	61171	31944
39	06902	16776	21397	08146	73465	63956	87943	31235	97172	70914
40	57681	54796	22462	45560	34145	59656	18743	94348	32680	58816
41	24337	02003	26936	84715	98741	25673	37411	33753	46384	74234
42	26555	73470	21623	23906	45569	03888	59514	61529	94441	96103
43	70567	85562	11128	33413	57527	20531	45974	96576	54797	02684
44	44524	17652	27751	17804	17156	97187	52187	29044	49815	02464
45	62114	24707	78445	69160	78153	49941	37079	85896	96680	01535
46	74315	72014	15216	05785	47610	18554	30990	44005	67003	83168
47	90764	96186	36767	74342	00245	47347	67799	35085	21945	42284
48	58730	80875	39991	32706	12408	66678	10777	63190	85167	10633
49	40823	54503	46751	42376	63223	69125	03024	50685	18715	26785

(Continued)

Table 9 *(Continued)*

	00 04	05 09	10 14	15 19	20 24	25 29	30 34	35 39	40 44	45 49
50	31873	63103	33599	00799	51351	64884	85207	91264	27670	45644
51	22141	67169	27315	45917	92367	51346	23708	94772	37345	41623
52	20176	25244	31528	87367	08660	71980	35091	21022	24238	98602
53	86857	16357	52997	44133	82103	97920	77414	66942	45924	16451
54	68431	03784	07503	24526	83641	03239	35932	50349	78312	75057
55	64444	17044	40855	78079	76515	26228	64714	26472	41745	99146
56	37958	67527	63925	82635	10178	53634	28970	73752	78325	73836
57	03450	47672	38615	01545	46890	19727	43973	49033	56603	16510
58	77478	08441	57783	11701	17595	23352	91164	52251	59561	02762
59	11630	48074	27362	38994	01700	25060	46894	40842	71689	15030
60	50345	37759	99525	95850	36904	04201	33443	08772	80510	22958
61	15545	52216	60389	44513	96894	95997	66498	12310	79376	02309
62	75671	63882	05927	12792	83398	29802	89115	10118	46505	31217
63	40202	36687	83265	67884	56847	05073	31739	22137	61449	98936
64	47395	53833	46829	99363	05413	74953	87500	40405	74203	50057
65	38032	76008	77570	89185	57691	89308	85968	18425	60416	23721
66	65318	15803	90517	28245	82349	48469	17427	07520	13453	76716
67	85665	70013	19250	33732	19745	67770	75614	44693	48689	45307
68	28921	91563	50990	02780	61185	85558	25429	30094	55534	12415
69	64175	93462	68845	34137	94198	01354	57694	81280	30352	42749
70	41240	62609	06737	45076	08233	44034	35386	48703	13211	61071
71	64496	77463	08766	78085	53339	69447	96232	51394	58080	88589
72	17844	41441	14366	32425	06192	08350	77025	25872	48495	68944
73	22948	73692	37784	87083	10172	03571	86819	60408	21056	94903
74	37689	82552	59799	46186	62844	77490	76754	52779	27190	21966
75	23751	14167	64954	53808	29673	35744	98232	78421	22068	85108
76	45205	23841	72105	67255	95039	72870	23214	62882	43578	59316
77	42840	44533	02017	14673	70603	09526	21732	65620	99966	82620
78	26212	75885	77783	23126	30795	25888	46847	38364	44512	84910
79	44823	83232	55931	81180	46713	42727	56818	96489	94023	39765
80	08793	86769	48388	32706	79538	41374	57418	14169	18847	90319
81	94587	00394	68422	02276	68836	05988	93623	47465	97556	42856
82	93934	08842	64523	46333	44672	73571	73373	90140	73528	62014
83	88415	26922	95473	50309	20381	30004	29348	85914	61808	84033
84	71370	48662	18574	96278	06234	49047	87648	57139	65430	96679
85	49110	27215	17965	99626	96707	83762	95758	14660	63418	68936
86	93213	40041	47045	05952	63920	06758	43643	30698	23045	21344
87	94271	35880	86985	31606	27257	26691	76062	60855	76167	11704
88	26612	75937	93902	01752	62766	98678	85743	59475	41106	12411
89	74290	42170	18191	44615	51154	23101	49296	35399	04607	87174
90	71603	28926	21487	49504	92783	41070	32582	18682	29190	48032
91	73789	41606	10012	44967	10310	65660	90853	40683	65991	53301
92	60855	93855	37427	40003	01689	98136	61685	23146	34671	86901
93	79606	91469	07831	90267	11023	61706	04001	01093	66401	65195
94	25460	34006	14941	48347	40090	38544	15037	67926	20062	87308
95	56903	92929	79676	89725	42631	58928	49449	70983	56168	71252
96	12146	00048	74810	14783	84569	75047	39573	93673	74035	72879
97	94531	47518	20983	76301	59893	97841	68230	62142	28772	78737
98	48028	73059	69015	33312	07943	56595	81233	78801	22998	10985
99	04156	82157	88026	25766	06715	89769	34550	08000	34492	41526

(Continued)

Table 9 (Continued)

	50 54	55 59	60 64	65 69	70 74	75 79	80 84	85 89	90 94	95 99
50	02728	78462	21117	34097	06938	20560	54318	91515	22133	53898
51	70510	45796	50202	51360	32592	88775	72756	27634	17536	51201
52	43571	56411	65856	39723	37566	33106	32106	78206	67602	40689
53	47992	37904	81681	64569	51039	13749	85192	50351	08854	26532
54	14278	49680	33320	62979	84817	97941	02195	95438	31765	90435
55	89773	68630	90107	17128	80218	01552	63681	18254	88040	89270
56	45236	08672	17720	66165	01252	12996	05783	76948	26478	81310
57	45520	53982	17196	10324	56195	18288	83355	01557	36814	48756
58	84823	85885	45142	31607	88748	18822	44706	45530	69631	21862
59	07075	89724	21446	87135	33187	20229	25424	59486	65706	27849
60	37697	66987	26463	83721	36432	45576	59886	13895	64199	44836
61	90323	22653	24817	12161	50484	45834	33548	92714	43098	22900
62	45474	99519	57172	06376	80336	27401	41960	82000	53489	29788
63	35094	38418	16217	94728	56071	08351	71264	66881	21338	06960
64	09194	58417	45344	93279	07897	69706	43388	63670	74888	74436
65	58937	71374	70193	65228	35686	69789	10749	61962	48474	24356
66	49962	00477	71365	00706	85145	69880	17792	04538	73235	00575
67	33548	11560	24275	59033	00309	17444	93429	14729	76163	61711
68	10937	60025	28105	88524	86107	48263	03710	82805	81092	08264
69	33515	45035	64827	41351	78496	00663	29145	38510	46758	40290
70	54792	53652	52923	91456	97647	38756	27928	37399	83458	23302
71	82729	01641	77468	66318	50038	16718	47088	83059	64063	17201
72	66975	12412	56415	11996	95954	36480	41624	87942	19104	52035
73	39723	14404	31325	34590	53869	91235	87235	43350	45277	48288
74	94960	08077	09195	18184	48417	64811	23529	78634	09240	92986
75	02951	95309	92881	23356	95263	58837	73851	60235	58032	42702
76	30084	87527	01013	17513	72556	94284	14728	86976	06627	49718
77	10042	38840	41439	81233	36368	84579	57035	81867	42359	35846
78	37780	00668	17776	38231	41445	16139	05693	74800	09062	26386
79	73933	08029	50260	49364	97712	91738	19418	69293	06370	02419
80	17921	85605	65998	27509	55353	58995	13325	46135	80608	10635
81	99051	63933	70332	10947	72064	94866	82567	11893	43212	17849
82	37356	81201	52109	95428	41325	19753	17668	72038	15020	04804
83	02069	22123	62648	44966	51172	18605	31843	18805	27380	22504
84	36995	97381	66660	96284	13375	80995	98497	76115	14289	86890
85	68847	45675	98858	57403	23525	65148	34182	48024	74106	37595
86	74386	62359	03569	98024	67622	19481	25255	63646	63770	04825
87	28538	37320	04709	66034	73976	96061	55204	80889	95058	37555
88	41442	21290	86564	12935	07735	56487	14019	36630	90154	44434
89	78174	33895	40566	84051	71371	02356	83374	61516	31752	01708
90	05916	54673	03478	88646	22400	02080	92922	30088	88212	38658
91	81634	54650	62367	76283	99403	99381	48368	90058	72755	33581
92	54499	17825	87417	22515	24149	80946	67546	87764	62636	71838
93	03970	89957	56078	91811	13864	04459	73573	82248	94740	25365
94	29046	04794	69304	96869	33321	87568	60292	26280	41665	65293
95	86251	13475	54071	18827	07339	28678	69810	38641	41581	45135
96	76593	87699	75105	78490	05760	91493	69730	83470	71347	59713
97	64350	59161	09034	91122	03314	81903	29083	55417	66056	34769
98	09819	08754	89759	90070	31090	26094	82655	67232	70423	63221
99	68159	10344	60298	26022	89162	51999	09008	17316	59998	96420

Index

A

Active groups, 2, 12
Addition principle, 115, 116
Alternative, two-sided, 270
Alternative hypothesis, 264, 266
Among groups, 411
Analysis-of-variance table, 409–413
Answers to odd-numbered problems, 465–478
Approximate tests, 333–360
    binomial data, 341–344
    contingency tables, 350–358
    goodness-of-fit test, 344–350
    hypotheses about multinomial parameters, 335–341
    multinomial distribution, 334
Arithmetic mean, 50
Association, 456–461
    measure of, 457
    negative, 458
    positive, 458
Assumptions, relaxing, 311–312
Asymmetrical distribution, 61
Average, 50

B

Bar charts, 24
Bayes' theorem, 144–146
Binomial, normal approximation to, 194–202
Binomial coefficients, 135
Binomial data, 341–344
Binomial distribution, 189–194
    approximation of, 194
    95 percent confidence intervals, 484
Binomial p:
    estimating, 252–258
    hypotheses on, 275–278
Binomial population, 190–191
    two samples from, 321–328
Binomial probabilities, 135–139
    cumulative, 480
Binomial random variable, 189
Bins, 12–13
    boundaries, 14–15, 19
    frequency, 15
    intervals, 15, 16–18
    limits, 14, 19
    mark, 15
Blocking principle, 431–432

C

Cause, correlation and, 395–398
Cause-and-effect relationship, 396
Central limit theorem, 221–225
Central tendency, 49
Change of scale, 70–75
Charts, bar, 24
Chi-square distribution, 336–337
    percentage points of, 485
Chi-square statistic, 335–336
Coding, 70–75
Coefficient:
    correlation, 393
    of determination, 385

Coefficient (*cont.*):
  rank correlation, 457–458
  sample correlation, 393
Combinations, 121, 125–128
Combinatorial analysis, 128
Comparisons:
  group, 312
  paired, 312–318
Complementary events, 106–107
Completely randomized designs, 295, 407–408
Compound bar chart, 24
Computation, 7–8
Conditional probability, 107–108
Confidence intervals, 236
  for μ_i, 300–304
  for $\mu_1 - \mu_2$, 300–304
Confidence level, 237
Contingency tables, 350–358
Continuous random variables, 156
Control groups, 2, 12
Correlation, 392–398
  cause and, 395–398
  negative, 393
  positive, 393
Correlation coefficient, 393
  rank, 457–458
  sample, 393
Counting, 115–133
CPI (Consumer Price Index), 85, 88
Critical region, 265
Cumulative binomial probabilities, 480
Cumulative distribution function, 164–165
Cumulative frequency distributions, 21–24
Curves, frequency, 161

D

Data, 2
Degrees of freedom, 241
Dependent variable, 364
Descriptive measures, 6–7, 44–92
  coding (change of scale), 70–75
  index numbers, 83–89
  the mean, 50–53
  means and variance for grouped data, 76–81
  measures of location, 49–50
  selecting, 61–62
  measures of variance, 62
  the median, 56–59
  the mode, 59–61
  quick measures of location and spread, 81–83
  the range, 63
  standard deviation, 63–69, 171–173
  summation notation, 44–49
  symbols, 44–49
  the variance, 63–69, 171–173
  the weighted mean, 53–56
Descriptive statistics, *see* Descriptive measures
Designs, 5
  completely randomized, 295, 407–408
Determination, coefficient of, 385
Deviation:
  mean, 64
  standard, 63–69, 171–173
Discrete random variables, 156
Disjoint events, 104, 111
Dispersion, measures of, 62
Distribution, 159
  binomial, 189–194
    approximation of, 194
    95 percent confidence intervals for, 484
  chi-square, 336–337
    percentage points of, 485
  frequency
    cumulative, 21–24
    empirical, 12–40
  geometric, 168
  hypergeometric, 174–175
  multimodal, 59
  multinomial, 334
  normal, 178–188
  percentage, 15
  Poisson, 202–206
  population, 37
  sampling, 210–226
    central limit theorem, 221–225
    expected values and variance, 212–216
    from normal populations, 217–218
    sampling and inference, 210–212
    standardized sample mean, 219–220
  standard normal, areas of, 481
  symmetrical, 61

INDEX

    t, 241–243
    theoretical, 37
    unimodal, 59
Distribution function, cumulative, 164–165
Double blind, 6

E

$e^{-\mu}$, values of, 482
Empirical frequency distributions, 12–40
Equal means, hypothesis of, 419–423
Errors, 411, 412
    experimental, 430–431
    standard, 232
    Type I, 281–285
    Type II, 281–285
Estimate, the, 231
Estimation, 230–260
    confidence intervals for normal means, 234–241, 243–249
    estimating binomial p, 252–258
    known variance, 234–241
    problem of, 230
    sample size, 249–251
    t-distribution, 241–243
    unknown variance, 243–249
Estimator, the, 231
    minimum-variance unbiased, 232–233
    pooled variance, 297
    unbiased, 231–232
Event, 99
    probability of, 101–102
Expected mean square (EMS), 420
Expected values, 169–176, 212–216, 335
Experimental error, 430–431

F

Factorial notation, 124
Favorable cases, 99
F-distributions, 421–422
    percentage points of, 487–494
Finite-population correction factor, 215
Finite populations, 152–153
Finite sampling, 141–144

Fixed model, 409
Fraction of successes, 252
Freedom, degrees of, 241
Frequency curve, 161
Frequency distributions, 12–21
    cumulative, 21–24
    empirical, 12–40

G

Generalization, inductive, 7
Geometric distribution, 168
Goodness-of-fit test, 344–350
Gosset, W. S., 245 n
Graphs, 24–35
    histograms, 24–27
    ogives, 24, 27–31
Group comparisons, 312
Grouped data, means and variances for, 76–81
Group effects, 409
Groups:
    among, 411
    into pairs, 312–313
    versus pairs, 318–320
    within, 411
Group sizes, unequal, effects of, 424–430

H

Histograms, 24–27
Homogeneity, 355
Horizontal bar chart, 24
Hypergeometric distribution, 174–175
Hypertension study, 2–8, 12, 230, 294, 295
Hypothesis:
    alternative, 264, 266
    null, 264, 266
Hypothesis $\mu_1 = \mu_2$, testing, 304–311
Hypothesis of equal means, 419–423
Hypothesis testing, 264–290
    on a binomial p, 275–278
    on a normal mean, 267–275
    p-values, 278–281
    theory of testing, 281–288

I

Independence, 110–111, 351–352, 355
Independent trials, repeated, 133–141
Independent variable, 364
Index numbers, 83–89
Index of summation, 45
Inductive generalization, 7
Inference, 210–212
  about A and B, 374–377
Infinite populations, 152–153
Intercept of a line, 365
Interquartile range, 82

K

Known variance, 234–241
Kruskal-Wallis test, 462–463

L

Laspeyres index number, 88
Least-squares estimates, 398–399
Least-squares principle, 367–372
Level of the test, 265–266
Line:
  intercept of, 365
  slope of, 365
Linear regression, 364–367
Linear transformation, 71
Location, measures of, 49–50
  quick, 81–83
  selecting, 61–62
Lower quartile, 81–82

M

Machine formulas, 68
Market basket, 86
Mathematical model, 408–409
Mean, the, 50–53
  arithmetic, 50
  of the mean, 214
  nonparametric, 448
  normal, hypotheses on, 267–275
  standardized sample, 219–220
  weighted, 53–56
Mean deviation, 64
Means:
  nonparametric difference of, 454
  normal, confidence intervals for, 234–241, 243–249
  of two different populations, 294–295
Measure of association, 457
Measures of dispersion, 62
Measures of location, 49–50
  quick, 81–83
  selecting, 61–62
Measures of variation, 62
Median, the, 56–59, 446
Minimum-variance unbiased estimator, 232–233
Modal class, 59
Mode, the, 59–61
Model, the, 264
  fixed, 409
  mathematical, 408–409
  random, 409
μ_i, confidence intervals for, 300–304
$\mu_1 - \mu_2$, confidence intervals for, 300–304
Multimodal distribution, 59
Multinomial distribution, 334
Multinomial parameters, hypothesis about, 335–341
Multinomial populations, 334
Multiple regression, 386–392
Multiplication principle, 115, 116–117
Mutually exclusive events, 104, 111

N

Negative association, 458
Negative correlation, 393
Nonparametric difference of means, 454
Nonparametric mean, 448
Nonparametric methods, 445–463
  association, 456–461
  rank sum test, 452–456
  sign test, 451–452
  Wilcoxon test, 446–450, 451, 452
Normal approximation to the binomial, 194–202

INDEX

Normal distributions, 178–188
  standard, areas of, 481
Normal means:
  confidence intervals for, 234–241, 243–249
    known variance, 234–241
    unknown variance, 243–249
  hypotheses on, 267–275
Normal populations, sampling from, 217–218
Null hypothesis, 264, 266

O

Objective statistical analysis, 5
Observed frequencies, 335
Odd-numbered problems, answers to, 465–478
Ogives, 24, 27–31
One-sided alternative, 270
Open-ended interval, 19
Outcomes, 97–98
Outcome space, *see* Sample space

P

Paasche index number, 88
Paired comparisons, 312–318
Pairs:
  groups into, 312–313
  groups versus, 318–320
Parameters, 211
Partitioning sum of squares, 382–386
Percentage distribution, 15
Percentage point, upper, 237–238
Percentiles, the, 83
Permutations, 121, 122–128
Pictographs, 24
Pie diagrams, 24
Placebo, 2, 5–6, 12
Poisson distribution, 202–206
Pooled sum of squares, 296
Pooled variance estimator, 297
Population distributions, 37
Populations, 37, 152–153
  binomial, 190–191
    two samples from, 321–328
  finite, 152–153
  infinite, 152–153

multinomial, 334
normal, sampling from, 217–218
symmetric, 446–447
two different, means of, 294–295
two normal, samples from, 294–300
Positive association, 458
Positive correlation, 393
Posterior probability, 145
Power, 281–287
Prediction equation, 367, 378–379
Probability, 95–149
  addition of, 105
  Bayes' theorem, 144–146
  binomial, 135–139
    cumulative, 480
  computing, 102–115
  conditional, 107–108
  counting, 115–133
  finite sampling, 141–144
  meaning of, 97–102
  posterior, 145
  repeated independent trials, 133–141
  subjective, 144–146
  unconditional, 108
Probability density, 161
Probability of the event, 101–102
Problems, odd-numbered, answers to, 465–478
Product moment correlation coefficient, 392–393
Proportion of successes, 252
p-values, 278–281

Q

Quartile:
  lower, 81–82
  upper, 81–82

R

r, relationship between z and, 486
Randomization, 406–408
Randomized complete block designs, 313, 432–440
Randomized design, completely, 407–408
Random model, 409

Random numbers, 495–498
Random samples, 154
Random variables, 155–157
   binomial, 189
   continuous, 156
   discrete, 156
   distribution of, 157–169
   expected values of, 169–176
   sets of, 176–178
Range, the, 63
Rank correlation coefficient, 457–458
Ranks, 447–448
Rank sum, 453
Rank sum test, 452–456
Regression, 364–392
   inferences about A and B, 374–378
   least squares principle, 367–372
   linear, 364–367
   multiple, 386–392
   partitioning sum of squares, 382–386
   uses of, 378–382
   variance estimates, 372–374
Regression line, 367
Rejection region, 265
Relative frequency, 15
Relaxing assumptions, 311–312
Repeated independent trials, 133–141
Replication, 431

S

Sample correlation coefficient, 393
Samples, 153–155
   from binomial populations, 321–328
   from two normal populations, 294–300
Sample size, 249–251
Sample space, 98–99
Sampling, 210–212
   finite, 141–144
   from the normal populations, 217–218
   with replacement, 139
   without replacement, 139
Sampling distributions, 210–226
   central limit theorem, 221–225
   expected values and variance, 212–216
   from normal populations, 217–218
   sampling and inference, 210–212
   standardized sample means, 219–220
Scale, change of, 70–75
Scatter diagram, 364–365
Significance level, 279
Sign test, 451–452
Size of the test, 265–266
Slope of a line, 365
Spread, quick measures of, 81–83
Square of the sum, 47
Standard deviation, 63–69, 171–173
Standard error, 232
Standardized sample mean, 219–220
Standardized variables, 173, 184
Standard normal distribution, areas of, 481
Standard normal variable, 184
Statistical analysis, objective of, 5
Statistics, 211
   t, 241
Stem-and-leaf plots, 35–37
Subjective probability, 144–146
Subset, 99
Summary table, 295–296
Summation, index of, 45
Summation notation, 44–49
Summation symbol, 45
Sum of squares, 47, 65–67
   partitioning, 382–386
   pooled, 296
Symbols, 44–49
Symmetrical distribution, 61
Symmetric population, 446–447
Systematic sample, 153–154

T

t, values for given probability levels, 483
t-distribution, 241–243
Test, 264–265
   size of, 265–266
Testing, theory of, 281–288
Theoretical distributions, 37
Total variation, partitioning, 406
Treatment effects, 409
Treatments, 295, 411, 412
Trials, independent, repeated, 133–141

Triangle test, 264
Trimean, the, 82
t-statistics, 241
Two normal populations, samples from, 294–300
Two-sided alternative, 270
Type I error, 281–285
Type II error, 281–285

U

Unbiased estimator, 231–232
  minimum-variance, 232–233
Unconditional probability, 108
Uncorrelated variables, 393
Unequal group sizes, effects of, 424–430
Unimodal distribution, 59
U.S. Bureau of Labor Statistics Consumer Price Index (CPI), 85, 88
U.S. Public Health Service, hypertension study, 2–8, 12, 230, 294, 295
U.S. Veterans Administration, 6
Unknown variance, 243–249
Upper percentage point, 237–238
Upper quartile, 81–82

V

Values:
  expected, 169–176, 212–216, 335
  p, 278–281
Variables:
  dependent, 364
  independent, 364
  random, 155–157
    binomial, 189
    continuous, 156
    discrete, 156
    distribution of, 157–169
    expected values of, 169–176
    sets of, 176–178
    standardized, 173, 184
    standard normal, 184
    uncorrelated, 393
Variance, 63–69, 171–173, 212–216
  known, 234–241
  measures of, 62
  unknown, 243–249
Variance analysis, 406–443
  constructing the table, 409–413
  effects of unequal group sizes, 424–430
  estimation of effects, 416–419
  hypotheses of equal means, 419–423
  the model, 408–409
  numerical example, 413–416
  principles of design, 430–432
  randomization, 406–408
  randomized complete block design, 313, 432–440
Variance estimates, 372–374
Variation:
  measures of, 62
  total, partitioning, 406

W

Weighted mean, 53–56
Wilcoxon statistics, 447
Wilcoxon test, 446–450, 451, 452
Within groups, 411
Words, 122

Z

z, relationship between r and, 486

Percentage Points of the Chi-Square Distribution

df	\.995	\.990	\.975	\.950	\.050	\.025	\.010	\.005
1	—	—	—	.004	3.84	5.02	6.63	7.88
2	.01	.02	.05	.10	5.99	7.38	9.21	10.60
3	.07	.11	.22	.35	7.81	9.35	11.34	12.84
4	.21	.30	.48	.71	9.49	11.14	13.28	14.86
5	.41	.55	.83	1.15	11.07	12.83	15.09	16.75
6	.68	.87	1.24	1.64	12.59	14.45	16.81	18.55
7	.99	1.24	1.69	2.17	14.07	16.01	18.48	20.28
8	1.34	1.65	2.18	2.73	15.51	17.53	20.09	21.96
9	1.73	2.09	2.70	3.33	16.92	19.02	21.67	23.59
10	2.16	2.56	3.25	3.94	18.31	20.48	23.21	25.19
11	2.60	3.05	3.82	4.57	19.68	21.92	24.72	26.76
12	3.07	3.57	4.40	5.23	21.03	23.34	26.22	28.30
13	3.57	4.11	5.01	5.89	22.36	24.74	27.69	29.82
14	4.07	4.66	5.63	6.57	23.68	26.12	29.14	31.32
15	4.60	5.23	6.26	7.26	25.00	27.49	30.58	32.80
16	5.14	5.81	6.91	7.96	26.30	28.85	32.00	34.27
17	5.70	6.41	7.56	8.67	27.59	30.19	33.41	35.72
18	6.26	7.01	8.23	9.39	28.87	31.53	34.81	37.16
19	6.84	7.63	8.91	10.12	30.14	32.85	36.19	38.58
20	7.43	8.26	9.59	10.85	31.41	34.17	37.57	40.00
21	8.03	8.90	10.28	11.59	32.67	35.48	38.93	41.40
22	8.64	9.54	10.98	12.34	33.92	36.78	40.29	42.80
23	9.26	10.20	11.69	13.09	35.17	38.08	41.64	44.18
24	9.89	10.86	12.40	13.85	36.42	39.36	42.98	45.56
25	10.52	11.52	13.12	14.61	37.65	40.65	44.31	46.93
26	11.16	12.20	13.84	15.38	38.89	41.92	45.64	48.29
27	11.81	12.88	14.57	16.15	40.11	43.19	46.96	49.64
28	12.46	13.56	15.31	16.93	41.34	44.46	48.28	50.99
29	13.12	14.26	16.05	17.71	42.56	45.72	49.59	52.34
30	13.79	14.95	16.79	18.49	43.77	46.98	50.89	53.67
40	20.71	22.16	24.43	26.51	55.76	59.34	63.69	66.77
50	27.99	29.71	32.36	34.76	67.50	71.42	76.15	79.49
60	35.53	37.48	40.48	43.19	79.08	83.30	88.38	91.95
70	43.28	45.44	48.76	51.74	90.53	95.02	100.43	104.22
80	51.17	53.54	57.15	60.39	101.88	106.63	112.33	116.32
90	59.20	61.75	65.65	69.13	113.14	118.14	124.12	128.30
100	67.33	70.06	74.22	77.93	124.34	129.56	135.81	140.17

Column header: Probability of a Larger Value

In the row for $d = 6$ degrees of freedom and the column for probability $h = .050$, the entry is $\chi^2_{h,d} = \chi^2_{.05,6} = 12.59$; if χ^2 has the chi-square distribution with 6 degrees of freedom, then $P(\chi^2 > 12.59) = .05$.

Abridged from Thompson, Catherine M., "Table of percentage points of the χ^2 distribution," *Biometrika*, Vol. 32 (1942), p. 187, by permission.